Lecture Notes in Computer Science 1671

Edited by G. Goos, J. Hartmanis and J. van Leeuwen

Springer

Berlin
Heidelberg
New York
Barcelona
Hong Kong
London
Milan
Paris
Singapore
Tokyo

Dorit Hochbaum   Klaus Jansen
José D.P. Rolim   Alistair Sinclair (Eds.)

Randomization, Approximation, and Combinatorial Optimization

Algorithms and Techniques

Third International Workshop on Randomization
and Approximation Techniques in Computer Science,
and Second International Workshop
on Approximation Algorithms
for Combinatorial Optimization Problems
RANDOM-APPROX'99
Berkeley, CA, USA, August 8-11, 1999
Proceedings

 Springer

Volume Editors

Dorit Hochbaum
University of California at Berkeley, Business School
Berkeley, CA 94720-1776, USA
E-mail: hochbaum@ieor.berkeley.edu

Klaus Jansen
IDSIA Lugano
Corso Elvezia 36, CH-6900 Lugano, Switzerland
E-mail: klaus@idsia.ch

José D.P. Rolim
University of Geneva, Computer Science Center
24, Rue Général Dufour, CH-1211 Geneva 4, Switzerland
E-mail: Jose.Rolim@cui.unige.ch

Alistair Sinclair
University of California at Berkeley, Computer Science Division
Soda Hall, Berkeley, CA 94720-1776, USA
E-mail: sinclair@cs.berkeley.edu

Cataloging-in-Publication data applied for
Die Deutsche Bibliothek - CIP-Einheitsaufnahme

Randomization, approximation, and combinatorial optimization :
algorithms and techniques ; proceedings / Third International
Workshop on Randomization and Approximation Techniques in
Computer Science, and Second International Workshop on
Approximation Algorithms for Combinatorial Optimization Problems,
Random-Approx '99, Berkeley, CA, August 8 - 11, 1999. Dorit
Hochbaum ... (ed.). - Berlin ; Heidelberg ; New York ; Barcelona ;
Hong Kong ; London ; Milan ; Paris ; Singapore ; Tokyo : Springer,
1999
  (Lecture notes in computer science ; Vol. 1671)
  ISBN 3-540-66329-0

CR Subject Classification (1998): F.2, G.1.2, G.1.6, G.2, G.3, E.1, I.3.5

ISSN 0302-9743
ISBN 3-540-66329-0 Springer-Verlag Berlin Heidelberg New York

© Springer-Verlag Berlin Heidelberg 1999
Printed in Germany

Typesetting: Camera-ready by author
SPIN: 10705458   06/3142 – 5 4 3 2 1 0   Printed on acid-free paper

Foreword

This volume contains the papers presented at the **3rd International Workshop on Randomization and Approximation Techniques in Computer Science** (RANDOM'99) and the **2nd International Workshop on Approximation Algorithms for Combinatorial Optimization Problems** (APPROX'99), which took place concurrently at the University of California, Berkeley, from August 8–11, 1999. RANDOM'99 is concerned with applications of randomness to computational and combinatorial problems, and is the third workshop in the series following Bologna (1997) and Barcelona (1998). APPROX'99 focuses on algorithmic and complexity issues surrounding the development of efficient approximate solutions to computationally hard problems, and is the second in the series after Aalborg (1998).

The volume contains 24 contributed papers, selected by the two program committees from 44 submissions received in response to the call for papers, together with abstracts of invited lectures by Uri Feige (Weizmann Institute), Christos Papadimitriou (UC Berkeley), Madhu Sudan (MIT), and Avi Wigderson (Hebrew University and IAS Princeton). We would like to thank all of the authors who submitted papers, our invited speakers, the external referees we consulted and the members of the program committees, who were:

RANDOM'99
Alistair Sinclair, UC Berkeley
Noga Alon, Tel Aviv U.
Jennifer Chayes, Microsoft
Monika Henzinger, Compaq-SRC
Mark Jerrum, U. of Edinburgh
Ravi Kannan, Yale U.
David Karger, MIT
Valerie King, U. of Victoria
Jon Kleinberg, Cornell U.
Andrzej Ruciński, U. Poznán
Raimund Seidel, U. Saarbrücken
Joel Spencer, Courant Institute
Amnon Ta-Shma, ICSI Berkeley
Emo Welzl, ETH Zürich

APPROX'99
Dorit Hochbaum, UC Berkeley
Sanjeev Arora, Princeton U.
Leslie Hall, Johns Hopkins U.
Samir Khuller, U. of Maryland
Phil Klein, Brown U.
Kurt Mehlhorn, MPI Saarbrücken
Joe Mitchell, SUNY Stony Brook
Seffi Naor, Bell Labs and Technion
David Peleg, Weizmann Institute
Vijay Vazirani, Georgia Tech.
David Williamson, IBM Yorktown
Gerhard Woeginger, TU Graz

We gratefully acknowledge support from the European agency INTAS, the Computer Science Department of the University of California at Berkeley, and the University of Geneva. We also thank Germaine Gusthiot and Thierry Zwissig for their help.

June 1999   Dorit Hochbaum, APPROX'99 Program Chair
                Klaus Jansen and José D. P. Rolim, Workshop Chairs
                        Alistair Sinclair, RANDOM'99 Program Chair

Contents

Session Random 1

Completeness and Robustness Properties of Min-Wise Independent   1
Permutations
Andrei Z. Broder and Michael Mitzenmacher

Low Discrepancy Sets Yield Approximate Min-Wise Independent   11
Permutation Families
Michael Saks, Aravind Srinivasan, Shiyu Zhou and David Zuckerman

Session Approx 1

Independent Sets in Hypergraphs with Applications to Routing via   16
Fixed Paths
Noga Alon, Uri Arad and Yossi Azar

Approximating Minimum Manhattan Networks   28
Joachim Gudmundsson, Christos Levcopoulos and Giri Narasimhan

Approximation of Multi-color Discrepancy   39
Benjamin Doerr and Anand Srivastav

A Polynomial Time Approximation Scheme for the Multiple   51
Knapsack Problem
Hans Kellerer

Session Approx 2

Set Cover with Requirements and Costs Evolving over Time   63
Milena Mihail

Multicoloring Planar Graphs and Partial k-Trees   73
Magnús M. Halldórsson and Guy Kortsarz

Session: Random 2

Testing the Diameter of Graphs   85
Michal Parnas and Dana Ron

Improved Testing Algorithms for Monotonicity   97
Yevgeniy Dodis, Oded Goldreich, Eric Lehman, Sofya Raskhodnikova,
Dana Ron and Alex Samorodnitsky

Linear Consistency Testing   109
Yonatan Aumann, Johan Håstad, Michael O. Rabin and Madhu Sudan

Improved Bounds for Sampling Contingency Tables   121
Benjamin James Morris

Invited Talk
Probabilistic and Deterministic Approximations of the Permanent   130
Avi Wigderson

Session Random 3
Improved Derandomization of BPP Using a Hitting Set Generator   131
Oded Goldreich and Avi Wigderson

Probabilistic Construction of Small Strongly Sum-Free Sets via   138
Large Sidon Sets
Andreas Baltz, Tomasz Schoen and Anand Srivastav

Session Approx 3
Stochastic Machine Scheduling: 144
Performance Guarantees for LP-based Priority Policies
Rolf H. Möhring, Andreas S. Schulz and Marc Uetz

Efficient Redundant Assignments under Fault-Tolerance Constraints   156
Dimitris A. Fotakis and Paul G. Spirakis

Scheduling with Machine Cost   168
Csanád Imreh and John Noga

A Linear Time Approximation Scheme for the Job Shop Scheduling Problem 177
Klaus Jansen, Roberto Solis-Oba and Maxim Sviridenko

Invited Talk
Randomized Rounding for Semidefinite Programs - Variations on the   189
MAX CUT Example
Uriel Feige

Session Approx 4
Hardness Results for the Power Range Assignment Problem in   197
Packet Radio Networks
Andrea E. F. Clementi, Paolo Penna and Riccardo Silvestri

A New Approximation Algorithm for the Demand Routing and   209
Slotting Problem with Unit Demands on Rings
Christine T. Cheng

Session Random 4

Algorithms for Graph Partitioning on the Planted Partition Model   221
Anne E. Condon and Richard M. Karp

A Randomized Time-Work Optimal Parallel Algorithm for Finding   233
a Minimum Spanning Forest
Seth Pettie and Vijaya Ramachandran

Fast Approximate PCPs for Multidimensional Bin-Packing Problems   245
Tuğkan Batu, Ronitt Rubinfeld and Patrick White

Pfaffian Algorithms for Sampling Routings on Regions with Free   257
Boundary Conditions
Russell A. Martin and Dana Randall

Minisymposium on Scheduling Talks
Organizer Klaus Jansen

Scheduling with Unexpected Machine Breakdowns   269
Susanne Albers and Günter Schmidt

Scheduling on a Constant Number of Machines   281
F. Afrati, E. Bampis, C. Kenyon and I. Milis

Author Index   289

Completeness and Robustness Properties of Min-Wise Independent Permutations

Andrei Z. Broder[1] and Michael Mitzenmacher[2,*]

[1] Compaq Systems Research Center
130 Lytton Avenue, Palo Alto, CA 94301, USA. broder@pa.dec.com
[2] Harvard University
29 Oxford St., Cambridge, MA 02138. michaelm@eecs.harvard.edu

Abstract. We provide several new results related to the concept of min-wise independence. Our main result is that any randomized sampling scheme for the relative intersection of sets based on testing equality of samples yields an equivalent min-wise independent family. Thus, in a certain sense, min-wise independent families are "complete" for this type of estimation.

We also discuss the notion of robustness, a concept extending min-wise independence to allow more efficient use of it in practice. A surprising result arising from our consideration of robustness is that under a random permutation from a min-wise independent family, any element of a fixed set has an equal chance to get any rank in the image of the set, not only the minimum as required by definition.

1   Introduction

A family of permutations $\mathcal{P} \subseteq S_n$ is called *min-wise independent* (abbreviated *MWI*) if for any set $X \subseteq [n] = \{1, \ldots, n\}$ and any $x \in X$, when π is chosen at random in $\mathcal{P}$ according to some specified probability distribution we have

$$\mathbf{Pr}\big(\min\{\pi(X)\} = \pi(x)\big) = \frac{1}{|X|} \ . \tag{1}$$

In other words we require that all the elements of any fixed set X have an equal chance to become the minimum element of the image of X under π.

When the distribution on $\mathcal{P}$ is non-uniform, the family is called *biased*, and it is called *unbiased* otherwise. In general in this paper we will not specify the probability distribution on $\mathcal{P}$ unless relevant, and from now on when we say "π chosen at random in (the min-wise independent family) $\mathcal{P}$" we mean "π chosen in $\mathcal{P}$ according to the probability distribution associated to $\mathcal{P}$ such that (1) holds."

Together with Moses Charikar and Alan Frieze, we introduced this notion in [4] motivated by the fact that such a family (under some relaxations) is essential

---

* Parts of this work were done while this author was at Compaq Systems Research Center.

to the algorithm used in practice by the AltaVista web index software to detect and filter near-duplicate documents. The crucial property that enables this application is the following: let X be a subset of $[n]$. Pick a "sample" $s(X) \in X$ by choosing at random a permutation π from a family of permutations $\mathcal{P}$ and letting

$$s(X) = \pi^{-1}(\min\{\pi(X)\}) . \tag{2}$$

Then, if $\mathcal{P}$ is a MWI-family, for any two nonempty subsets A and B, we have

$$\mathbf{Pr}\big(s(A) = s(B)\big) = \frac{|A \cap B|}{|A \cup B|} . \tag{3}$$

Hence such samples can be used to estimate the relative size of the intersection of sets, a quantity that we call the *resemblance* of A and B, defined as

$$R(A, B) = \frac{|A \cap B|}{|A \cup B|} . \tag{4}$$

We estimate resemblance by first picking, say, 100 permutations from a MWI-family, and then computing samples for each set of interest. Then the resemblance of any two sets can be estimated simply by determining the fraction of samples that coincide.

In practice we can allow small relative errors. We say that $\mathcal{P} \subseteq S_n$ is *approximately min-wise independent with relative error ϵ* (or just approximately min-wise independent, where the meaning is clear) if for any set $X \subseteq [n]$ and any $x \in X$, when π is chosen at random in $\mathcal{P}$ we have

$$\left| \mathbf{Pr}\big(\min\{\pi(X)\} = \pi(x)\big) - \frac{1}{|X|} \right| \le \frac{\epsilon}{|X|} . \tag{5}$$

For further details about the use of these ideas to estimate document similarity see [6, 1, 2]. An optimal (size-wise) construction for a MWI-family was obtained by Takei, Itoh, and Shinozaki [13]. Explicit constructions of approximately MWI-families were obtained by Indyk [8] and by Saks & al. [11]. For an application of these families to derandomization see [5].

We also note that concepts similar to min-wise independence have appeared prior to our work [4] as well. For example, the monotone ranged hash functions described in [9] have the min-wise independence property; Cohen [7] uses the property that the minimum element of a random permutation is uniform to estimate the size of the transitive closure, as well as to solve similar related problems; and Mulmuley [10] uses what we call approximate min-wise independence to use fewer random bits for several randomized geometric algorithms.

The main result of this paper, presented in Sect. 2, is that, rather surprisingly, *any* sampling scheme that has property (3) is equivalent to a scheme derived via equation (2) from a min-wise independent family of permutations. More precisely we have the following theorem:

Theorem 1. *Let $\mathcal{F}$ be a family of functions from nonempty subsets of $[n]$ to some arbitrary set Ω. Assume there exists a probability distribution on $\mathcal{F}$ such that for any two nonempty subsets, A and B,*

$$\mathbf{Pr}\big(f(A) = f(B)\big) = \frac{|A \cap B|}{|A \cup B|} \ .$$

Then there exists a min-wise independent family of permutations $\mathcal{P}$ such that every $f \in \mathcal{F}$ is defined by

$$f(X) = f\left(\left\{\pi_f^{-1}(\min\{\pi_f(X)\})\right\}\right)$$

for some $\pi_f \in \mathcal{P}$.

We note here some immediate consequences of the theorem:

(a) The induced family of permutations has the same size as the initial family of functions, that is $|\mathcal{P}| = |\mathcal{F}|$.
(b) Each $f \in \mathcal{F}$ takes exactly n distinct values $f(\{x_1\}), \ldots, f(\{x_n\})$. (A priori each f can take $2^n - 1$ values.)
(c) Assume that we add the condition that for every $X \subseteq [n]$, each $f \in \mathcal{F}$ satisfies $f(X) \in X$; in other words, the "sample" must belong to the set being sampled. Then for every $x \in [n]$ each f satisfies $f(\{x\}) = x$, and hence each f has the form

$$f(X) = \pi_f^{-1}(\min\{\pi_f(X)\}) \ .$$

(The converse of the assumption is also true: if for every $x \in [n]$ we have $f(\{x\}) = x$ then $f(X) \in X$ follows. See Corollary 1 below.)
(d) Thus every estimation scheme that has property (3) is equivalent under renaming to a sampling scheme derived via equation (2) from a min-wise independent family of permutations. (For each f, $f(\{x_1\})$ is the "name" of x_1, $f(\{x_2\})$ is the "name" of x_2, etc.)

Of course in practice it might be more convenient to represent $\mathcal{F}$ directly rather than via $\mathcal{P}$. (See [3] for an example.) But the fact remains that any method of sampling to estimate resemblance via equation (3) is equivalent to sampling with min-wise independent permutations.

To develop some intuition, before plunging into the proof, we start by observing that the choice of "min" in the definition (1) is somewhat arbitrary. Clearly if we replace "min" with "max" both in (1) and in (2), property (3) holds. More generally, we can fix a permutation $\sigma \in S_n$ (think of it as a total order on $[n]$), and require $\mathcal{P}$ to satisfy the property

$$\mathbf{Pr}\big(\min\{\sigma(\pi(X))\} = \sigma(\pi(x))\big) = \frac{1}{|X|} \ . \tag{6}$$

Then we can choose samples according to the rule

$$s(X) = \pi^{-1}\left(\sigma^{-1}\left(\min\{\sigma(\pi(X))\}\right)\right) \ .$$

(We obtain "max" by taking $\sigma(i) = n + 1 - i$.)

Is there any advantage to choosing a particular σ? A moment of reflection indicates that there is nothing to be gained since we can simply replace the family $\mathcal{P}$ by the family $\mathcal{P} \circ \sigma$. This is, in fact, a very simple instance of Theorem 1. However, it could be of interest if a family $\mathcal{P}$ satisfies condition (6) with respect to more than one order σ. One reason is that, in practice, computing $\pi(X)$ is expensive (see [3] for details). If a family has the min-wise independence property with respect to several orders, then we can extract a sample for each order. Obviously these samples are correlated, but if the correlation can be bounded, these samples are still usable.

Takei, Itoh, and Shinozaki [13] presented an optimal (size-wise) construction for a MWI-family under the uniform distribution. Their family has size $\text{lcm}(1, \ldots, n)$, matching the lower bound of [4]. They observed that their construction produces a family that is simultaneously min-wise independent and max-wise independent. In Sect. 3 we show that this is not a fluke; in fact, any min-wise independent family is also max-wise independent. Moreover, if $\mathcal{P} \subseteq S_n$ is min-wise independent, then for any set $X \subseteq [n]$, any $x \in X$, and any fixed $r \in \{1, \ldots, |X|\}$, when π is chosen at random in $\mathcal{P}$ we have

$$\mathbf{Pr}\big(\text{rank}(\pi(x), \pi(X)) = r\big) = \frac{1}{|X|} , \tag{7}$$

where $\text{rank}(x, X)$ for $x \in X$ is the number of elements in X not greater than x. Hence the max-wise independence property follows by taking $r = |X|$.

In Sect. 4 we discuss families that have the min-wise independence property with respect to *all* possible orders σ. We call such families *robust*. We show that although not every min-wise independent family is robust, there are non-trivial robust families. On the other hand, robust families under the uniform distribution of size $\text{lcm}(1, \ldots, n)$ do not necessarily exist for every n.

2 Any Sampling Scheme is a MWI-Family

In this section we prove the following:

Theorem 1 *Let $\mathcal{F}$ be a family of functions from nonempty subsets of $[n]$ to some arbitrary set Ω. Assume there exists a probability distribution on $\mathcal{F}$ such that for any two nonempty subsets, A and B,*

$$\mathbf{Pr}\big(f(A) = f(B)\big) = \frac{|A \cap B|}{|A \cup B|} .$$

Then there exists a min-wise independent family of permutations $\mathcal{P}$ such that every $f \in \mathcal{F}$ is defined by

$$f(X) = f\left(\left\{\pi_f^{-1}(\min\{\pi_f(X)\})\right\}\right)$$

for some $\pi_f \in \mathcal{P}$.

Proof. Assume the premises of the Theorem. We start with two Lemmas.

Lemma 1. *Let X be a nonempty subset of $[n]$. Then for any $x \in X$*

$$\mathbf{Pr}(f(X) = f(\{x\})) = \frac{|X \cap \{x\}|}{|X \cup \{x\}|} = \frac{1}{|X|} \ .$$

Corollary 1. *Let $X = \{x_1, x_2, \ldots, x_k\}$ be a nonempty subset of $[n]$. Then for each $f \in \mathcal{F}$*

$$f(X) \in \{f(\{x_1\}), f(\{x_2\}), \ldots, f(\{x_k\})\} \ .$$

Proof.

$$\mathbf{Pr}\left(f(X) \in \{f(\{x_1\}), f(\{x_2\}), \ldots, f(\{x_k\})\}\right)$$

$$= \sum_{i=1}^{k} \mathbf{Pr}\left(f(X) = f(\{x_i\})\right) = 1.$$

□

Lemma 2. *Let $X = \{x_1, x_2, \ldots, x_k\}$ and Y be a nonempty subsets of $[n]$. If $X \subseteq Y$, then for every $f \in \mathcal{F}$, if $f(Y) \in \{f(\{x_1\}), f(\{x_2\}), \ldots, f(\{x_k\})\}$, then $f(Y) = f(X)$.*

Proof. By hypothesis

$$\mathbf{Pr}(f(X) = f(Y)) = \frac{|X \cap Y|}{|X \cup Y|} = \frac{k}{|Y|} \ .$$

On the other hand,

$$\mathbf{Pr}(f(X) = f(Y))$$

$$= \mathbf{Pr}(f(X) = f(\{x_1\}) \wedge f(Y) = f(\{x_1\})) + \cdots$$
$$\cdots + \mathbf{Pr}(f(X) = f(\{x_k\}) \wedge f(Y) = f(\{x_k\}))$$

$$= \mathbf{Pr}(f(X) = f(\{x_1\}) \mid f(Y) = f(\{x_1\})) \, \mathbf{Pr}(f(Y) = f(\{x_1\})) + \cdots$$
$$\cdots + \mathbf{Pr}(f(X) = f(\{x_k\}) \mid f(Y) = f(\{x_k\})) \, \mathbf{Pr}(f(Y) = f(\{x_k\}))$$

$$= \mathbf{Pr}(f(X) = f(\{x_1\}) \mid f(Y) = f(\{x_1\}))(1/|Y|) + \cdots$$
$$\cdots + \mathbf{Pr}(f(X) = f(\{x_k\}) \mid f(Y) = f(\{x_k\}))(1/|Y|).$$

(The last equality follows from Lemma 1.) Hence for every $x_i \in X$

$$\mathbf{Pr}(f(X) = f(\{x_i\}) | f(Y) = f(\{x_i\})) = 1 \ ,$$

and therefore for every $f \in F$, if $f(Y) = f(\{x_i\})$ then $f(X) = f(\{x_i\})$ as well.

□

Lemma 3. *For any two distinct elements $x_1, x_2 \in [n]$ and each $f \in \mathcal{F}$.*

$$f(\{x_1\}) \neq f(\{x_2\}) \ .$$

Proof. By hypothesis $\mathbf{Pr}(f(\{x_1\}) = f(\{x_2\})) = 0$. ☐

Returning to the proof of the Theorem, we show now how to construct for each $f \in \mathcal{F}$ a permutation π_f such that for every nonempty set X

$$f(X) = f\left(\left\{\pi_f^{-1}(\min\{\pi_f(X)\})\right\}\right) . \tag{8}$$

Note that the family $\mathcal{P}$ given by the π_f above are clearly min-wise independent by Lemma 1.

Fix f and let $g : \{f(\{x_1\}), \ldots, f(\{x_n\})\} \to [n]$ be the function defined by $g(f(\{x_i\})) = x_i$. In view of Lemma 3 g is well-defined. Now define a sequence $y_1, y_2, \ldots, y_n$ as follows:

$$y_1 = g(f([n]))$$
$$y_2 = g(f([n] \setminus \{y_1\}))$$
$$y_3 = g(f([n] \setminus \{y_1, y_2\}))$$
$$\vdots$$

In view of Corollary 1 g is correctly used and we have

$$f([n]) = f(\{y_1\})$$
$$f([n] \setminus \{y_1\})) = f(\{y_2\})$$
$$f([n] \setminus \{y_1, y_2\})) = f(\{y_3\})$$
$$\vdots$$

Furthermore $y_1, y_2, \ldots, y_n$ is a permutation of $[n]$. Finally we take π_f to be the inverse of the permutation determined by the y_i; that is, π_f maps y_1 to 1, y_2 to 2, etc. We need to show that f satisfies equation (8) for every nonempty set X.

Fix X and consider the sets $Y_1 = [n]$, $Y_2 = [n] \setminus \{y_1\}$, $Y_3 = [n] \setminus \{y_1, y_2\}$, $\ldots$, $Y_n = \{y_n\}$. Let k be the largest index such that Y_k still includes X. This implies that

(a) $y_k \in X$ since otherwise we could have taken Y_{k+1}.
(b) $\{y_1, y_2, \ldots, y_{k-1}\} \cap X = \emptyset$ since none of these elements belong to Y_k.

By definition $f(Y_k) = f(\{y_k\})$. But $y_k \in X \subseteq Y_k$ and therefore Lemma 2 implies that $f(X) = f(\{y_k\})$ as well. On the other hand property (a) above implies that $\min\{\pi_f(X)\} \le k$ and property (b) implies that $\min\{\pi_f(X)\} > k - 1$. Hence $\min\{\pi_f(X)\} = k$ and $\pi_f^{-1}(\min\{\pi_f(X)\}) = y_k$ as required. ☐

3   Rank Uniformity for MWI-Families

In this section, we show that any min-wise independent family actually has the property that every item in any fixed set is equally likely to have any rank in the image of the set – not just the minimum rank as required by definition. Our analysis is based on the following lemma, proven in [12]. (Alternatively, the "only if" part follows also from Theorem 6 of [4] and the "if" part follows from the proof of Theorem 2 below.)

Lemma 4. *A family of permutations $\mathcal{P}$ is min-wise independent if and only if for any set $X \subset [n]$ of size k and any element $x \in [n] \setminus X$*

$$\mathbf{Pr}\left(\pi(X) = [k] \wedge \pi(x) = k + 1\right) = \frac{1}{\binom{n}{k}(n - k)},$$

when π is chosen at random in $\mathcal{P}$.

In other words, if we fix a set X of size k and an extra element x, the probability that x maps to $k + 1$ and X maps to $\{1, \ldots, k\}$ in some arbitrary order is exactly what "it should be" if we were sampling uniformly from the entire set of permutations S_n.

Theorem 2. *If $\mathcal{P}$ is min-wise independent, and π is chosen at random from $\mathcal{P}$, then*

$$\mathbf{Pr}\left(\mathrm{rank}(\pi(x), \pi(X)) = r\right) = \frac{1}{|X|} . \tag{9}$$

Proof. We sum over all the possible ways such that $\mathrm{rank}(\pi(x), \pi(X)) = r$ and $\pi(x) = s$ and consider which elements map to $[s - 1]$. Note that we must have $r \leq s \leq n - (|X| - r)$. There must be $r - 1$ other elements of X, call them $\{x_1, x_2, \ldots, x_{r-1}\}$, such that $\pi(x_i) \in [s - 1]$, and there are $\binom{|X|}{r-1}$ ways to choose them. Similarly, there must be $s - r$ elements of $[n] \setminus X$, call them $\{y_1, y_2, \ldots, y_{n-r}\}$, such that $\pi(y_i) \in [s-1]$ and there are $\binom{n-|X|}{s-r}$ ways to choose these elements. For each possible combination of choices, we have from Lemma 4 that the probability that these elements are mapped to $[s - 1]$ and x is mapped to s is

$$\frac{1}{\binom{n}{s-1}(n - s + 1)} .$$

Hence

$$\mathbf{Pr}\left(\mathrm{rank}(\pi(x), \pi(X)) = r\right) = \sum_{s=r}^{n-|X|+r} \frac{\binom{|X|-1}{r-1}\binom{n-|X|}{s-r}}{\binom{n}{s-1}(n - s + 1)}$$

$$= \frac{1}{|X|\binom{n}{|X|}} \sum_{s=r}^{n-|X|+r} \binom{s-1}{r-1}\binom{n-s}{|X|-r}$$

$$= \frac{1}{|X|\binom{n}{|X|}} \binom{n}{|X|} = \frac{1}{|X|} .$$

(The second equality is obtained by expanding binomials into factorials and regrouping. The third equality is obtained by counting the ways of choosing $|X|$ elements out of $[n]$ by summing over all possible values s for the r'th largest element among those chosen.) $\qquad \square$

4 Robust Families

We now consider *robust* families. As described in the introduction, robustness is an extension of min-wise independence. Formally, a family $\mathcal{P}$ is robust if for *every* possible permutation σ, when π is chosen at random in $\mathcal{P}$

$$\mathbf{Pr}\big(\min\{\sigma\big(\pi(X)\big)\} = \sigma\big(\pi(x)\big)\big) = \frac{1}{|X|} \ . \tag{10}$$

Trivially, S_n is a robust family. We first demonstrate that there exist non-trivial robust families. To this end, we extend the condition for min-wise independent families given in Lemma 4 to the equivalent condition for robust families. Since robust families are min-wise independent under any order σ we obtain the following:

Lemma 5. *A family of permutations $\mathcal{P}$ is robust if and only if for any set $X \subset [n]$ of size k and any element $x \in [n] \setminus X$, and any other set $A \subset [n]$ of size also k and any element $a \in [n] \setminus A$*

$$\mathbf{Pr}\big(\pi(X) = A \wedge \pi(x) = a\big) = \frac{1}{\binom{n}{k}(n-k)} \ . \tag{11}$$

Theorem 3. *There exist biased robust families of size at most*

$$n^2 \binom{2(n-1)}{n-1} \ .$$

Proof. Following an idea used in [4], we establish a linear program for determining a robust family of the required size. There are $n!$ variables x_{π_i}, one for each possible permutation π_i. The variable x_{π_i} represent the probability that π_i is chosen within our family; if $x_{\pi_i} = 0$, we may exclude π_i from the family.

Our linear program is based on Lemma 5. We set up an equation for each pair (a, A) and (x, X) with $|A| = |X|$, with each equation representing the constraint that (a, A) maps to (x, X) with the required probability. Hence there are

$$\sum_{i=0}^{n-1} n^2 \binom{n-1}{i}^2 = n^2 \binom{2(n-1)}{n-1}$$

equations. We know there exists a solution to the linear program, since if each permutation is chosen with probability $1/n!$ we have a robust family. Hence there must be a basic feasible solution with at most $n^2 \binom{2(n-1)}{n-1}$ variables taking non-zero values. This solution yields a biased robust family. $\square$

It is also worthwhile to ask if there are any non-trivial unbiased robust families. We demonstrate that in fact there are non-trivial families for $n \geq 4$.

Recall that the permutations S_n can be split into two groups, each of size $n!/2$, as follows: a permutation is called *even* if it can be obtained by an even number of transpositions from the identity, and odd *odd* otherwise.

Theorem 4. *For $n \geq 4$, the even permutations and the odd permutations of $[n]$ both yield robust families.*

Proof. We use Lemma 5. That is, we must show that for each pair (x, X) with $x \in [n]$, $X \subseteq [n]$, $x \notin X$, the probability that $\pi(x) = a$ and $\pi(X) = A$ is correct for every (a, A) with $a \in [n]$, $A \subseteq [n]$, $|A| = |X|$, and $a \notin A$.

Equivalently, since the odd permutations and even permutations divide the set of all permutations into two equal-sized families, it suffices to show that the number of even permutations mapping (x, X) into (a, A) is the same as the number of odd permutations that do so. Note that as $n \geq 4$, either $|X| \geq 2$ or $|[n] - X - \{x\}| \geq 2$. In the first case, we can determine a one-to-one mapping of even permutations to odd permutations that map (x, X) into (a, A) by choosing two particular elements of X (say the two smallest) and transposing them. In the second case, we may do the same by transposing two elements of $[n] - X - \{x\}$. □

From the lower bound in [4], we know that unbiased min-wise independent families (and hence robust families) have size at least $\mathrm{lcm}(1, \ldots, n)$. As $\mathrm{lcm}(1, \ldots, n) = n!/2$ for $n = 4$ and $n = 5$, the result of Theorem 4 is optimal for these cases. We suspect that Theorem 4 is in fact optimal for all $n \geq 4$; that is, there is no unbiased robust family of size less than $n!/2$. While we cannot yet show this, we can show that for $n = 6$, there is no unbiased robust family of size $\mathrm{lcm}(1, \ldots, n) = 60$.

Theorem 5. *All the unbiased robust families of permutations of $\{1, 2, 3, 4, 5, 6\}$ have size greater than 60.*

Proof. The proof uses an exhaustive search, where the search for a robust family is reduced by using symmetry and Lemma 5. Details will appear in the full paper. □

Given the development of approximate min-wise independent families of permutations developed in [4], it is natural to ask about approximate robust families of permutations as well. A family of permutations is said to be *approximately robust with relative error ϵ* if and only if for every permutation order σ,

$$\left| \mathbf{Pr}\left(\min\{\sigma(\pi(X))\} = \sigma(\pi(x)) \right) - \frac{1}{|X|} \right| \leq \frac{\epsilon}{|X|} . \tag{12}$$

That is, regardless of σ, the probability over the choice of π that an element x is the minimum of a set $|X|$ is within a factor of $(1 \pm \epsilon)$ of the natural probability $\frac{1}{|X|}$. It is straightforward to show that there must be small approximate robust families.

Theorem 6. *There are approximate robust families of size $O(n^2 \log(n)/\epsilon)$.*

Proof. The proof follows Theorem 3 of [4]. We simply choose a random set of permutations of the appropriate size, and show that with some probability, we obtain an unbiased approximate robust family. Details will appear in the full paper. □

Acknowledgment

We wish to thank Uri Feige for his help.

References

[1] A. Z. Broder. On the resemblance and containment of documents. In *Proceedings of Compression and Complexity of Sequences 1997*, pages 21–29. IEEE Computer Society, 1988.

[2] A. Z. Broder. Filtering near-duplicate documents. In *Proceedings of FUN 98*, 1998. To appear.

[3] A. Z. Broder, M. Burrows, and M. S. Manasse. Efficient computation of minima of random functions. Manuscript.

[4] A. Z. Broder, M. Charikar, A. Frieze, and M. Mitzenmacher. Min-wise independent permutations. In *Proceedings of the 30th Annual ACM Symposium on Theory of Computing (STOC-98)*, pages 327–336, New York, May 1998. ACM Press.

[5] A. Z. Broder, M. Charikar, and M. Mitzenmacher. A derandomization using min-wise independent permutations. In *Proceedings of Random 98*, pages 15–24, 1998. Available as *Lecture Notes in Computer Science*, vol. 1518.

[6] A. Z. Broder, S. C. Glassman, M. S. Manasse, and G. Zweig. Syntactic clustering of the Web. In *Proceedings of the Sixth International World Wide Web Conference*, pages 391–404, April 1997.

[7] E. Cohen. Estimating the size of the transitive closure in linear time. In *35th Annual Symposium on Foundations of Computer Science*, pages 190–200, Santa Fe, New Mexico, 20–22 Nov. 1994. IEEE.

[8] P. Indyk. A small approximately min-wise independent family. In *Proceedings of the Tenth Annual ACM-SIAM Symposium on Discrete Algorithms*, pages 454–456, 1999.

[9] D. Karger, E. Lehman, T. Leighton, M. Levine, D. Lewin, and R. Panigrahy. Consistent hashing and random trees: Distributed caching protocols for relieving hot spots on the World Wide Web. In *Proceedings of the Twenty-Ninth Annual ACM Symposium on Theory of Computing*, pages 654–663, El Paso, Texas, 4–6 May 1997.

[10] K. Mulmuley. Randomized geometric algorithms and pseudorandom generators. *Algorithmica*, 16(4/5):450–463, Oct./Nov. 1996.

[11] M. Saks, A. Srinivasan, S. Zhou, and D. Zuckerman. Discrepant Sets Yield Approximate Min-Wise Independent Permutation Families. In these Proceedings.

[12] Y. Takei and T. Itoh. A characterization of min-wise independent permutation families. In *Proceedings of the Language and Automata Symposium*, Kyoto-Univ, Japan, Feb 1-3 1999. To appear.

[13] Y. Takei, T. Itoh, and T. Shinozaki. An optimal construction of exactly min-wise independent permutations. Technical Report COMP98-62, IEICE, 1998.

Low Discrepancy Sets Yield Approximate Min-Wise Independent Permutation Families

Michael Saks[1]* Aravind Srinivasan[2]** Shiyu Zhou[3] David Zuckerman[4]***

[1] Department of Mathematics, Rutgers University, Hill Center,
110 Frelinghuysen Road, Piscataway, NJ 08854.
saks@math.rutgers.edu
[2] Bell Laboratories, Lucent Technologies,
700 Mountain Ave., Murray Hill, NJ 07974-0636.
srin@research.bell-labs.com
[3] Department of Computer & Information Science, University of Pennsylvania.
shiyu@cis.upenn.edu
[4] Computer Science Division, University of California, Berkeley, CA 94720.
diz@cs.berkeley.edu

Abstract. Motivated by a problem of filtering near-duplicate Web documents, Broder, Charikar, Frieze & Mitzenmacher defined the following notion of ϵ-approximate min-wise independent permutation families [2]. A multiset $\mathcal{F}$ of permutations of $\{0, 1, \ldots, n-1\}$ is such a family if for all $K \subseteq \{0, 1, \ldots, n-1\}$ and any $x \in K$, a permutation π chosen uniformly at random from $\mathcal{F}$ satisfies

$$\mid \Pr[\min\{\pi(K)\} = \pi(x)] - \frac{1}{|K|} \mid \leq \frac{\epsilon}{|K|}.$$

We show connections of such families with *low discrepancy sets for geometric rectangles*, and give explicit constructions of such families $\mathcal{F}$ of size $n^{O(\sqrt{\log n})}$ for $\epsilon = 1/n^{\Theta(1)}$, improving upon the previously best-known bound of Indyk [4]. We also present polynomial-size constructions when the min-wise condition is required only for $|K| \leq 2^{O(\log^{2/3} n)}$, with $\epsilon \geq 2^{-O(\log^{2/3} n)}$.

Key words and phrases: Min-wise independent permutations, document filtering, pseudorandom permutations, explicit constructions.

* Research supported in part by NSF grant CCR-9700239.
** Parts of this work were done: (i) while at the School of Computing, National University of Singapore, Singapore 119260, and (ii) while visiting DIMACS (Center for Discrete Mathematics and Theoretical Computer Science). DIMACS is an NSF Science and Technology Center, funded under contract STC-91-19999; and also receives support from the New Jersey Commission on Science and Technology.
*** On leave from the University of Texas at Austin. Supported in part by NSF NYI Grant No. CCR-9457799, a David and Lucile Packard Fellowship for Science and Engineering, and an Alfred P. Sloan Research Fellowship. Part of this research was done while the author attended the DIMACS Workshop on Randomization Methods in Algorithm Design.

1 Introduction

Constructing pseudorandom permutation families is often more difficult than constructing pseudorandom function families. For example, there are polynomial size constructions of k-wise independent function families for constant k [5, 6, 1, 8]. On the other hand, although there are polynomial-size 3-wise independent permutation families (see, e.g. [10]), there are only exponential size constructions known for higher k. In fact, the only subgroups of the symmetric group that are 6-wise independent are the alternating group and the symmetric group itself; for 4-wise and 5-wise independence there are only finitely many besides these (see [3]). There are constructions of almost k-wise independent permutation families with error $\epsilon = O(k^2/n)$ [9], again not as good as is known for function families.

We address a different type of pseudorandom permutation family, called a *min-wise independent permutation family*. Motivated by a problem of filtering near-duplicate Web documents, Broder, Charikar, Frieze & Mitzenmacher [2] defined them as follows:

Definition 1.1 ([2]) *Let* $[n]$ *denote* $\{0, 1, \ldots, n-1\}$, *and* S_n *denote the set of permutations of* $[n]$. *A multiset* $\mathcal{F}$ *contained in* S_n *is called* min-wise independent *if for all* $K \subseteq [n]$ *and any* $x \in K$, *when a permutation* π *chosen uniformly at random from* $\mathcal{F}$ *we have that* $\Pr[\min\{\pi(K)\} = \pi(x)] = \frac{1}{|K|}$. *(*$\pi(K)$ *denotes the set* $\{\pi(y) : y \in K\}$.)

While $\mathcal{F} = S_n$ of course satisfies the above, even indexing from such an $\mathcal{F}$ is difficult, as some applications have n of the order of magnitude of 2^{64} [2]. Furthermore, it is shown in [2] that any min-wise independent family must have exponential size: more precisely, its cardinality is at least $\text{lcm}(1, 2, \ldots; n) \geq e^{n - o(n)}$. (This lower bound of $\text{lcm}(1, 2, \ldots, n)$ is in fact tight [11].) This motivates one to study families that are only *approximately* min-wise independent; moreover, in practice, we may have an upper bound d on the cardinality of the sets K of Definition 1.1, such that $d \ll n$. Thus, the following notion is also introduced in [2]; we use slightly different terminology here.

Definition 1.2 ([2]) *Suppose a multi-set* $\mathcal{F}$ *is contained in* S_n; *let* π *be as in Definition 1.1.* $\mathcal{F}$ *is called an* (n, d, ϵ)-mwif *(for d-wise ϵ-approximate min-wise independent family) if for all* $K \subseteq [n]$ *with* $|K| \leq d$ *and any* $x \in K$, *we have*

$$\left| \Pr[\min\{\pi(K)\} = \pi(x)] - \frac{1}{|K|} \right| \leq \frac{\epsilon}{|K|}.$$

Using a random construction, Broder *et. al.* showed the *existence* of an (n, d, ϵ)-mwif of cardinality $O(d^2 \log(2n/d)/\epsilon^2)$ [2]. Indyk presented an explicit construction of an (n, n, ϵ)-mwif of cardinality $n^{O(\log(1/\epsilon))}$ in [4]. In this paper, we show a connection between the construction of approximate min-wise independent families and the construction of low discrepancy sets for geometric rectangles, and use this connection to give a new construction of an (n, d, ϵ)-mwif.

To state our main result we first need some definitions. Let m, d and n be integers with $d \leq n$. We denote by $\mathcal{GR}(m, d, n)$ the set of *(geometric) rectangles* $[a_1, b_1) \times [a_2, b_2) \times \cdots [a_n, b_n)$ such that:

- For all i, $a_i, b_i \in \{0, 1, \ldots, m-1\}$ with $a_i \leq b_i$;
- $a_i = 0$ and $b_i = m - 1$ simultaneously hold for at least $n - d$ indices i (i.e., the rectangle is "nontrivial" in at most d dimensions).

Given such a rectangle $R \in \mathcal{GR}(m, d, n)$, its *volume* vol$(R)$ is defined to be $(\prod_{i=1}^{n}(b_i - a_i))/m^n$. A set $D \subseteq [0, m)^n$ is called a δ-*discrepant set* for $\mathcal{GR}(m, d, n)$ if:

$$\forall R \in \mathcal{GR}(m, d, n), \ | \frac{|D \cap R|}{|D|} - \text{vol}(R) \ | \leq \delta. \tag{1}$$

For an element $r = (r_1, r_2, \ldots, r_n) \in [0, m)^n$, define $\Gamma(r)$ to be the induced permutation $\pi_r \in S_n$ such that for any $0 \leq i, j \leq n - 1$, $\pi_r(i) < \pi_r(j)$ if and only if $r_i < r_j$, or $r_i = r_j$ but $i < j$. For a subset $D \subseteq [0, m)^n$, $\Gamma(D)$ is defined to be the multiset of $\Gamma(r)$ where $r \in D$.

Our main theorem is the following:

Theorem 1. *Let m be arbitrary. Suppose $D \subseteq [0, m)^n$ is any δ-discrepant set for $\mathcal{GR}(m, d, n)$. Then for any $\frac{1}{m} \leq \alpha < 1$, $\Gamma(D)$ is an (n, d, ϵ)-mwif, where $\epsilon = (\alpha + \frac{\delta}{\alpha})d^2$.*

Lu [7] gave an explicit construction of δ-discrepant sets for $\mathcal{GR}(m, d, n)$ of cardinality

$$(mn)^{O(1)} \cdot (1/\delta)^{O(\sqrt{\log(\max\{2, d/\log(1/\delta)\})})}.$$

Therefore, setting $m = 2d^2/\epsilon$, $\alpha = 1/m$ and $\delta = 1/m^2$ in the main theorem and invoking Lu's construction, we obtain the following corollary:

Corollary 2. *There exists an explicit construction of an (n, d, ϵ)-mwif of cardinality*

$$n^{O(1)} \cdot (d/\epsilon)^{O(\sqrt{\log(\max\{2, d/\log(1/\epsilon)\})})}.$$

Note that this size is poly(n) if $d \leq 2^{O(\log^{2/3} n)}$ and $\epsilon \geq 2^{-O(\log^{2/3} n)}$. Also, when $d = n$, our bound is better than that of [4] if $\epsilon \leq 2^{-c_0 \sqrt{\log n}}$, where $c_0 > 0$ is a certain absolute constant.

2 Proof of Main Theorem

Fix an arbitrary set $K \subseteq [n]$ of any size $k \leq d$, and choose any $x \in K$. We want to show that

$$| \Pr[\min\{\pi(K)\} = \pi(x)] - \frac{1}{k} \ | \leq \frac{\epsilon}{k},$$

where π is chosen uniformly at random from $\Gamma(D)$.

Assume without loss of generality that $t = 1/\alpha$ and αm are integers. Given x and K, we will define a sequence of pairwise disjoint rectangles $\{R_i = R_i(K, x) : 1 \leq i \leq t-1\}$ such that the permutations corresponding to points in $R = \cup_i R_i$ all satisfy $\min\{\pi(K)\} = \pi(x)$, and such that $\text{vol}(R)$ is approximately $\frac{1}{k}$. Using the fact that D is a good discrepant set for each R_i we will conclude that $\Gamma(D)$ has the required property.

We define R_i as follows.

$$R_i = \{(r_1, \ldots, r_n) \mid (i-1)\alpha m \leq r_x < i\alpha m; \ \forall y \in (K - \{x\}), \ i\alpha m \leq r_y < m;$$
$$\text{and } 0 \leq r_z < m \text{ for } z \notin K\}.$$

The following facts are easily seen:

1. For any $1 \leq i < j \leq t-1$, $R_i \cap R_j = \phi$.
2. $\text{vol}(R_i) = \alpha(1 - i\alpha)^{k-1}$.
3. For any $\pi \in \Gamma(R_i)$, $\min\{\pi(K)\} = \pi(x)$.

Define $R = \cup_{i=1}^{t-1} R_i$. Using the first two facts, we can lower bound the volume of R as follows:

$$\text{vol}(R) = \sum_{i=1}^{t-1} \text{vol}(R_i)$$

$$= \sum_{i=1}^{t-1} \alpha(1 - i\alpha)^{k-1}$$

$$\geq \int_1^t \alpha(1 - \alpha x)^{k-1} dx$$

$$= -\frac{1}{k}(1 - \alpha x)^k \mid_1^{1/\alpha}$$

$$= (1 - \alpha)^k / k$$

$$\geq \frac{1}{k} - \alpha.$$

Since D is a δ-discrepant set for $\mathcal{GR}(m, d, n)$, (1) shows that for each $1 \leq i \leq t-1$,

$$\mid \frac{|D \cap R_i|}{|D|} - \text{vol}(R_i) \mid \leq \delta.$$

Therefore,

$$\frac{|D \cap R|}{|D|} = \sum_{i=1}^{t-1} \frac{|D \cap R_i|}{|D|}$$

$$\geq \sum_{i=1}^{t-1} (\text{vol}(R_i) - \delta)$$

$$= \text{vol}(R) - (t-1)\delta$$

$$\geq \frac{1}{k} - (\alpha + \frac{\delta}{\alpha}).$$

Thus,

$$\Pr[\min\{\pi(K)\} = \pi(x)] \geq \frac{|D \cap R|}{|D|}$$

$$\geq \frac{1}{k} - (\alpha + \frac{\delta}{\alpha}).$$

Since this holds for any $x \in K$, an upper bound on this probability can be derived as follows:

$$\Pr[\min\{\pi(K)\} = \pi(x)] \leq 1 - (k-1)(\frac{1}{k} - (\alpha + \frac{\delta}{\alpha}))$$

$$\leq \frac{1}{k} + k(\alpha + \frac{\delta}{\alpha}).$$

Since $k \leq d$, this completes the proof of the theorem.

Acknowledgments. We thank Andrei Broder and Michael Mitzenmacher for helpful discussions. We also thank Leonard Schulman and Monica Vazirani for the references about k-wise independent permutations and for interesting discussions about them. Our thanks also to the RANDOM '99 referees for their helpful comments.

References

1. Alon, N., Babai, L., Itai, A.: A fast and simple randomized parallel algorithm for the maximal independent set problem. Journal of Algorithms **7** (1986) 567–583.
2. Broder, A. Z., Charikar, M., Frieze, A., Mitzenmacher, M.: Min-wise independent permutations. In Proc. ACM Symposium on Theory of Computing, pages 327–336, 1998.
3. Cameron, P. J.: Finite permutation groups and finite simple groups. Bull. London Math. Soc. **13** (1981) 1–22.
4. Indyk, P.: A small approximately min-wise independent family of hash functions. In Proc. ACM-SIAM Symposium on Discrete Algorithms, pages 454–456, 1999.
5. Joffe, A.: On a set of almost deterministic k-independent random variables. Annals of Probability **2** (1974) 161–162.
6. Karp, R.M., Wigderson, A.: A fast parallel algorithm for the maximal independent set problem. Journal of the ACM **32** (1985) 762–773.
7. Lu, C.-J.: Improved pseudorandom generators for combinatorial rectangles. In Proc. International Conference on Automata, Languages and Programming, pages 223–234, 1998.
8. Luby, M.: A simple parallel algorithm for the maximal independent set problem. SIAM J. Comput. **15** (1986) 1036–1053.
9. Naor, M., Reingold, O.: On the construction of pseudo-random permutations: Luby-Rackoff revisited. J. of Cryptology **12** (1999) 29–66.
10. Rees, E. G.: Notes on Geometry. Springer Verlag, 1983.
11. Takei, Y., Itoh, T., Shinozaki, T.: An optimal construction of exactly min-wise independent permutations. Technical Report COMP98-62, IEICE, 1998.

Independent Sets in Hypergraphs with Applications to Routing Via Fixed Paths

Noga Alon[1], Uri Arad[2], and Yossi Azar[3]

[1] Department of Mathematics and Computer Science, Tel-Aviv University.
noga@math.tau.ac.il [†]
[2] Dept. of Computer Science, Tel-Aviv University, Tel-Aviv, 69978, Israel.
uria@math.tau.ac.il
[3] Dept. of Computer Science, Tel-Aviv University, Tel-Aviv, 69978, Israel.
azar@math.tau.ac.il [‡]

Abstract. The problem of finding a large independent set in a hypergraph by an online algorithm is considered. We provide bounds for the best possible performance ratio of deterministic vs. randomized and non-preemptive vs. preemptive algorithms. Applying these results we prove bounds for the performance of online algorithms for routing problems via fixed paths over networks.

1   Introduction

The problem of finding the maximum independent set in a graph is a fundamental problem in Graph Theory and Theoretical Computer Science. It is well known that the problem is NP-hard ([18]), and that even the task of finding a rough approximation for the size of the maximum independent set is NP-hard ([4]). The intensive study of this problem includes the design and analysis of approximation algorithms ([10], [3]) and the investigation of online algorithms. The performance ratio of such an algorithm is the (worst case) ratio between the size of the maximum independent set, and the size (or expected size, when dealing with a randomized algorithm) of the independent set found by the algorithm.

In the online version of the maximum independent set problem the input graph is not known in advance, and the vertices arrive online. Here each vertex arrives with its incident edges towards previously presented vertices and the algorithm has to make an online decision if to add the vertex to the independent set. The adversary has the freedom to build the graph in any way he chooses.

The online algorithms can be deterministic or randomized. In addition, they can be non-preemptive or preemptive, where a preemptive algorithm may discard previously selected vertices (but may not pick a vertex that has already been discarded). This results in four basic online variants of the problem.

[†] Supported in part by the Israel Science Foundation, and by a USA-Israel BSF grant.
[‡] Research supported in part by the Israel Science Foundation and by the US-Israel Binational Science Foundation (BSF).

Here we extend the study of the online independent set problem from the domain of graphs to that of hypergraphs. We consider the case of k-uniform hypergraphs, where the hypergraph is not known in advance, and vertices are presented along with their edges. The first part of the paper contains lower and upper bounds for the performance ratio (usually called the competitive ratio) of online algorithms for these problems.

Besides being interesting in their own rights, the results on the performance of online algorithms for the hypergraph maximum independent set problem have nice applications in obtaining lower bounds for the performance of online algorithms for routing over networks via fixed paths. These applications are obtained by an on-line reduction, a notion that differs from the usual reduction and works in the online setting.

In the online routing problems considered here a network graph is given in advance and the algorithm is presented with requests for calls over given paths in the network. We refer to the throughput version of the problem in which each call is accompanied by a required bandwidth, and must be either allocated this bandwidth, or rejected. The goal is to maximize the weighted number of calls (that is, the total bandwidth) accepted by the network. Routing received a lot of attention recently with various results. We explore the relation between the hypergraph independent set problem and the routing problem, obtaining lower bounds for the performance of online algorithms for both. This relation also captures randomized and preemptive algorithms.

1.1 Independent sets in graphs and hypergraphs

The offline version of the problem is defined as follows. Given a hypergraph $G = (V, E)$, find a maximum subset of V such that the vertex induced subgraph on it does not contain any edge.

In the *online* version of the problem, vertices are presented one by one along with edges which connect them to previously presented vertices. The online algorithm must decide for each vertex, as it is presented, whether to accept it or not. The accepted set must induce an independent set at all times. The goal is to maximize the size of the selected set.

We consider deterministic or randomized algorithms. Our discussion will allow both preemptive and non-preemptive algorithms. In the preemptive version of the problem, the algorithm may discard previously selected vertices. However, a vertex which has been discarded, at any time, can not be returned to the set.

Deterministic, non-preemptive algorithms for the online graph independent set problem have been considered before. A well known folklore result states that any deterministic algorithm has a competitive ratio $\Omega(n)$ when the graph is not known in advance. Here we provide tight bounds for the competitive ratio of deterministic non-preemptive, randomized non-preemptive and deterministic preemptive algorithms for graphs as well as for hypergraphs. We also obtain upper and lower bounds for the randomized preemptive case. Note that our upper bound for the randomized preemptive case for hypergraphs is obtained using a polynomial time algorithm, and its performance bound matches the bound of the best known polynomial time approximation off-line algorithm that can be obtained using the methods of [9] (see also [17], [3]). To the best of our knowledge, this is the first online algorithm for the hypergraph independent set problem that achieves sub-linear competitive ratio. Note that by the result of [16] following [4], one cannot hope to obtain a much better bound even by an off-line polynomial time algorithm, unless NP have polynomial time randomized algorithms.

It is interesting to note that our polynomial time algorithm does not rely on the special properties of independent sets in uniform hypergraphs, but on the fact that being an independent set is a hereditary property. A property of subsets of a universe U is hereditary, if for every subset $A \subset U$ that satisfies it, every subset of A has the property as well. Hence, the same algorithm and upper bound hold for any hereditary property. In particular the upper bound holds for independent set in arbitrary hypergraphs, which are not necessarily uniform.

A related version of the online independent set problem deals with the model in which a graph (or a hypergraph) is known in advance to the algorithm, and a subset of vertices of it is presented by the adversary in an online manner. The goal here is also that of finding a large independent set, and the performance is measured by comparing it to the maximum size of an independent set in the induced subgraph on the presented vertices. It is quite easy to show that an $\Omega(n^\epsilon)$ lower bound holds for deterministic algorithms when the graph is known in advance where $\epsilon < 1$ is some fixed positive number. Bartal, Fiat and Leonardi [8] showed that an $\Omega(n^\epsilon)$ lower bound still holds when randomization and preemption are allowed.

1.2 Routing via fixed paths

Our results for the online hypergraph independent set problem can be applied in the study of the problem of (virtual circuit) routing over networks. Here the network graph is known in advance, and each edge has a known capacity. The algorithm is presented with requests for calls over the network with a certain required bandwidth. The algorithm either allocates this bandwidth on a path or rejects the call. The goal is to maximize the throughput (total bandwidth) of the accepted calls. Clearly, one may allow to use randomization and preemption. In the preemptive case accepted calls may be preempted, but preempted or rejected calls cannot be re-accepted. Obviously, calls preempted by an algorithm are not counted for the value of the throughput of this algorithm.

Two different versions for routing via fixed paths can be considered. In the first, the algorithm is presented with a request consisting of a source and a destination node, and must assign a route with the required bandwidth over the network to accept the call, while in the second version, each request includes a path to be routed over the network, and the algorithm may only decide to accept or reject the call.

There are numerous results for the virtual circuit routing problem for both versions (for surveys see [11, 15]). The competitive ratio of any deterministic (non-preemptive) algorithm has been shown to have an $\Omega(n)$ lower bound when the bandwidth request could be as large as the capacity. On the other hand, an $O(\log n)$-competitive deterministic routing algorithm has been shown for general networks when all bandwidth requirements are bounded by the network capacity over $\log n$ [2].

A lot of research has been invested to overcome the small capacity requirements for special networks such as lines, tress, meshes [13, 14, 7, 5, 6, 12, 1]. However, the problem of deciding whether randomized or preemptive algorithms can achieve poly-logarithmic bound for large bandwidth requests over general networks remained open. A major step has been taken by [8] that showed an $\Omega(n^\epsilon)$ lower bound for randomized preemptive online routing algorithms on general networks. Their lower bound holds for requests of maximal bandwidth, i.e. unit bandwidth for each request in a unit capacity network. The lower bound was proved by a reduction from the online maximum independent set problem in a known graph to the problem of routing calls over a network. The reduction does not extend for unit bandwidth and capacity k networks. In fact, it is still a major open problem to show a lower bound even for capacity 2.

Interestingly, we show a reduction between the independent set problem with an unknown graph and the fixed paths routing problem. Our reduction does extend for the case of capacity k. Specifically, we show a reduction from the independent set problem in a k uniform hypergraph to the fixed paths routing problem in a network of capacity $k - 1$. This enables us to obtain lower bounds for the latter problem by using our lower bounds for the performance of online algorithms for the hypergraph independent set problem. The reduction holds also for randomized and preemptive algorithms.

Our result covers the gap between the known results for unit bandwidth and logarithmic bandwidth by giving a lower bound that approaches the known results as the bandwidth grows from 1 to $\log n$.

1.3 The presented results

We show the following,

- For the Independent Set problem in k-uniform hypergraphs with n vertices,
  - A $\Theta(\frac{n}{k})$ tight lower bound for the competitive ratio of deterministic, deterministic preemptive or randomized non-preemptive algorithms.
  - An $\Omega(\frac{n^{1/2}}{k})$ lower bound for the competitive ratio of randomized preemptive algorithms.
  - An $O(\frac{n}{\log n})$ upper bound for the competitive ratio of randomized preemptive algorithms.
- For the fixed paths routing problem over a network of N vertices with capacity $k - 1$,
  - An $\Omega(\frac{N^{1/k}}{k})$ lower bound for the competitive ratio of deterministic, deterministic preemptive or randomized non-preemptive algorithm.
  - An $\Omega(\frac{N^{1/(2k)}}{k})$ lower bound for the competitive ratio of randomized preemptive algorithms.

2 Independent sets in k-uniform hypergraphs

As mentioned in the introduction, the algorithmic problem discussed in this section is the following. Given a k-uniform hypergraph $G = (V, E)$, with $V = \{v_1, v_2, \ldots, v_n\}$ and $E \subseteq 2^V$ ($\forall e \in E$, $|e| = k$), find a subset $V' \subseteq V$ of maximum cardinality such that for all $v_{i_1}, v_{i_2}, \ldots, v_{i_k} \in V'$: $(v_{i_1}, v_{i_2}, \ldots, v_{i_k}) \notin E$.

In the online version, the vertices are presented one by one, along with the edges which connect them to previously presented vertices.

2.1 A tight lower bound for online deterministic or randomized algorithms

Since G is a k-uniform hypergraph, any set of $k-1$ vertices forms an independent set. Therefore an upper bound of $\frac{n}{k-1} = O(\frac{n}{k})$ is trivially achievable. We now prove an $\Omega(\frac{n}{k})$ lower bound.

Theorem 1. *Any deterministic or randomized non-preemptive algorithm for the hypergraph independent set problem in a k-uniform hypergraph on n vertices has a competitive ratio $\Omega(\frac{n}{k})$.*

Proof. We use the online version of Yao's lemma by evaluating the performance of deterministic algorithms on a probability distribution on the inputs. Define the following probability distribution on the input sequences:

- Vertices are presented in pairs.
- One vertex of each pair will be selected randomly and marked as a "good" vertex, the other vertex will be marked as "bad".
- A set of k vertices containing a vertex from the current pair is an edge iff it contains at least one "bad" vertex from a previous pair.

Clearly, once the online algorithm picked one "bad" vertex, it can no longer pick more than $k - 2$ additional vertices. Note that, crucially, the two vertices in each pair are indistinguishable when they are presented. Therefore, whenever the online algorithm picks a vertex, the probability it is "bad" is $\frac{1}{2}$, regardless of the history. The expected number of vertices the algorithm picked until the first "bad" vertex is picked, is 2. Hence the expected size of the independent set it finds is at most $2 + (k - 2) = k$. The offline algorithm, on the other hand, can always pick all "good" vertices, yielding a competitive ratio of $\Omega(\frac{n}{k})$.

2.2 A tight lower bound for online deterministic preemptive algorithms

Theorem 2. *Any deterministic preemptive algorithm for the hypergraph independent set problem for k-uniform hypergraphs on n vertices has a competitive ratio $\Omega(\frac{n}{k})$.*

Proof. We define the following input sequence:

- Vertices are presented in steps. In each step there are $2k - 2$ vertices such that any subset of k of them is an edge.
- At most $k - 1$ vertices from each step will be selected as "bad" vertices, all the other vertices will be marked as "good".
- A set of k vertices that contains vertices from the current step and previous steps is an edge if it contains at least one "bad" vertex from a previous step.

The deterministic algorithm may choose at most $k-1$ vertices from each step. The adversary will mark them as "bad" and all other vertices (at least $k - 1$) as "good". Therefore all the vertices which may be selected by the online algorithm are "bad", and may be replaced, by preemption, only by other "bad" vertices. By the construction of the sequence the online algorithm may hold a maximum of $k - 1$ vertices at any time without having an edge (at most $k - 1$ from one step or at most $k - 1$ from several steps). However, the optimal algorithm will collect all "good" vertices, thus creating an independent set of at least $\frac{n}{2}$ vertices.

2.3 A lower bound for online randomized preemptive algorithms

We prove a lower bound of $\Omega(\frac{\sqrt{n}}{k})$ for the competitive ratio of any randomized preemptive on-line algorithm. We make use of Yao's lemma to establish a lower bound for any deterministic algorithm on a given probability distribution, thus yielding a lower bound for the randomized case.

Theorem 3. *Any randomized preemptive on-line algorithm for the online independent set problem for k-uniform hypergraphs on n vertices has competitive ratio $\Omega(\frac{\sqrt{n}}{k})$.*

Proof. Define the following probability distribution on the input sequences. Each sequence will be constructed of vertices, presented in steps. Each step consists of l vertices, with a total of n vertices in all steps. Each step will be generated according to the following distribution:

- At step j, l vertices are presented such that any subset of k of them is an edge.
- One vertex chosen uniformly at random will be marked as a "good" vertex, while all others will be marked as "bad".
- A set of k vertices that contains vertices from the current step and previous steps is an edge iff it contains at least one "bad" vertex from a previous step.

For the proof, we reveal at the end of each step, which is the "good" vertex, thus giving the algorithm the opportunity to immediately discard all "bad" vertices, at the beginning of the next step. Note that all the vertices in each step look indistinguishable given all the history since they participate in exactly the same edges. Thus, there is no way for the algorithm to distinguish between the "good" and the "bad" vertices before the step ends. Therefore, at the end of each step, the algorithm may hold any number of "good" vertices from previous steps, plus a set of at most $k - 1$ additional vertices. Some of these additional vertices may be "bad" vertices from previous steps, and some may belong to the current step. The probability of the algorithm to select each "good" vertex in a step is at most $\frac{k-1}{l}$, regardless of previous selections. The expected benefit of the algorithm is thus:

$$E(\text{ON}) \le \frac{n}{l} \cdot \frac{k-1}{l} + k - 1 \le \frac{nk}{l^2} + k$$

On the other hand, the optimum algorithm OPT may pick all the "good" vertices, giving a benefit of at least $\frac{n}{l}$. Choosing, optimally, $l = \sqrt{n}$ we get a competitive ratio of $\Omega(\frac{\sqrt{n}}{k})$.

2.4 A sublinear upper bound

Here we present a randomized, preemptive algorithm for the independent set problem in an arbitrary (not necessarily uniform) hypergraph and show that its competitive ratio is $O(n/\log n)$. The algorithm also runs in polynomial time.

Given an input sequence of n vertices, the algorithm divides the sequence into groups of y vertices each. Each of these groups will be called a phase. At the beginning of each phase we uniformly select at random β distinct vertices of that phase. During the phase we pick all selected vertices, as long as they induce an independent set. If they do not induce an independent set, then the phase fails, and we drop all the vertices but one. If the phase succeeds we stop. Otherwise, we start another phase and replace the one remaining vertex with the first selected vertex of the next phase. We assume that the portion of the maximal independent set size, x, is known in advance (i.e. the set contains n/x vertices). Later we use an appropriate weighted version of classify and randomly select to relieve this restriction.

Claim. For each $4 \le x \le \log n$ define $\beta = \frac{\log n}{4 \log x}$ and $y = 4\beta x$. Then our algorithm picks an independent set of size β with high probability, or there is no independent set of size $\frac{n}{x}$ in the graph.

Proof. We assume that there is an independent set in the graph, consisting of at least $\frac{n}{x}$ vertices. We distinguish between phases with a lot of vertices from the set, and those with few vertices from the independent set. Phases with more than $\frac{y}{2x}$ vertices are good phases. There are at least $\frac{n}{2xy}$ good phases, otherwise the total number of vertices in the independent set is less than $\frac{n}{2xy} \cdot y + \frac{n}{x} \cdot \frac{y}{2x} = \frac{n}{x}$ in contradiction to the size of the independent set. From each good set we select β vertices at random. Since $y = 4\beta x$, each of these vertices has a conditional probability greater than $\frac{y/2x - \beta}{y} = \frac{1}{4x}$ of being a vertex from the independent set, given that all the previously selected vertices are from the independent set. Therefore the probability of failure is less than $1 - (\frac{1}{4x})^\beta$, for each good phase. Since we have $\frac{n}{2xy}$ good phases, the total probability of failure is bounded by $\left(1 - (\frac{1}{4x})^\beta\right)^{\frac{n}{2xy}}$. As $\beta = \frac{\log n}{4 \log x}$ and $y = 4\beta x$, we get that the probability of failure is less than

$$\left(1-(\frac{1}{4x})^\beta\right)^{\frac{n}{2\pi y}} \le \left(1-\frac{1}{n^{1/2}}\right)^{\frac{n}{2\pi y}} \le e^{\frac{-n^{1/2}}{2\pi y}} \le e^{-n^{0.49}}$$

Theorem 4. *There exists a randomized preemptive algorithm which achieves a competitive ratio of $O(\frac{n}{\log n})$ for the independent set problem in an arbitrary hypergraph on n vertices.*

Proof. We use classify and randomly select. Divide the range of x from 4 to $\log n$ into classes by powers of two, and assign a probability to each such class. For the class of $2^{i-1} \le x < 2^i$ we assign probability proportional to $\frac{i}{2^i}$.
Using the above algorithm for the chosen class we get an algorithm for which:

- If $\text{OPT}(\sigma) \le \frac{n}{\log n}$, then $\text{ON}(\sigma) \ge 1$ and the competitive ratio is at most $\frac{n}{\log n}$.
- If $\text{OPT}(\sigma) = \frac{n}{x} > \frac{n}{\log n}$, then $E(\text{ON}(\sigma)) \ge \Omega(\frac{\log x}{x}) \cdot \Theta(\frac{\log n}{\log x}) = \Omega(\frac{\log n}{x})$, and again the competitive ratio is at most $O(\frac{n}{\log n})$.

Note that if the length of the sequence is unknown we may use a technique similar to the standard doubling techniques by selecting an initial value for n and then squaring its value if the sequence turns out to be too long. Squaring ensures that only a small portion of the independent set will be processed using a wrong value of n, while having the wrong value for n (by at most a power of 2) will only add a constant factor to the competitive ratio. To avoid the sequence from ending just as we update the value of n, we use a simple boundary smoothing technique, such as randomly selecting a multiplicative factor between 1 and 2, and multiplying the updated value by this factor.

3 Routing via fixed paths in constant capacity networks

We next show a reduction from the independent set problem for k-uniform hypergraphs to routing with fixed paths over a $k-1$ bandwidth network. The reduction step translates vertices into paths over the given graph, while making sure that any hyperedge results in an inconsistent set of calls. The reduction yields lower bounds for the routing problem.

Note that while the hypergraph was unknown to the algorithm in the independent set problem, the network structure is known in advance in the case of routing. The process of adding a new (unknown) vertex of the hypergraph while revealing the edges which connect it to previously presented vertices, corresponds to the process of presenting a new path, to be allocated over the network.

A vertex v is called the *completing vertex* of an edge e, if all the other vertices of e were presented before v, and thus the edge e was revealed at the appearance of v.

3.1 The reduction step

Let $G = (V, E)$ be a k-uniform hypergraph, with $V = \{c_1, c_2, \ldots, c_n\}$, and assume the vertices are presented in this order. We construct a graph $G' = (V', E')$, where each edge has capacity $k - 1$ and a set of paths $P = \{p_1, p_2, \ldots, p_n\}$, such that

1. Each vertex $c_i \in V$ corresponds to the path p_i.
2. For every set of paths $p_{i_1}, p_{i_2}, \ldots, p_{i_k}$, there exists an edge $e \in E'$ such that $e \in p_{i_j}, \forall j$ if and only if $(c_{i_1}, c_{i_2}, \ldots, c_{i_k}) \in E$.

Note that the reduction we present is an on-line reduction, and not a standard reduction. In the on-line reduction the network, and the paths are built as the algorithm advances, without knowing what are the actions of the algorithm, and how the sequence will continue, while in a regular "offline" reduction the whole sequence in known in advance. Moreover, in a standard reduction, any input sequence might result in a completely different image, even if the sequences have a common prefix. In an on-line reduction, on the other hand, we must allow any prefix to be completed in every possible way without restricting it by the reduction process itself.

G' consists of $2n + 2\binom{n}{k}$ vertices, and $2n\binom{n}{k} + \binom{n}{k}^2$ edges. First we construct a graph with $\binom{n}{k}$ independent edges. Each edge has a unique label $(i_1, i_2, \ldots, i_k)$ where $i_j < i_{j+1} \forall j$, i_k is called the last coordinate in the edge label. We refer to these edges as *restricting* edges. We assign a left and a right vertex to each restricting edge. The right end of each edge is connected to the left end of every other restricting edge, we refer to these edges as *connecting* edges. We then add two sets of vertices s_i and t_i $i = 1 \ldots n$ with s_i connected to all left ends, and t_i connected to all right ends (see Figure 1).

For each vertex c_i we assign the following path p_i; starting from s_i, we pass through all edges containing i in their label not as their last coordinate, and through all the edges labeled $(i_1, i_2, \ldots, i)$ with i as the last coordinate of the label if and only if $(c_{i_1}, c_{i_2}, \ldots, c_i) \in E$. Note that c_i is the completing vertex of the edge $(c_{i_1}, c_{i_2}, \ldots, c_i)$. Finally we connect the last edge to the vertex t_i. Starting from s_i, the path will enter each edge through its left vertex, and leave through its right vertex.

To complete the reduction we prove the following lemma,

Lemma 1. *In the resulting graph, for every set of paths $p_{i_1}, p_{i_2}, \ldots, p_{i_k}$, there exists an edge $e \in E'$ such that $e \in p_{i_j}, \forall j$ if and only if $(c_{i_1}, c_{i_2}, \ldots, c_{i_k}) \in E$.*

Proof. We first show that each connecting edge is used by at most $k - 1$ different paths. Each path passing through a connecting edge must pass through the restricting edges connected to it. A path p_i uses a certain restricting edge only if its index is one of the coordinates in the label of that edge, but two restricting edges may share at most $k - 1$ coordinates. Thus no connecting edge is used more than $k - 1$ times.

Therefore, we limit our attention to restricting edges. Consider the paths going through the restricting edge $e \in E'$ whose label is $(i_1, i_2, \ldots, i_k)$. All the

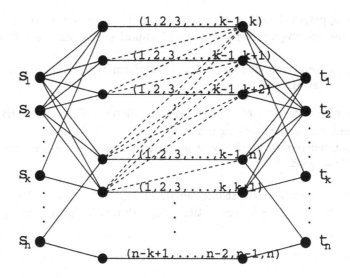

Fig. 1. The structure of G'. Labeled edges are the restricting edges, dashed lines represent connecting edges.

paths $p_{i_1}, p_{i_2}, \ldots, p_{i_{k-1}}$ pass through the edge, as their index is not the last coordinate in the edge label. Therefore, all edges are used to their maximum capacity by the paths corresponding to the first $k-1$ coordinates in their label. According to our construction, the last path p_{i_k} goes through this edge if and only if $(c_{i_1}, c_{i_2}, \ldots, c_{i_k}) \in E$, thus creating an inconsistency.

3.2 The resulting lower bounds

Theorem 5. *The following lower bounds hold for the online routing problem with fixed paths, over a network with N vertices and constant capacity $k-1$.*

- *Any deterministic, deterministic preemptive or randomized non-preemptive on-line algorithm has competitive ratio $\Omega(\frac{N^{1/k}}{k})$.*
- *Any randomized preemptive on-line algorithm has competitive ratio $\Omega(\frac{N^{1/(2k)}}{k})$.*

Proof. By the above lemma, any online algorithm for the fixed paths routing problem over a network with capacity $k-1$, is also an algorithm for the independent set problem over a k-uniform hypergraph. Each path selected matches a vertex in the hypergraph, and vice-versa. Moreover, any independent set in the hypergraph defines a set of consistent paths in the network, and any set of consistent paths defines an independent set. Therefore, any algorithm(online or offline) which achieves a value of $A(\sigma)$ for the network routing problem, may be used to build an independent set of the same size in the hypergraph. Thus, any lower bound on the competitive ratio for the independent set problem for a

k-uniform hypergraph with n vertices, is also a lower bound for the competitive ratio for the routing problem on a $k - 1$ capacity network with $N = \Theta(n^k)$ vertices.

Using the lower bounds found for the k-uniform hypergraph problem, we get the following set of lower bounds:

- The competitive ratio of any deterministic, deterministic preemptive or randomized non-preemptive algorithm is $\Omega(\frac{n}{k}) = \Omega(\frac{N^{1/k}}{k})$.
- The competitive ratio of any randomized preemptive on-line algorithm, is $\Omega(\frac{n^{1/2}}{k}) = \Omega(\frac{N^{1/(2k)}}{k})$.

Note that the lower bound becomes smaller than $\log N$ for $k = \Theta(\log N)$. This conforms with the $O(\log N)$ competitive algorithm of [2] for $k = \Theta(\log n)$.

References

[1] R. Adler and Y. Azar, *Beating the logarithmic lower bound: randomized preemptive disjoint paths and call control algorithms*, Proc. 10th ACM-SIAM Symp. on Discrete Algorithms, 1999, pp. 1–10.

[2] B. Awerbuch, Y. Azar, and S. Plotkin, *Throughput-competitive online routing*, 34th IEEE Symposium on Foundations of Computer Science, 1993, pp. 32–40.

[3] N. Alon and N. Kahale, *Approximating the independence number via the θ-function*, Math. Programming 80 (1998), 253-264.

[4] S. Arora, C. Lund, R. Motwani, M. Sudan, and M. Szegedy, *Proof verification and intractability of approximation problems*, Proc. of the 33rd IEEE FOCS, IEEE (1992), pages 14–23.

[5] B. Awerbuch, Y. Bartal, A. Fiat, and A. Rosén, *Competitive non-preemptive call control*, Proc. of 5th ACM-SIAM Symposium on Discrete Algorithms, 1994, pp. 312–320.

[6] B. Awerbuch, R. Gawlick, T. Leighton, and Y. Rabani, *On-line admission control and circuit routing for high performance computation and communication*, Proc. 35th IEEE Symp. on Found. of Comp. Science, 1994, pp. 412–423.

[7] A. Bar-Noy, R. Canetti, S. Kutten, Y. Mansour, and B. Schieber, *Bandwidth allocation with preemption*, Proc. 27th ACM Symp. on Theory of Computing, 1995, pp. 616–625.

[8] Y. Bartal, A. Fiat, and S. Leonardi, *Lower bounds for on-line graph problems with application to on-line circuit and optical routing*, Proc. 28th ACM Symp. on Theory of Computing, 1996, pp. 531–540.

[9] B. Berger and J. Rompel, *A better performance guarantee for approximate graph coloring*, Algorithmica 5(1990),459–466.

[10] R. Boppana and M. M. Halldorsson, *Approximating maximum independent sets by excluding subgraphs*, BIT 32 (1992), 180–196.

[11] A. Borodin and R. El-Yaniv, *Online computation and competitive analysis*, Cambridge University Press, 1998.

[12] R. Canetti and S. Irani, *Bounding the power of preemption in randomized scheduling*, Proc. 27th ACM Symp. on Theory of Computing, 1995, pp. 606–615.

[13] J.A. Garay and I.S. Gopal, *Call preemption in communication networks*, Proceedings of INFOCOM '92 (Florence, Italy), vol. 44, 1992, pp. 1043–1050.

[14] J. Garay, I. Gopal, S. Kutten, Y. Mansour, and M. Yung, *Efficient on-line call control algorithms*, Journal of Algorithms **23** (1997), 180–194.

[15] S. Leonardi, *On-line network routing*, Online Algorithms - The State of the Art (A. Fiat and G. Woeginger, eds.), Springer, 1998, pp. 242–267.

[16] J. Håstad, *Clique is hard to approximate within $n^{1-\epsilon}$*, Proc. 37th IEEE FOCS, IEEE (1996), 627 – 636.

[17] T. Hofmeister and H. Lefmann, *Approximating maximum independent sets in uniform hypergraphs*, Proc. of the 23rd International Symposium on Mathematical Foundations of Computer Science, Lecture Notes in Computer Science 1450, Springer Verlag (1998), 562-570.

[18] R. Karp, *Reducibility among combinatorial problems*, Plenum Press, New York, 1972, Miller and Thatcher (eds).

Approximating Minimum Manhattan Networks

(Extended Abstract)

Joachim Gudmundsson[1] *, Christos Levcopoulos[1] †, and Giri Narasimhan[2] ‡

[1] Dept. of Computer Science, Lund University, Box 118, 221 00 Lund, Sweden.
[2] Dept. of Mathematical Sciences, The Univ. of Memphis, Memphis, TN 38152, USA.

Abstract. Given a set S of n points in the plane, we define a *Manhattan Network* on S as a rectilinear network G with the property that for every pair of points in S, the network G contains the shortest rectilinear path between them. A *Minimum Manhattan Network* on S is a Manhattan network of minimum possible length. A Manhattan network can be thought of as a graph $G = (V, E)$, where the vertex set V corresponds to points from S and a set of steiner points S', and the edges in E correspond to horizontal or vertical line segments connecting points in $S \cup S'$. A Manhattan network can also be thought of as a 1-spanner (for the L_1-metric) for the points in S.

Let R be an algorithm that produces a rectangulation of a staircase polygon in time $R(n)$ of weight at most r times the optimal. We design an $O(n \log n + R(n))$ time algorithm which, given a set S of n points in the plane, produces a Manhattan network on S with total weight at most $4r$ times that of a minimum Manhattan network. Using known rectangulation algorithms, this gives us an $O(n^3)$-time algorithm with approximation factor four, and an $O(n \log n)$-time algorithm with approximation factor eight.

1 Introduction

A *rectilinear path* connecting two points in the plane is a path consisting of only horizontal and vertical line segments. A rectilinear path of minimum possible length connecting two points will be referred to as a *Manhattan path*, where the length of a rectilinear path is equal to the sum of the lengths of its horizontal and vertical line segments. Manhattan paths are monotonic. The *Manhattan distance* (or L_1-distance) between two points in the plane is the length of the Manhattan path connecting them. In this paper we introduce the concept of geometric networks that guarantee Manhattan paths between every pair of points from a given set of points.

Consider a set S of points in the plane. A *geometric network* on S can be modeled as an undirected graph $G = (V, E)$. The vertex set V corresponds to

---

* email: Joachim.Gudmundsson@cs.lth.se
† email: Christos.Levcopoulos@cs.lth.se
‡ email: giri@msci.memphis.edu; Funded by grants from NSF (CCR-940-9752) and Cadence Design Systems, Inc.

the points in $S \cup S'$, where S' is a set of newly added Steiner points; the edge set E corresponds to line segments joining points in $S \cup S'$. If all the line segments are either horizontal or vertical, then the network is called a *rectilinear geometric network*. Each edge $e = (a, b) \in E$ has length $|e|$ that is defined as the Euclidean distance $|ab|$ between its two endpoints a and b. The total length of a set of edges is simply the sum of the lengths of the edges in that set. The *total length* of a network $G(V, E)$ is denoted by $|E|$. For $p, q \in S$, a pq-path in G is a path in G between p and q.

For a given set S of n points in the plane, we define a *Manhattan Network* on S as a rectilinear geometric network G with the property that for every pair of points $p, q \in S$, the network G contains a Manhattan pq-path connecting them. A *Minimum Manhattan Network* on S is a Manhattan network of minimum possible length.

The complete grid on the point set S is clearly a Manhattan network. In other words, the network obtained by drawing a horizontal line and a vertical line through every point in S and by considering only the portion of the grid inside the bounding box of S is a network that includes the Manhattan path between every pair of points in S. It is easy to show that the minimum Manhattan network on S need not be unique and that the complete grid on the point set S can have total weight $O(n)$ times that of a minimum Manhattan network on the same point set. Figures 1(a) and (b) below show examples of a Manhattan

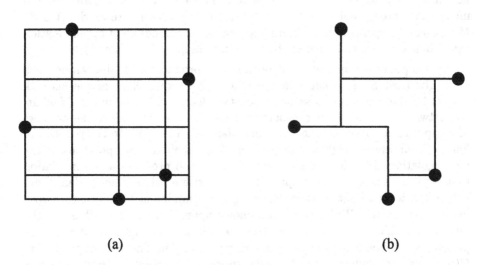

(a)   (b)

Fig. 1. (a) A Manhattan network, and (b) A minimum Manhattan network.

network and a minimum Manhattan network on the same set of points. In fact, the network in Figure 1(a) is also a complete grid on the input points.

Many VLSI circuit design applications require that a given set of terminals in the plane be connected by networks of small total length. Rectilinear steiner minimum trees were studied in this context. Manhattan networks impose additional constraints on the distance between the terminals in the network. The concept of Manhattan networks seems to be a very natural concept; it is surprising that this concept has not been previously studied. Manhattan networks are likely to have many applications in geometric network design and in the design of VLSI circuits.

Manhattan networks are also closely connected to the concept of spanners. Given a set S of n points in the plane, and a real number $t>1$, we say that a geometric network G is a *t-spanner* for S for the L_p-norm, if for each pair of points $p, q \in S$, there exists a pq-path in G of length at most t times the L_p-distance between p and q. In this connection, a minimum Manhattan network can be thought of as a sparsest 1-spanner for S for the L_1-norm, assuming that Steiner points are allowed to be added. 1-spanners are also interesting since they represent the network with the most stringent distance constraints. However, note that the sparsest 1-spanner for S in the L_p-norm (for $p \geq 2$) is the trivial complete graph on S. It is also interesting to note that a Manhattan network can be thought of as a $\sqrt{2}$-spanner (with Steiner points) for the L_2 norm. Although complete graphs represent ideal communication networks, they are expensive to build; sparse spanners represent low cost alternatives. The weight of the spanner network is a measure of its sparseness; other sparseness measures include the number of edges, maximum degree and the number of steiner points. Spanners have applications in network design, robotics, distributed algorithms, and many other areas, and have been a subject of considerable research [1, 3–6]. More recently, spanners have found applications in the design of approximation algorithms for problems such as the traveling salesperson problem [2, 9].

In this paper we present an algorithm that produces a Manhattan network for a given set S of n points in the plane. The total weight of the network output by the algorithm is within a constant factor of the minimum Manhattan network. It is interesting to note that in this paper we reduce the problem of computing an approximate minimum Manhattan network to the problem of finding a minimum-weight rectangulation of a set of staircase polygons. If the rectangulation algorithm runs in time $O(R(n))$ and produces a rectangulation that is within a factor r of the optimal, our algorithm will produce a Manhattan network of total weight $4r$ times the weight of a minimum Manhattan network in time $O(n \log n + R(n))$. Using two known approximation algorithms for the minimum-weight rectangulation problem, we obtain two algorithms for the approximate minimum Manhattan network problem. The first algorithm runs in $O(n^3)$ time and produces a Manhattan network of total weight at most four times that of a minimum Manhattan network. The second algorithm runs in $O(n \log n)$ time and has an approximation factor of eight. It is unknown whether the problem of computing the minimum Manhattan network is a NP-hard problem. It is also unknown whether a polynomial-time approximation scheme exists for this problem.

A noteworthy feature of our result is that unlike most of the results on *t*-spanners, we compare the output of our algorithm to that of minimum Manhattan networks and our results involve small constants (4 or 8). Most results on sparse t-spanners prove weight bounds that compare it to the length of a minimum spanning tree; the constants involved in those results are usually very large.

In Section 3, we present the approximation algorithm. In Section 3.2 we prove that the algorithm produces a Manhattan network; in Section 3.3 we prove that the network produced is of weight at most $4r \times |E_{opt}|$, where E_{opt} is the set of edges in a minimum Manhattan network on S.

2 Definitions

Let u and v be two points in S, where u lies to the left of and above v. A L-path between u and v is a rectilinear path consisting of one vertical line segment with upper endpoint at u, and one horizontal line segment with right endpoint at v. Note that such a path is a minimum-weight, minimum-link path connecting u and v. The ⌐-path is defined in a similar fashion, as shown in Fig. 2. If u lies to the left of and below v, we define a ⌐-path and a ⌐-path in a similar fashion. Note that each of the four paths described above introduces one steiner point at the bend. We will denote by $[v, u]$ the closed region described by a rectilinear rectangle with corners in v and u. The coordinates of a vertex $v \in V$ are denoted $v.x$ respectively $v.y$.

Fig. 2. (left) An L-path, and (right) an ⌐-path.

If two different edges of the graph intersect, we will assume that their intersection defines a Steiner point.

3 The approximation algorithm

In this section we present an approximation algorithm to construct a small length Manhattan network $G' = (V', E')$. The algorithm will construct the edge set in four independent steps. In each step, for each vertex, the algorithm constructs a (possibly empty) local network connecting that vertex to a set of chosen "neighboring" vertices. Since the four steps are symmetrical the first step is explained in more detail than the later steps.

1. Sweep the points from left to right. As shown in Fig. 3(a), for each point $v' \in V$ let v be the leftmost point below and not to the left of v' (if several, take uppermost). We say that v' *1-belongs to* v. For each vertex v let $B_1(v)$

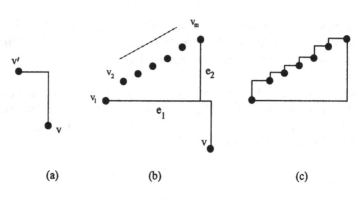

Fig. 3.

denote the set of vertices in V that 1-belong to v. Let $v_1, \ldots, v_m$ be the vertices in $B_1(v)$ ordered from left to right, as shown in Fig. 3(b). First, construct an $\urcorner$-path, denoted by e_1, connecting v_1 and v. If $m > 1$, draw a vertical edge e_2, with top endpoint at v_m and bottom endpoint on e_1. Now, we construct a "local" Manhattan network such that there is a Manhattan path from each vertex v_i, $2 \le i \le v_{m-1}$, to e_1 or e_2; this step is explained in more detail in Section 3.1. Add the edges constructed in this step to the set of edges E_1'.

2. Sweep the points from left to right. For each point $v' \in V$ let v be the leftmost point above and not to the left of v' (if several, take bottommost). We say that v' *2-belongs to* v. For each vertex v let $B_2(v)$ denote the set of vertices in V that 2-belong to v. Perform the symmetrical procedure as performed in step 1 on the set $B_2(v)$ for every $v \in V$, to obtain the set of edges E_2'.

3. Sweep the points from bottom to top. For each point $v' \in V$ let v be the bottommost point to the left and not below v' (if several, take leftmost). We say that v' *3-belongs to* v. For each vertex v let $B_3(v)$ denote the set of vertices in V that 3-belong to v. Perform the symmetrical procedure as performed in step 1 on the set $B_3(v)$ for every $v \in V$, to obtain the set of edges E_3'.

4. Sweep the points from top to bottom. For each point $v' \in V$ let v be the topmost point to the left and not above v' (if several, take leftmost). We say that v' *4-belongs to* v. For each vertex v let $B_4(v)$ denote the set of vertices in V that 4-belong to v. Perform the symmetrical procedure as performed in step 1 on the set $B_4(v)$ for every $v \in V$, to obtain the set of edges E_4'.

After building the appropriate "local" networks, we say that every vertex v is *directly connected* to the vertices in $B_1(v) \cup \ldots \cup B_4(v)$. From these four sweeps we get four edge sets, $E_1', \ldots, E_4'$. Now the Manhattan network is defined as $G = (V', E')$, where $E' = E_1' \cup \ldots \cup E_4'$, and V' includes the points in S and all the Steiner points that are generated when adding the edges in E'.

Also note that there is some asymmetry in the above construction. This is deliberate; the asymmetric cases are required in the proof of correctness of the algorithm (see Lemma 3).

3.1 Constructing the local networks

We will explain the construction of the local networks involved in step 1 of the algorithm in detail. The constructions involved in the other steps are symmetrical. Let v be an arbitrary vertex of V and let $v_1, \ldots, v_m$ be the vertices in $B_1(v)$. By step 1 of the algorithm, we note that v_i lies below and to the left of v_{i+1}, $1 \leq i < m$; thus, $v.x \geq v_m.x$ and $v.y \leq v_1.y$.

In this section we describe how to construct a local network connecting v with the vertices $v_1, \ldots, v_m$; the local network is a Manhattan network on $v, v_1, \ldots, v_m$. We assume that $m > 2$, otherwise we are done. Recall that v_1 and v are connected by an ⌐-path, e_1, and that v_m is connected to this edge by a vertical segment corresponding to edge e_2, as shown in Fig. 3(b). Let e'_1 be the horizontal part of e_1 between v_1 and e_2. Our aim is to produce a set of edges of minimum total weight such that there is a Manhattan path between v_i and v, $1 \leq i \leq m$. This already holds for v_1 and v_m. Consider the following staircase polygon obtained by adding ⌐-paths between v_i and v_{i+1}, $1 \leq i < m$, to the base e'_1 and the right side e_2, as shown in Fig. 3(c). This polygon will be referred to as the C-hull of the set of vertices $\{v_1, \ldots, v_m\}$. We claim that a rectangulation of this polygon would give us a rectilinear network that guarantees Manhattan paths from v to every vertex in the set $\{v_1, \ldots, v_m\}$. This claim is easily proved by observing that every vertex $v_i, i = 1, \ldots, m$, must lie on the corner of a distinct rectangle (hence you should be able to proceed either down or to the right from v_i), and there always exists a monotonic rectilinear path from v_i to v that follows the borders of the rectangles encountered along the way. Denote by $E'_1(v)$ the set of edges constructed in the rectangulation of the C-hull plus the two edges e_1 and e_2. That is, $E'_1(v)$ is the set of edges produced in step 1 to connect the vertices in $B_1(v)$ to v. It should be noted that $E'_1(v)$ only consists of the edges in the interior of the C-hull of the set of vertices $\{v_1, \ldots, v_m\}$ and that the ⌐-paths from v_i to v_{i+1} are not included in it. While a rectangulation of these C-hulls guarantees Manhattan paths, we show later that a minimum-weight rectangulation of the C-hulls gives us an approximation algorithm for the problem. The following results on minimum-weight rectangulations were previously known:

Theorem 1. *([8]) An optimal rectangulation of a staircase polygon can be computed in time $O(n^3)$.*

Theorem 2. *([7]) A thickest-first rectangulation of a staircase polygon can be computed in linear time such that the weight of the added edges in the rectangulation is at most twice the weight of an optimal rectangulation.*

Since a sweep takes $O(n \log n)$, if the optimal rectangulation procedure is used the time-complexity of our approximation algorithm is $O(n^3)$, and if the thickest-first rectangulation algorithm is used then our algorithm runs in time $O(n \log n)$.

It remains to prove that the graph $G'=(V',E')$ is a Manhattan network for the points in S and that $|E'| \leq 4r \times |E_{opt}|$, where r is the approximation factor of the rectangulation algorithm.

3.2 The Algorithm outputs a Manhattan network

To show that the algorithm outputs a Manhattan network, it suffices to prove the following lemma:

Lemma 3 *For each pair of points $s,t \in V$ there is a Manhattan path in G' connecting s and t.*

Proof. Without loss of generality, either t lies to the right and above s, or t lies to the right and below s (if not, switch s and t). We first assume that t lies to the right and above s. Without loss of generality, we may assume that no two points have the same x- or y-coordinate, since the algorithm always connects two such points by a line segment.

Consider $[s,t]$, the rectangle with corners at s and t. If $s \in B_2(t)$ or $t \in B_4(s)$, then s is directly connected to t and we are done. Otherwise, we know from the algorithm that s is directly connected to a point $s_1 \in V$ above and to the right of s and to the left of t (Case 4 of construction), and that t is directly connected to a point $t_1 \in V$ below and to the left of t and above s (Case 2 of construction). We consider the following two cases:

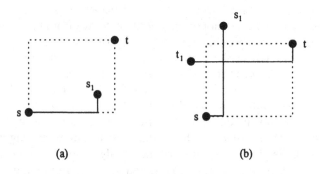

(a)   (b)

Fig. 4.

Case 1 [s_1 or t_1 lies within $[s,t]$]: Without loss of generality, assume that s_1 lies within $[s,t]$, as shown in Fig. 4(a). In this case, we let s_1 be the new s and continue recursively.

Case 2 [s_1 and t_1 lie outside $[s,t]$]: Then we know that s_1 lies above and to the left of t and that t_1 lie above and to the left of s, as shown in Fig. 4(b). The Manhattan path connecting s and s_1 must intersect the Manhattan path connecting t and t_1 within $[s, t]$. Hence there is a Manhattan path connecting s and t.

The case when t lies to the right and below s uses Cases 1 and 3 of the construction, and is otherwise similar to the proof above. Hence the lemma. □

3.3 The total length of the output network is $\leq 4r \times |E_{opt}|$

For the length analysis, once again, it is sufficient to consider only one sweep, i.e., step 1 of the algorithm. The approximation factor for this sweep is then multiplied by four to obtain the approximation factor for the algorithm. Let v be an arbitrary vertex of V. Let $B_1(v) = \{v_1, \ldots, v_m\}$, let $V' = B_1(v) \cup \{v\}$. We define a *charging area* of v (with respect to this sweep) as the region $\cup_{i=1}^{m}[v, v_i]$, as shown by the shaded region in Fig. 5(a). The charging area is denoted by $C_1(v)$. Note that the interior of the charging area for any vertex must be devoid of input points. We start with the following observation:

Lemma 4 *For every pair of vertices $v_i, v_j \in V$, the charging areas $C_1(v_i)$ and $C_1(v_j)$ are disjoint, except possibly for the point v_i or v_j.*

Proof. Since no vertex can 1-belong to more than one vertex, the staircase parts of the two charging areas cannot share any vertices. Thus either v_i is on the staircase part of $C_1(v_j)$ or vice versa. But then, the rest of the charging areas cannot overlap because of its shape and orientation. □

It is important to point out that the charging areas may share a point, but cannot share an edge of the boundary. Also note that all edges of $E_1'(v)$ that were added in step 1 must lie entirely within $C_1(v)$. Hence, the edges produced in step 1 connecting vertices in V' cannot be used to connect vertices in any other charging areas. We will prove that $|E_1'(v)|$ is at most equal to the length of the edges in E_{opt} lying within the charging area of v. Since the charging areas are disjoint this implies that the total length of the edges produced in step 1 of the algorithm, $|E_1'| = \sum_{v \in V} |E_1'(v)|$, is at most $|E_{opt}|$.

First we partition the charging area $C_1(v)$ into three regions, where $R_1 = [v, v_1]$, $R_2 = [v, v_m] \setminus R_1$ and R_3 is the remaining region of the charging area. The three regions are shown in Fig. 5(b). Before we continue, we recall that $E_1'(v)$ consists of a ⌐-path, e_1, connecting v_1 and v, a vertical edge e_2 connecting v_m with e_1, and a rectangulation of the C-hull of $B_1(v)$ (which is meant to connect v_i, $2 \leq i < m$, with e_1 or e_2).

Consider a minimum Manhattan network connecting the vertices in V. What do we know about the edges E_{opt} of such a network?

1. E_{opt} must include a Manhattan path between v_1 and v within R_1. Note that this path has the same length as e_1.
2. E_{opt} must include a Manhattan path between v_m and v within $[v, v_m]$. Note that within R_2 this path has at least the same weight as e_2.

3. If $m>2$ then there must be a network N_{opt} connecting v_i, $2\leq i<m$, with e_1 or e_2 within R_3. Hence, it remains to prove that a minimum weight network connecting v_i, $2\leq i<m$, to the right or bottom side of R_3 has weight equal to a minimum weight rectangulation of R_3.

Lemma 5 *A minimum-weight network connecting v_i, $2\leq i<m$, with e_1 or e_2 has length at least equal to the length of a minimum-weight rectangulation of the C-hull.*

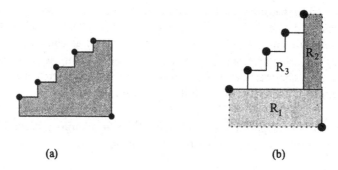

(a)   (b)

Fig. 5. (a) The charging area. (b) Partitioning the charging area into three regions.

Lemma 5 is a direct consequence of the following two lemmas. Consider a grid induced by the vertices of V'. Let N_{opt} be an optimal network within R_3 connecting v_i, $2\leq i<m$, to the right or bottom side of R_3. Let N_{rect} be a minimum-weight network, connecting v_i, $2\leq i<m$, to the right or bottom side of R_3, whose segments lie (only) on the grid induced by the vertices in V'.

Lemma 6 $|N_{rect}|=|N_{opt}|$.

Proof. Let P be a Manhattan path between vertices v_i and v. Assume that the path is moving right on the grid and that it changes direction downwards without reaching an intersection in the grid. Denote by t the last intersection of the grid that the path passed. Firstly, it is obvious that a path from v_i to v lying on the grid is of equal weight. Hence, the only reason to change direction "outside" the grid is that other paths may use this segment. Since all paths that may use this segment start at vertices to the left of t, they must start from vertices v_j, where $j < i$, which cannot intersect till the next horizontal grid line below it. Thus the path can be straightened out within that grid cell to follow the grid lines without decreasing its length. Going step by step, all paths can be modified to follow grid lines. Hence, the observation follows. □

Lemma 7 *There exists a minimum-weight rectangulation of the interior of the C-hull of length $|N_{rect}|$.*

Proof. Every path between an interior point v_i and v moves (seen from v_i) only in two directions, right and down. The only case when a path would induce a non-rectangular network of the C-hull is when it turns right or down without meeting another path. Assume we follow a path from v_i going down and then turning right, without meeting a horizontal segment. In this case the path could have been shortened by not changing direction, see Fig. 6(b), or by starting going right from the beginning. This is easy to see since the horizontal distance between a vertical segment and the turning point of the path is equal to the horizontal distance to the vertical segment and v_i. Hence, we do not gain anything by going downwards if the path is not meeting a horizontal segment. This means that there exists a rectangulation of the interior of the C-hull of weight $|N_{rect}|$.
□

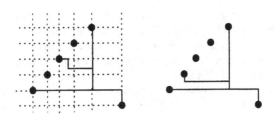

Fig. 6.

Putting these results together we obtain the following lemma.

Lemma 8 $(\sum_{v \in V} |E'_1(v)|) \leq r \times |E_{opt}|$.

To obtain the approximation factor for the algorithm just multiply the approximation factor for each step by four, since there are four sweeps. We summarize this paper by giving the main theorem.

Theorem 9. *Given a set of n points in the plane, and given an r-approximate, $R(n)$-time algorithm to compute a minimum-weight rectangulation of a staircase polygon, there exists an $O(n \log n + R(n))$-time algorithm that outputs a Manhattan network of length at most $4r$ times that of a minimum Manhattan network.*

4 Conclusions and Open Problems

We introduce the concept of Manhattan networks on a given set of points; these networks guarantee Manhattan paths between every pair of points from the input set. We present efficient 4- and 8-approximation algorithms for the problem of computing minimum Manhattan networks. We conjecture that an efficient 2-approximate algorithm and a polynomial-time approximation scheme does exist for the problem. The following problems remain open:

1. Determine the complexity of the problem of computing minimum Manhattan networks.
2. Design a PTAS for the problem of computing minimum Manhattan networks.
3. Design a 2-approximate algorithm for the problem of computing minimum Manhattan networks.

Acknowledgments

The last author would like to thank Dr. Lubomir Soltes for useful discussions and an implementation of the algorithm presented in this paper.

References

1. I. Althöfer, G. Das, D. P. Dobkin, D. Joseph, and J. Soares. On sparse spanners of weighted graphs. *Discrete Comput. Geom.*, 9:81–100, 1993.
2. Sanjeev Arora, Michelangelo Grigni, David Karger, Philip Klein, and Andrzej Woloszyn. A polynomial-time approximation scheme for weighted planar graph TSP. In *Proc. 9th ACM-SIAM Sympos. Discrete Algorithms*, pages 33–41, 1998.
3. S. Arya, G. Das, D. M. Mount, J. S. Salowe, and M. Smid. Euclidean spanners: short, thin, and lanky. In *Proc. 27th Annu. ACM Sympos. Theory Comput.*, pages 489–498, 1995.
4. B. Chandra, G. Das, G. Narasimhan, and J. Soares. New sparseness results on graph spanners. *Internat. J. Comput. Geom. Appl.*, 5:125–144, 1995.
5. G. Das and G. Narasimhan. A fast algorithm for constructing sparse Euclidean spanners. *Internat. J. Comput. Geom. Appl.*, 7:297–315, 1997.
6. C. Levcopoulos, G. Narasimhan, and M. Smid. Efficient algorithms for constructing fault-tolerant geometric spanners. In *Proc. 30th Annu. ACM Sympos. Theory Comput.*, pages 186–195, 1998.
7. C. Levcopoulos and A. Östlin. Linear-time heuristics for minimum weight rectangulation. In *5th Scandinavian Workshop on Algorithm Theory*, volume 1097 of *LNCS*, pages 271–283, 1996.
8. A. Lingas, R. Pinter, R. Rivest, and A. Shamir. Minimum edge length partitioning of rectilinear polygons. In *Proc. 20th Allerton Conf. Commun. Control Comput.*, pages 53–63, 1982.
9. S. B. Rao and W. D. Smith. Approximating geometrical graphs via "spanners" and "banyans". In *Proceedings of the 30th Annual ACM Symposium on Theory of Computing*, 1998.

Approximation of Multi-color Discrepancy

Benjamin Doerr[1]* and Anand Srivastav[2]

[1] Mathematisches Seminar, Christian–Albrechts–Universität zu Kiel,
Ludewig–Meyn–Str. 4, D–24098 Kiel, Germany,
bed@numerik.uni-kiel.de,
WWW home page: http://www.numerik.uni-kiel.de/~bed/
[2] Mathematisches Seminar, Christian–Albrechts–Universität zu Kiel,
Ludewig–Meyn–Str. 4, D–24098 Kiel, Germany,
asr@numerik.uni-kiel.de,
WWW home page: http://www.numerik.uni-kiel.de/~asr/

Abstract. In this article we introduce (combinatorial) multi–color discrepancy and generalize some classical results from 2–color discrepancy theory to c colors. We give a recursive method that constructs c–colorings from approximations to the 2–color discrepancy. This method works for a large class of theorems like the six–standard–deviation theorem of Spencer, the Beck–Fiala theorem and the results of Matoušek, Welzl and Wernisch for bounded VC–dimension. On the other hand there are examples showing that discrepancy in c colors can not be bounded in terms of two–color discrepancy even if c is a power of 2. For the linear discrepancy version of the Beck–Fiala theorem the recursive approach also fails. Here we extend the method of floating colors to multi–colorings and prove multi–color versions of the the Beck–Fiala theorem and the Barany–Grunberg theorem.

1 Introduction

Combinatorial discrepancy theory deals with the problem of partitioning the vertices of a hypergraph (set–system) in such a way that all hyperedges are split into about equal parts by the partition classes. Discrepancy measures the deviation of an optimal partition to an ideal one, that is one where all edges contains the same number of vertices in any partition class. As discrepancy is a NP–hard problem, efficient methods for constructing a good coloring can only approximate the discrepancy.

Usually one represents the partition by a coloring, that is a mapping from the vertices into some set, such that the classes of equal images form the partition classes. In this language, most results known so far only deal with two colors. Recent results from communication complexity (e. g. [BHK]) motivate the study of multi–color discrepancy.

---

* supported by the graduate school 'Effiziente Algorithmen und Multiskalenmethoden', Deutsche Forschungsgemeinschaft

1.1 Basic Definitions of Multi–Color Discrepancy

Let $\mathcal{H} = (V, \mathcal{E})$ denote a finite hypergraph, i. e. V is a finite set and $\mathcal{E} \subseteq 2^V$. A c–coloring of $\mathcal{H}$ is a mapping $\chi : X \to M$, where M is any set of cardinality c. For convenience, we may take $M = [c] := \{1, \ldots, c\}$. Sometimes — as we will see in this paper — a different set M will be of advantage. We define the *discrepancy of an edge $E \in \mathcal{E}$ for color $i \in M$ with respect to χ* by

$$\text{disc}_{\chi,i}(E) := \left| |E \cap \chi^{-1}(i)| - \frac{|E|}{c} \right|.$$

This measures the deviation of the number of i-colored points in X from the average $\frac{|E|}{c}$. The *discrepancy of $\mathcal{H}$ with respect to χ* is

$$\text{disc}(\mathcal{H}, \chi) := \max_{i \in M, E \in \mathcal{E}} \text{disc}_{\chi,i}(E),$$

and the *discrepancy of $\mathcal{H}$ in c colors* is defined by

$$\text{disc}(\mathcal{H}, c) := \min_{\chi : X \to [c]} \text{disc}(\mathcal{H}, \chi). \tag{1}$$

For a definition of multi–color discrepancy of a matrix, the representation of the colors by suitable vectors is useful. For a color $i \in [c]$ let $m^{(i)} \in \mathbb{R}^c$ be a vector with components defined by

$$m_j^{(i)} := \begin{cases} \frac{c-1}{c} & \text{if } i = j \\ -\frac{1}{c} & \text{otherwise} \end{cases},$$

for $j = 1, \ldots, c$. Put $M_c := \{m^{(i)} | i \in [c]\}$. For a c–coloring $\chi : X \to M_c$ we define

$$\text{disc}(\mathcal{H}, \chi) = \max_{E \in \mathcal{E}} \left\| \sum_{x \in E} \chi(x) \right\|_\infty,$$

and

$$\text{disc}(\mathcal{H}, c) := \min_{\chi : X \to M_c} \text{disc}(\mathcal{H}, \chi). \tag{2}$$

It is straightforward to see that the definitions in (1) and (2) are equivalent.

Let A be the incidence matrix of $\mathcal{H}$ and let $\bar{A}$ be the matrix which results from replacing every a_{ij} in A by $a_{ij} I_c$, where I_c denotes the unit matrix of dimension c. By identifying a $\chi : X \to M_c$ with a $c|X|$–dimensional vector in the natural way, we get

$$\text{disc}(\mathcal{H}, c) = \min_{\chi : X \to M_c} \left\| \bar{A} \chi \right\|_\infty.$$

This motivates the matrix notion of multi–color discrepancy for arbitrary matrices $A \in \mathbb{R}^{m \times n}$:

$$\text{disc}(A, c) = \min_{\chi : [n] \to M_c} \left\| \bar{A} \chi \right\|_\infty.$$

Let $\overline{M}_c$ be the the convex hull of M_c in $\mathbb{R}^c$. For $p \in \overline{M}_c$ set $\overline{p} : X \to \mathbb{R}^c; x \mapsto p$. We define the *weighted discrepancy of $\mathcal{H}$ with weight p* by

$$\mathrm{wd}(\mathcal{H}, c, p) := \min_{\chi : X \to M_c} \left\| \overline{A}(\overline{p} - \chi) \right\|_\infty \left(= \min_{\chi : X \to M_c} \max_{j \in [c], E \in \mathcal{E}} \left| |E \cap \chi^{-1}(j)| - p_j |E| \right| \right)$$

and the *weighted discrepancy of $\mathcal{H}$* by

$$\mathrm{wd}(\mathcal{H}, c) := \max_{p \in \overline{M}_c} \mathrm{wd}(\mathcal{H}, c, p).$$

There is an equivalent way to define weighted discrepancy which puts more emphasis on the aspect of weights: Denote by E_c the standard basis of $\mathbb{R}^c$ and by $\overline{E}_c$ its convex hull, that is all $p \in [0, 1]^c$ such that $\|p\|_1 = 1$. We have

$$\mathrm{wd}(\mathcal{H}, c) = \max_{p \in \overline{E}_c} \min_{\chi : X \to E_c} \left\| \overline{A}(\overline{p} - \chi) \right\|_\infty .$$

We define the *linear discrepancy* of a matrix $A \in \mathbb{R}^{m \times n}$ in c colors by

$$\mathrm{lindisc}(A, c) := \max_{p : [n] \to \overline{M}_c} \min_{\chi : [n] \to M_c} \left\| \overline{A}(p - \chi) \right\|_\infty .$$

The linear discrepancy of a hypergraph is of course the linear discrepancy of its incidence matrix. Let us write $A_0 \leq A$ to indicate that the matrix A_0 consists of some columns of the matrix A. For hypergraphs we write $\mathcal{H}_0 \leq \mathcal{H}$ if $\mathcal{H}_0$ is an induced subgraph of $\mathcal{H}$. Finally, the hereditary discrepancy in c colors is defined by

$$\mathrm{herdisc}(\mathcal{H}, c) := \min_{\mathcal{H}_0 \leq \mathcal{H}} \mathrm{disc}(\mathcal{H}_0, c).$$

Replace $\mathcal{H}$ by A for the matrix version. It is easy to show that the different notions of c–color discrepancy in the case $c = 2$ are identical with the usual notions of discrepancy (cf. the survey of Beck and Sós [BSó]) up to the constant factor of $1/2$. When citing 2–color results we will use the conventional notation (which has no parameter c in it, e. g. $\mathrm{herdisc}(\mathcal{H})$, so there is no danger of confusion).

1.2 Results

In this paper we give two methods for approximating the discrepancy in c colors, a recursive approach and the extension of the floating–color technique to multi–colorings. The recursive approach uses 2–color discrepancy theory. It turns out that under some not too strong assumptions we have an upper bound for the discrepancy in any number of colors that is roughly twice the bound for 2 colors. We show $\mathrm{disc}(\mathcal{H}, c) \leq 2.0005 \Delta(\mathcal{H})$[1] (Beck–Fiala theorem for c colors), prove $\mathrm{disc}(\mathcal{H}, c) \leq 12\sqrt{n}$ for $|V| = |\mathcal{E}| = n$ (Spencer's bound generalized to c colors)

---

[1] $\Delta(\mathcal{H})$ is the degree of the hypergraph.

and derive bounds for c–color discrepancy of hypergraphs with bounded VC–dimension extending theorems of Matoušek, Welzl and Wernisch [MWW].

We give an example which shows the limit of the recursive approach: There are hypergraphs having 2–discrepancy zero, but arbitrarily large c–discrepancy even if c is a power of 2. Furthermore, the recursive method does not apply to the stronger linear discrepancy version of the Beck–Fiala theorem. For this situation and in the setting of the Barany–Grunberg theorem we present an approximation algorithm using vector–colorings and an extension of the floating–color technique to multi–colorings.

2 Recursive Coloring

As a warming up exercise let us fix the bound for discrepancy in c colors by the basic probabilistic method.

Theorem 1. *Let $\mathcal{H} = (V, \mathcal{E})$ be any hypergraph. Set $m := |\mathcal{E}|$ and $s = \max_{E \in \mathcal{E}} |E|$. Then $\mathrm{disc}(\mathcal{H}, c) \leq \sqrt{\frac{1}{2} s \ln(2mc)}$. With probability greater than $\frac{1}{2}$ we can find a c-coloring χ with discrepancy at most $\sqrt{\frac{1}{2} s \ln(4mc)}$ (this leads to a randomized algorithm with arbitrarily small error probability).*

Proof. Define a random c–coloring χ by independently picking a random color uniformly distributed from $[c]$ for every vertex $v \in V$. Define random variables $X_{i,v}$ by

$$X_{i,v} := \begin{cases} \frac{c-1}{c} & \text{if } \chi(v) = i \\ -\frac{1}{c} & \text{else} \end{cases}$$

for all $v \in V$, $i \in [c]$. For fixed i these are independent random variables. Set $X_{i,E} := \sum_{v \in E} X_{i,v}$ for all $E \in \mathcal{E}$, $i \in [c]$. From [ASE, Theorem A.4] we know $P(|X_{i,E}| > \alpha) < 2e^{-2\alpha^2/|E|}$ for all real $\alpha > 0$. With $\alpha = \sqrt{\frac{1}{2} s \ln(2mc)}$ it is easy to see that the random coloring χ fulfills $\mathrm{disc}(\mathcal{H}, \chi) \leq \sqrt{\frac{1}{2} s \ln(2mc)}$ with non-zero probability. Choosing $\alpha = \sqrt{\frac{1}{2} s \ln(4mc)}$ we get this probability below $\frac{1}{2}$.

For 2–colors we get $\mathrm{disc}(\mathcal{H}) \leq \sqrt{2s \ln(4m)}$, while the best bound in the classical approach is $\mathrm{disc}(\mathcal{H}) \leq \sqrt{2s \ln(2m)}$. Note that with the method of conditional probabilities the existence result of Theorem 1 can be derandomized.

The basic idea of recursive coloring is to find a suitable 2–coloring of X with color classes X_1, X_2 and then to iterate this process on the subhypergraphs induced by X_1 and X_2. If the 2–color discrepancy of all induced subhypergraphs is uniformly bounded, such a recursive method can be analyzed, even for n not a power of 2. This will lead to a generalization of the six–standard–deviation theorem of Spencer, the discrepancy bound of Beck–Fiala and the bound using the primal and dual shatter function of Matoušek, Welzl and Wernisch. At the end of this section we will show the limits of the recursive approach. For example, for the linear discrepancy in c–colors recursive methods fail, and we need other methods, which will be introduced in the next section.

The following lemma analyses a single step in the recursion.

Lemma 1. *Let C be a set of colors with $c = |C|$ and let C_1, C_2 be a partition of C. Let p be a weight of C, i. e. $p \in [0,1]^c$ such that $\|p\|_1 = 1$. Denote by $p_{|C_j}$ the vector taking the components of p with indices corresponding to the colors in the set C_j, and set $q_j = \|p_{|C_j}\|_1$, $j \in [2]$. Let χ_0 be a 2–coloring of X, set $X_1 := \chi_0^{-1}(1)$, $X_2 := \chi_0^{-1}(-1)$. Let $\chi_j : X_j \to C_j$ be any colorings. Set $\chi := \chi_1 \cup \chi_2$. For all $E \in \mathcal{E}$, $j \in [2]$ and $i \in C_j$ the discrepancy of E with respect to the color i, the coloring χ and the weight p is*

$$\big||E \cap \chi^{-1}(i)| - p_i|E|\big| \leq \tfrac{p_i}{q_j} \big||E \cap X_j| - q_j|E|\big| + \big||E \cap X_j \cap \chi_j^{-1}(i)| - \tfrac{p_i}{q_j}|E \cap X_j|\big|.$$

In particular $\mathrm{wd}(\mathcal{H}, c, p) \leq \tfrac{p_i}{q_j}\,\mathrm{wd}(\mathcal{H}, 2, (q_1, q_2)) + \max_{\mathcal{H}_0 \leq \mathcal{H}} \mathrm{wd}(\mathcal{H}_0, |C_j|, \tfrac{1}{q_j}p_{|C_j})$.

The proof is straightforward and will appear in the full version. We now investigate the case where we all induced subgraphs have a common bound on the weighted discrepancy in two colors. This is an important case for two reasons: Firstly, the proof of some results on two–color discrepancy actually gives an information about the weighted discrepancy of the induces subgraphs (e. g. in the Beck–Fiala setting). Secondly, the linear discrepancy (and thus also the weighted discrepancies of all subgraphs, note $\mathrm{wd}(\mathcal{H}, c) \leq \tfrac{1}{2}\,\mathrm{lindisc}(\mathcal{H})$) are bounded by the hereditary discrepancy: From [BSp,LSV] we know $\mathrm{lindisc}(\mathcal{H}) \leq 2\,\mathrm{herdisc}(\mathcal{H})$ (for a recent improvement see also [D]).

We represent the iterated partitioning of C by a binary tree. We call a binary tree $T = (V_T, E_T)$ a *partition tree* for C, if the following conditions are satisfied: the nodes are subsets of C, the root is C, all leaves are singletons of C and the two son nodes form a partition of their common father node. For every color $i \in C$ there is a unique path of type $C = C_0^{(i)} \subset C_1^{(i)} \subset \ldots \subset C_{k(i)}^{(i)} = \{i\}$ in the partition tree.

Theorem 2. *Let $\mathrm{wd}(\mathcal{H}_0, 2) \leq K$ for all induced subgraphs $\mathcal{H}_0$ of $\mathcal{H}$. Let C be a set of colors with $c = |C|$ and let p be a weight of C, i. e. $p \in [0,1]^c$ such that $\sum_{i=1}^c p_i = 1$. Let $T = (V_T, E_T)$ be a partition tree of C. Then there is a coloring $\chi : X \to C$ such that for all colors $i \in C$ and all $E \in \mathcal{E}$ we have*

$$\big||E \cap \chi^{-1}(i)| - p_i|E|\big| \leq p_i \sum_{l=1}^{k(i)} \frac{1}{\|p_{|C_l^{(i)}}\|_1} K.$$

Proof. By induction on the height of T. For $h(T) = 0$ we have just one color and both sides of the inequality become zero. So let T be of height greater than zero and assume that the theorem is true for smaller heights. Let C_1 and C_2 be the sons of C in T. Set $q_j := \sum_{k \in C_j} p_k$, $j = 1, 2$. By the assumption of the theorem there is a 2–coloring $\chi_0 : X \to [2]$ such that $\big||E \cap \chi_0^{-1}(j)| - q_j|E \cap X_j|\big| \leq K$ for all colors $j \in [2]$ and edges $E \in \mathcal{E}$. Put $X_j := \chi_0^{-1}(j)$, $j = 1, 2$. Denote by T_j the subtree having C_j as its root. Then $\mathcal{H}_{|X_j}$, the set of colors C_j with weight $\frac{1}{q_j}p_{|C_j}$ together with the T_j fulfill the assumption of this theorem. Hence by induction hypothesis there are $\chi_j : X_j \to C_j$, $j \in 1,2$ such that

$$\left| |E \cap X_j \cap \chi_j^{-1}(i)| - \frac{1}{q_j}p_i|E \cap X_j| \right| \le \frac{p_i}{q_j} \sum_{l=2}^{k(i)} \frac{1}{\frac{1}{q_j}\|p_{|C_l^{(i)}}\|_1} K$$

for all $E \in \mathcal{E}$, $j \in [2]$, $i \in C_j$. This and lemma 1 prove that $\chi = \chi_1 \cup \chi_2$ is as desired.

The c–color discrepancy problem is simply the case where all weights are equal. In this case only the size of the partitioning sets is of importance. Hence the following simpler structure can be investigated:

A *partition tree* for a positive integer n is binary tree $T = (V_T, E_T)$ together with a labeling $V_T \to [n]$ such that the following conditions are satisfied: the root is labeled n, all leaves are labeled 1 and for all non–leaf nodes the labels of the two sons sum up to the label of the node itself.

For every path $P : r = v_0, v_1, v_2, \ldots, v_k$ connecting the root r and a leaf v_k we call $v_T(P) = \sum_{i=1}^{k} \frac{1}{l(v_i)}$ the value of P and $v(T)$ the maximum $v_T(P)$ over all these paths P. Finally $v(n)$ is the minimum $v(T)$ over all partition trees T of n.

Theorem 3. *Let* $\mathrm{wd}(\mathcal{H}_0, 2) \le K$ *for all induced subgraphs* $\mathcal{H}_0$ *of* $\mathcal{H}$. *Then* $\mathrm{disc}(\mathcal{H}, c) \le v(c)K$.

Proof. Let $T = (V_T, E_T)$ together with l be a partition tree for c such that $v(T) = v(c)$. From (V_T, E_T) we build a partition tree of $[c]$ with same tree structure. Define $f : V_T \to 2^{[c]}$ recursively: Set $f(r) = [c]$ for the root r of T. For every node v with sons s_1 and s_2 such that $f(v)$ is already defined choose $f(s_1)$ to be any subset of $f(v)$ of size $l(s_1)$ and $f(s_2) = f(v) \setminus f(s_1)$. Note that f is injective, and by replacing every $v \in V_T$ by $f(v)$ we get a partition tree T^* for $[c]$. Set $p = \frac{1}{c}\mathbf{1}_c$. For every path P connecting the root with a leaf $\{i\}$ in T^* we get that the bound in theorem 2 for the discrepancy in color i is equal to $n_T(P)K$, hence the discrepancy is bounded by $v(T)K = v(c)K$.

What remains is the calculation of $v(c)$. Set $\lfloor n \rfloor_2 = 2^{\lfloor \log_2 n \rfloor}$, the largest power of 2 that is not larger than n, and $\lceil n \rceil_2 = 2^{\lceil \log_2 n \rceil}$, the smallest one not smaller than n. Denote by $n_1(n)$ the number of 1s in the binary representation of n (e.g. $n_1(9) = 2$). We give a lower and an upper bound for $v(n)$. If n is a power of 2, both bounds coincide.

Lemma 2. *For all* $n \in \mathbb{N}$, $n \ge 2$ *we have*

(i) $v(n) \ge 2 - \frac{1}{\lceil n \rceil_2}$.

(ii) $v(n) \le 2 + (n_1(n) - 2)\frac{1}{\lfloor n \rfloor_2} \le 2 + (\log_2(\lfloor n \rfloor_2) - 1)\frac{1}{\lfloor n \rfloor_2}$.

(iii) $v(n) \le 2.0005$.

Proof. (*i*): For the proof of the lower bound let $T = (V_T, E_T)$ together with l be any partition tree of n. Then there is a path $v_0, \ldots, v_k$ of length $k \geq \log_2\lceil n\rceil_2$ such that v_k is a leaf and $l(v_{i-1}) \leq 2l(v_i)$ for all $i \in [k]$. Thus

$$\sum_{i=1}^{k} \frac{1}{l(v_i)} \geq \sum_{i=0}^{k-1} 2^{-i} = 2 - \frac{1}{2^{k-1}} \geq 2 - \frac{1}{\lceil n\rceil_2}.$$

(*ii*): For the proof of the upper bound we give a strategy how to construct a partition tree of n. We do so recursively: For a vertex v labeled $\sum_{i\in[k]} a_i 2^k \neq 1$, $a_i \in \{0, 1\}$, we add sons $s_1(v)$ and $s_2(v)$ labeled $l(s_1(v)) = 2^{\min\{i\in[k]\,|\,a_i=1\}}$ and $l(s_2(v)) = l(v) - l(s_1(v))$. Immediately we see that we only need to investigate the path $P : r, s_2(r), s_2(s_2(r)), \ldots$ — if r denotes the root of T —, because the labels of all other paths occur also on this path. Thus $v(P)$ is maximal. The labels of the first $n_1(n)$ vertices of P are greater than or equal to $\lfloor n\rfloor_2$, so their contribution to $v(P)$ is not greater than $n_1(n)\frac{1}{\lfloor n\rfloor_2}$. The rest of the vertices are labeled by $\frac{2}{\lfloor n\rfloor_2}, \frac{4}{\lfloor n\rfloor_2}, \ldots$ up to 1. This sums up to $2 - \frac{2}{\lfloor n\rfloor_2}$ and the the first inequality is proven. For the second inequality note that $n_1(n) - 1 \leq \log_2(\lfloor n\rfloor_2)$.

(*iii*): For $n \geq n_0 := 2^{17} - 1$ the last inequality follows from (*ii*), as $(n_1(n) - 2)\frac{1}{\lfloor n\rfloor_2} \leq \frac{\log_2(\lfloor n\rfloor_2)-1}{\lfloor n\rfloor_2} \leq \frac{\log_2(\lfloor n_0\rfloor_2)-1}{\lfloor n_0\rfloor_2}$. For the remaining small numbers, $v(n)$ can be computed in $\mathcal{O}(n^2)$–time and attains its maximum value for $n = 909$, namely $v(909) \approx 2.000450$.

From this we derive a c–color version of the Beck–Fiala theorem:

Theorem 4. *For any hypergraph $\mathcal{H}$ we have*

$$\text{disc}(\mathcal{H}, c) < v(c)\Delta(\mathcal{H}) \leq 2.0005\Delta(\mathcal{H}).$$

The complexity to construct a c–coloring respecting this bound is less than $2\log_2\lfloor c\rfloor_2$ times the complexity for the 2–color case.

Proof (sketch). The bound is a direct consequence of theorem 3 and Lemma 2 and the Beck–Fiala theorem for 2 colors. For the complexity estimation let C be the constant such that the construction of a 2–coloring as in the theorem of Beck–Fiala has complexity bounded by Cn^4, where n shall denote the number of vertices of $\mathcal{H}$. Then the partition tree of Lemma 2 (*ii*) yields the bound of $n_1(c) - 1 + \log_2\lfloor c\rfloor_2 \leq 2\log_2\lfloor c\rfloor_2$.

The generalization of the 'Six Standard Deviation' theorem of Spencer [Sp1] is:

Theorem 5. *For any hypergraph $\mathcal{H} = (V, \mathcal{E})$ such that $n := |V| = |\mathcal{E}|$ we have*

$$\text{disc}(\mathcal{H}, c) < 5.32v(c)\sqrt{n} \leq 12\sqrt{n}.$$

Proof. Spencer's proof shows that $\text{herdisc}(\mathcal{H}) \leq 5.32\sqrt{n}$.

Without proof we may remark that this bound is tight (apart from the constant).

The recursive approach also generalizes a result of Matoušek, Welzl and Wernisch [MWW] connecting discrepancy with the primal shatter function $\pi_{\mathcal{H}}$ and dual shatter function $\pi_{\mathcal{H}}^*$ of a hypergraph (the shatter functions are closely related to the VC–dimension of $\mathcal{H}$).

Theorem 6. *Let $\mathcal{H} = (V, \mathcal{E})$ be a hypergraph on n points. Let $d > 1$. If $\pi_{\mathcal{H}} = \mathcal{O}(m^d)$, then $\operatorname{disc}(\mathcal{H}, c) = \mathcal{O}(n^{\frac{1}{2} - \frac{1}{2d}} (\log n)^{1 + \frac{1}{2d}})$. If $\pi_{\mathcal{H}}^* = \mathcal{O}(m^d)$, then $\operatorname{disc}(\mathcal{H}, c) = \mathcal{O}(n^{\frac{1}{2} - \frac{1}{2d}} \log n)$. In both cases the implicit constants can be chosen independent of c.*

Summary of the recursive method: We see that the recursive method is very effective in situations where we can bound the weighted discrepancy of the induced subgraphs. This is always the case if we know the hereditary discrepancy of $\mathcal{H}$. There are cases where the recursive approach is the only result we have. We do not have a direct proof for a result like theorem 5. We feel that the original proof relies heavily on the fact that only two colors are considered. On the other hand the recursive approach clearly has its limitations: We can get results on weighted discrepancy, but we do not get any on linear discrepancy (e. g. in the Beck–Fiala setting). To apply recursion, we need a 2–color result for the weighted discrepancy, even in the case that c is a power of two. The following example illustrates this limitation.

Example: Let $n \in \mathbb{N}$. Set $\mathcal{H}_n = ([2n], \{X \subseteq [2n] | |X \cap [n]| = |X \setminus [n]|\})$. Obviously, $\operatorname{disc}(\mathcal{H}_n) = 0$. On the other hand it is not difficult to show $\operatorname{disc}(\mathcal{H}_n, 4) = \frac{1}{8} n$ for all $n \in \mathbb{N}$.

3 Floating Vector–Colors

In this section we give analogous results to the Beck–Fiala theorem and the Barany–Grunberg theorem. In the 2–color case both are proved using a rounding strategy. We show how this strategy can be extended to the multi–color case. The key in both cases is the representation of the colors by the vectors $m^{(i)}$ defined in section 1.

3.1 Beck–Fiala Theorem for Linear Discrepancy

The maximum degree $\Delta(\mathcal{H}) := \max_{x \in X} |\{E \in \mathcal{E} | x \in E\}|$ is one of the few parameters of a hypergraph which give a good bound on the discrepancy. The Beck–Fiala theorem [BF] states $\operatorname{disc}(\mathcal{H}) < 2\Delta(\mathcal{H})$. Actually Beck and Fiala proved a stronger result. For any matrix $A = (a_{ij}) \in \mathbb{R}^{m \times n}$ denote by $\|A\|_1 := \max_{j \in [n]} \sum_{i \in [m]} |a_{ij}|$ the operator norm induced by the 1–norm on $\mathbb{R}^n$. Then $\operatorname{lindisc}(A) < 2\|A\|_1$. We were not able to generalize this theorem to c–colors by the recursive method, and this might also not be possible. The difficulty is that in the phase of coloring a subhypergraph in the recursive method some information on the weights $p_j(v)$, j a color, v a vertex, is not available anymore.

The method of floating colors though can be extended to multi–colorings. We have

Theorem 7. *For any matrix A, $\mathrm{lindisc}(A, c) < 2\|A\|_1$. The problem of computing a $\chi : [n] \rightarrow M_c$ for a given $p : [n] \rightarrow \overline{M}_c$ such that $\|\overline{A}(p - \chi)\|_\infty < 2\|A\|_1$ has time–complexity $\mathcal{O}(c^4)$ times the complexity of the 2–color problem.*

Proof. Set $\Delta := \|A\|_1$ and $\overline{A} = (\overline{a}_{ij})$ the matrix resulting from A by replacing every entry a_{ij} by $a_{ij}I_c$ as introduced in section 1. Note that $\Delta = \|\overline{A}\|_1$. Let $p : [n] \rightarrow \overline{M}_c$. Set $\chi = p$. Successively we will change χ to a $\chi : [n] \rightarrow M_c$.

Set $J := \{j \in [cn] | \chi_j \notin \{-\frac{1}{c}, \frac{c-1}{c}\}\}$ and call these columns floating (the others fixed). Set $I := \{i \in [cm] | \sum_{j \in J} |\overline{a}_{ij}| > 2\Delta\}$ and call these rows active (the others ignored). We will ensure that during the rounding process the following conditions are fulfilled (this is clear for the start because $\chi = p$):

(i) $(\overline{A}(p - \chi))|_I = 0$, i. e. all active rows have discrepancy zero, and
(ii) all colors are in $\overline{M}_c$, in particular $\sum_{k=0}^{c-1} \chi_{cj-k} = 0$ for all $j \in [n]$.

Note that (ii) is the crucial difference to the 2–color case, where we only need a condition of type (i). This will increase the number of equations investigated below, and is thus the reason why the multi–color bound is off the classical result by a factor of 2.

We have

$$|J|\Delta \geq \sum_{j \in J} \sum_{i \in I} |\overline{a}_{ij}| = \sum_{i \in I} \sum_{j \in J} |\overline{a}_{ij}| > |I| 2\Delta.$$

Note further that for every vertex it cannot happen that exactly one color is floating, so

$$\sum_{k=0}^{c-1} \chi_{cj-k} = 0, j \in [n] \text{ such that } c(j-1) + k \in J \text{ for some } k \in [c]$$

is a system of at most $\frac{1}{2}|J|$ equations. Hence the system

$$\overline{A}_{|I \times J} \chi_J = 0$$

$$\sum_{k=0}^{c-1} \chi_{cj-k} = 0, j \in [n] \text{ such that } c(j-1) + k \in J \text{ for some } k \in [c]$$

is under–determined (taking just the $\chi_j, j \in J$ as variables). Thus there is a non–trivial solution $x \in \mathbb{R}^J$. Expand x to $x_E \in \mathbb{R}^{cn}$ by

$$(x_E)_j := \begin{cases} x_j & \text{if } j \in J \\ 0 & \text{else} \end{cases}.$$

Note that for any such x we can replace χ by $\chi + x_E$ in (i), (ii). Choose $\lambda \in \mathbb{R}$ such that at least one component of $\chi + \lambda x_E$ becomes fixed and all colors are still in $\overline{M}_c$, i. e. $\chi + \lambda x_E \in \overline{M}_c^n$. Set $\chi := \chi + \lambda x_E$. Since (i), (ii) are fulfilled, we can continue this rounding process until all χ_j, $j \in [cn]$ are in $\{-\frac{1}{c}, \frac{c-1}{c}\}$. We show $\|\overline{A}(p-\chi)\|_\infty < 2\Delta$. Let $i \in [cm]$. Denote by $\chi^{(0)}$ and $J^{(0)}$ the values of χ and J when the row i first became ignored. Note that $\sum_{j \in J^{(0)}} |\bar{a}_{ij}| < 2\Delta$ by definition of I. Thus

$$|(A(p-\chi))_i| = |(A(p-\chi^{(0)}))_i + (A(\chi^{(0)} - \chi))_i| = 0 + \sum_{j \in J^{(0)}} \bar{a}_{ij}(\chi^{(0)} - \chi)(i) < 2\Delta.$$

3.2 Theorem of Barany–Grunberg

The theorem of Barany–Grunberg [BG] for 2 colors states:

Theorem 8. *Let $\|\cdot\|$ be any norm on $\mathbb{R}^n$ and $v_1, v_2, \ldots, v_k$ be a finite sequence of arbitrary length of vectors of norm at most 1 in $\mathbb{R}^n$. Then there are signs $\varepsilon_i, i = 1, \ldots, k$ such that for all $l \in [k]$ we have*

$$\left\| \sum_{i=1}^l \varepsilon_i v_i \right\| < 2n.$$

As in the proof of the Beck–Fiala theorem we describe the colors by vectors from the set M_c. As above let $v_1, v_2, \ldots, v_k$ be a finite sequence of vectors in $\mathbb{R}^n$ and $\|\cdot\|$ a norm on $\mathbb{R}^n$. A mapping $\chi : [k] \to M_c, i \mapsto \chi^{(i)}$ is called a coloring for these vectors. Since Barany–Grunberg works for any norm, we need to lift our norm to a suitable norm on $\mathbb{R}^{cn}$: Define a norm $\|\cdot\|_c$ on $\mathbb{R}^{cn}$ by $\|v\|_c := \max_{j \in [c]} \|v_{|\{j, j+c, \ldots, j+(n-1)c\}}\|$.

We need a calculus for substituting vectors into each other. For any two vectors $v \in \mathbb{R}^n$, $w \in \mathbb{R}^m$, we define $v * w$ to be the vector $u \in \mathbb{R}^{nm}$ such that $u_{(i-1)n+j} = v_i w_j$ for all $i \in [n]$, $j \in [m]$. So u is obtained by replacing every entry v_i of v by $v_i w$. The following lemma follows from direct calculations.

Lemma 3. $\left\| \sum_{i \in [k]} v_i * \chi^{(i)} \right\|_c = \max_{j \in [c]} \left\| \sum_{i \in [k], \chi^{(i)} = m_j} v_i - \frac{1}{c} \sum_{i \in [k]} v_i \right\|.$

The latter expression in the lemma measures the maximal deviation (over the colors) of the sum of vectors in this color from the average $\frac{1}{c} \sum_{i \in [k]} v_i$ with respect to the norm $\|\cdot\|$. This is the c–color analogue of the discrepancy term $\left\| \sum_{i=1}^l \varepsilon_i v_i \right\|$ in the Barany–Grunberg theorem.

The multi–color version of Barany–Grunberg is:

Theorem 9. *Let $\|\cdot\|$ be any norm on $\mathbb{R}^n$ and $v_1, v_2, \ldots, v_k$ be a finite sequence of vectors of norm at most 1 in $\mathbb{R}^n$. Then there is a c–partition $I_1, \ldots, I_c$ of $[k]$ such that for all $l \in [k]$ and $j \in [c]$ we have*

$$\left\| \sum_{i \in I_j \cap [l]} v_i - \frac{1}{c} \sum_{i=1}^l v_i \right\| < (c-1)n.$$

The complexity for computing this partition is $\mathcal{O}(c^3)$ times the complexity of the 2–color case.

Proof. By Lemma 3 it is enough to construct a coloring $\chi : [k] \to M_c$ such that

$$(*) \qquad \left\| \sum_{i \in [l]} v_i * \chi^{(i)} \right\|_c \leq (c-1)n \text{ for all } l \in [k].$$

We sketch an algorithm for this task: To start with put $A := [n]$ and $\chi_j^{(i)} := 0$ for all $i \in [k], j \in [c]$. We repeat the following rounding process: Let us call those $\chi_j^{(i)}$ where $i \in A$ and $\chi_j^{(i)} \notin \{\frac{c-1}{c}, -\frac{1}{c}\}$ *variables*. We try to find a nontrivial solution of the system of equations

$$\sum_{i \in A} v_i * \chi^{(i)} = 0 \quad \text{for all } j \in [c-1].$$

$$\sum_{j \in [c]} \chi_j^{(i)} = 0 \quad \text{for all } i \in A.$$

If one exists, change χ in the way that one variable becomes $\frac{c-1}{c}$ or $-\frac{1}{c}$ and all variable stay in $[-\frac{1}{c}, \frac{c-1}{c}]$. If not, then increase the number of vectors under consideration, i. e. set $A := A \cup \{\max A + 1\}$, if $A \neq [k]$, and stop, if $A = [k]$. After the rounding process stopped, change the remaining variables to $\frac{c-1}{c}$ or $-\frac{1}{c}$ in such a way that all $\chi^{(i)}$ are in M_c. For the correctness proof we calculate the number of indices in A such that the color of v_i is not completely determined to be at most $(c-1)n$. This together with the properties of $*$ and $\|\cdot\|_c$ shows that χ fulfills the sufficient condition $(*)$. The journal version of this paper will contain the missing details.

Note that this time the number of colors influences our bound in a much stronger way than in the theorems before: In the Beck–Fiala situation, the multi-color bound is off the 2–color one by a factor of 2 for the simple reason that in the 2–color case one can actually ignore one color (the discrepancy in both colors is the same). In the Barany–Grunberg theorem the bound contains a factor of $c - 1$ and both the original result (translated to our notion) and our result yield the same bound. This is due to the fact that we use an arbitrary norm. Thus in the analysis we can 'ignore' a vector v_i only from the point on when its color $\chi^{(i)}$ is completely determined. Compare this to the proof of Beck–Fiala, where the fixing of a single χ_k (which is equivalent to a $\chi_j^{(i)}$ here) improves the situation.

4 Conclusion

This paper presents two types of results. Firstly, some important discrepancy theorems can be lifted to c–colorings via 2–color results. Secondly, using the right vector–representation of the colors and an appropriate matrix calculus some results of the 2–color discrepancy theory can be generalized to any number of colors. Nevertheless c–color discrepancy theory is more than a generalization for generalization's sake as the recent applications in communication complexity might indicate. We hope that our paper can spur further research in this part of discrepancy theory.

References

[ASE]   N. Alon, J. Spencer, P. Erdős: *The Probabilistic Method*, John Wiley & Sons, Inc., 1992.

[BHK]   L. Babai, T. P. Hayes and P. G. Kimmel, *The cost of the Missing Bit: Communication Complexity with Help*, 30th STOC, 1998, 673–682.

[BF]   J. Beck and T. Fiala, *"Integer making" Theorems*, Discrete Applied Mathematics **3** (1981), 1–8.

[BG]   I. Barany and V. S. Grunberg, *On some combinatorial questions in finite dimensional spaces*, Linear Algebra Appl. **41** (1981), 1–9.

[BSó]   J. Beck and V. Sós, *Discrepancy Theory* in R. Graham, M. Grötschel and L. Lovász, Handbook of Combinatorics, 1995, Chapter 26.

[BSp]   J. Beck and J. Spencer, *Integral approximation sequences*, Math. Programming **30** (1984), 88–98.

[D]   B. Doerr, *Linear and Hereditary Discrepancy*, 1999, accepted for publication in Combinatorics, Probability and Computing.

[LSV]   L. Lovász, J. Spencer and K. Vesztergombi, *Discrepancies of set–systems and matrices*, European J. Combin. **7** (1986), 151–160.

[Sp1]   J. Spencer, *Six Standard Deviation Suffice*, Trans. Amer. Math. Soc. **289** (1985), 679–706.

[Sp2]   J. Spencer, *Ten Lectures on the Probabilistic Method*, SIAM, Philadelphia, 1987.

[MWW]   J. Matoušek, E. Welzl and L. Wernisch, *Discrepancy and approximations for bounded VC–Dimension*, Combinatorica **13** (1984), 455–466.

A Polynomial Time Approximation Scheme for the Multiple Knapsack Problem

Hans Kellerer

Institut für Statistik und Operations Research, Universität Graz,
Universitätsstraße 15, A-8010 Graz, Austria

Abstract. The *Multiple Knapsack Problem* (MKP) (with equal capacities) can be defined as follows: Given a set of n items with positive integer weights and profits, a subset has to be selected such that the items in this subset can be packed into m knapsacks of equal capacities and such that the total profit of all items in the knapsacks is maximized. For $m = 1$ (MKP) reduces to the classical 0-1 single knapsack problem. It is known that (MKP) admits no fully polynomial-time approximation scheme even for $m = 2$ unless $\mathcal{P} = \mathcal{NP}$. In this paper we present a polynomial time approximation scheme for (MKP) even if m is part of the input. This solves an important open problem in the field of knapsack problems.

1 Introduction

Knapsack problems belong to the basic and most well-known problems in combinatorial optimization (for an excellent introduction into knapsack theory see [4]). This paper concerns the so-called *Multiple Knapsack Problem* (MKP) (with equal capacities): We are given a set $N = \{1, \ldots, n\}$ of n items, each item i having positive integer *weight* w_i and *profit* p_i and m *knapsacks* of equal *capacity* c. A subset of N has to be selected such that the items in this subset can be packed into the knapsacks without exceeding the capacities so that the total profit of all items in the knapsacks is maximized. Formally (MKP) reads as follows.

$$\text{maximize} \quad \sum_{j=1}^{m} \sum_{i=1}^{n} p_i x_{ij}$$

$$\text{subject to} \quad \sum_{i=1}^{n} w_i x_{ij} \leq c \quad j = 1, \ldots, m$$

$$\sum_{j=1}^{m} x_{ij} \leq 1 \quad i = 1, \ldots, n$$

$$x_{ij} \in \{0, 1\}, \quad i = 1, \ldots, n, \, j = 1, \ldots, m$$

W.l.o.g. we assume $w_i \leq c$ for all $i = 1, \ldots, n$ and $n \geq m$. (MKP) can be considered as a generalization both of the classical 0-1 (single) knapsack problem and of the so-called Multiple Subset Sum Problem [1] which is (MKP) with profits identical to weights.

Of course, (MKP) is strongly NP-hard. While for 0-1 single knapsack problems efficient fully polynomial time approximation schemes (FPTAS) are known (see e.g. [2]), it can be easily shown that there is no FPTAS for (MKP) even for the multiple subset sum problem with two knapsacks unless $\mathcal{P} = \mathcal{NP}$ [1]. The status of approximability of (MKP), i.e. the existence of a PTAS for (MKP), was an important open problem in the field of knapsack problems. In this paper we will present a PTAS for (MKP).

A straightforward approximation algorithm for (MKP) can be obtained by generalizing the Greedy approach for the single knapsack problem. Assume the items to be sorted according to non-increasing profit-weight ratio. Then the continuous relaxation of (MKP) assigns the items to the knapsacks in this order, possibly "breaking" at most m critical items. The continuous solution is then transformed into a feasible solution for (MKP) by removing all these critical items. We call this approach in the following *generalized Greedy algorithm*. An alternative solution can be attained by assigning the m items with maximal profit to the m knapsacks. Let LB denote the maximum profit of both approaches and let z^* denote the optimal solution value of (MKP), then it can be easily seen that $LB \leq z^* \leq 2LB$. Therefore, we will consider in the rest of the paper only $(1 - \varepsilon)$-approximation algorithms for (MKP) with $\varepsilon < 1/2$.

Let us introduce some further notations. z^H denotes a heuristic solution value. Let $I \subseteq N$. Then $w(I) := \sum_{i \in I} w_i$ defines the *weight of set I* and $p(I) := \sum_{i \in I} p_i$ the *profit of set I*. A *feasible assignment* is an assignment of items to the knapsacks so that the total weight of the items in each knapsack does not exceed the knapsack capacity c. A set of items is called *feasible* if a feasible assignment for these items exists. Note that we will often identify knapsacks with the items put into them.

The paper is organized as follows: Section 2 contains an appropriate partition of the items into different classes and properties of the classes are formulated in Lemmas 1 to 4. Then in Section 3 these classes are transformed into item sets with only finitely many elements. An integer program is formulated for the new item sets. Finally, Section 4 contains the algorithm and it is shown that this algorithm is indeed a PTAS.

2 Partitioning of the Items

In this section we will give an appropriate partition of the item set N into four different classes I_1, I_2, I_3, I_4 which will be useful to construct a PTAS for (MKP).

The properties of these classes are formulated in Lemmas 1 to 4.

$$I_1 := \left\{ i \,\middle|\, p_i \leq \frac{4LB\varepsilon}{m} \right\},$$

$$I_2 := \left\{ i \,\middle|\, p_i > \frac{4LB\varepsilon}{m}, w_i \leq \varepsilon c \right\},$$

$$I_3 := \left\{ i \,\middle|\, \frac{4LB}{m} \geq p_i > \frac{4LB\varepsilon}{m}, w_i > \varepsilon c \right\},$$

$$I_4 := \left\{ i \,\middle|\, p_i > \frac{4LB}{m}, w_i > \varepsilon c \right\}.$$

Depending on the item profits the set $S := I_1$ denotes the set of *small items* and $B := I_2 \cup I_3 \cup I_4$ denotes the set of *big items*. For the profit weight ratio we use another partition. $G := \{ i \,|\, \frac{p_i}{w_i} > \frac{4LB}{mc} \}$ contains the *good ratio items* and $\bar{G} := \{ i \,|\, \frac{p_i}{w_i} \leq \frac{4LB}{mc} \}$ the *bad ratio items*.

It can be easily seen that I_4 is part of the optimal solution set. This is formulated in Lemma 1.

Lemma 1. $I_4 \subseteq I^*$. Moreover, $|I_4| < \frac{m}{2}$.

The next three lemmas show that there are solution sets which contain the items of I_2 such that the corresponding solution values do not differ "too much" from the optimal solution value. Let us remark that we mean with *First Fit* a procedure which takes items arbitrarily from a set and assigns them to the first knapsack in which they fit without exceeding capacity c.

Lemma 2. Let I be feasible and $w(I \cup I_2) \leq m(1 - \varepsilon)c$, then $I \cup I_2$ is feasible by assigning the remaining items of I_2 by First Fit.

Lemma 3. Let $G' \subseteq G$ be feasible. Then, $G' \cup I_2$ is feasible with $w(G' \cup I_2) < \frac{mc}{2} < m(1 - \varepsilon)c$. Especially, $I_2 \cup I_4$ is feasible with $w(I_2 \cup I_4) < \frac{mc}{2}$.

Lemma 4. Let I be feasible with $I_4 \subseteq I$. Then there is a feasible set $\tilde{I}$ with $I_2 \cup I_4 \subseteq \tilde{I}$ such that $p(\tilde{I}) \geq p(I) - 4LB(\frac{1}{m} + \varepsilon)$.

3 Redefinition of the Item Sets and Formulation of an Integer Linear Program

In this section we will do a further splitting of item sets I_3 and I_4 into subsets and from these sets we will obtain after two redefinitions of the item sets a finite number of subsets which contain each only identical elements. For these finitely many different elements an integer program is formulated.

The partitioning of class I_3 into sets I_{3j} will now be done in the following way:

$$I_{3j} := \{4\tfrac{LB}{m}\varepsilon + j\tfrac{LB}{m}\varepsilon^2 < p_i \leq 4\tfrac{LB}{m}\varepsilon + (j+1)\tfrac{LB}{m}\varepsilon^2\}$$
$$(j = 0,\ldots,f(\tfrac{1}{\varepsilon^2}) := \lceil \tfrac{4}{\varepsilon^2} - \tfrac{4}{\varepsilon}\rceil - 1)$$

Then, reduce I_{3j} to the $\min\{m\lceil\tfrac{1}{\varepsilon} - 1\rceil, |I_{3j}|\}$ items with minimal weight and assign all these items the profit $\tilde{p}_j := 4\tfrac{LB}{m}\varepsilon + j\tfrac{LB}{m}\varepsilon^2$. This defines new sets of items R_j $(j = 0,\ldots,f(\tfrac{1}{\varepsilon^2}))$ with profits $\tilde{p}_j$ and original weights. Set $r_j := |R_j|$ and denote the elements of R_j by $(1,j),(2,j),\ldots,(r_j,j)$. Assume the elements of R_j to be sorted in non-decreasing weight order, i.e. $w_{1,j} \leq w_{2,j} \leq \ldots \leq w_{r_j,j}$.

From now on we will assume that $m > \tfrac{1}{\varepsilon^4}$. The simpler special case $m \leq 1/\varepsilon^4$ will be treated in the end of the paper. Set $k := \lfloor m\varepsilon^3 \rfloor$ and let π_j and q_j be nonnegative integers such that $r_j = \pi_j k + q_j$ and $0 \leq q_j \leq k-1$. This is possible, since $k \geq 1$. We conclude with $m > 1/\varepsilon^4$ and $\varepsilon < 1/2$ that

$$\pi_j \leq \frac{r_j}{k} \leq \frac{m\left(\lceil\tfrac{1}{\varepsilon} - 1\rceil\right)}{\lfloor m\varepsilon^3 \rfloor} \leq \frac{\tfrac{m}{\varepsilon}}{m\varepsilon^3 - 1} < \frac{1}{\varepsilon^5}.$$

Each set R_j is partitioned into π_j+1 subsets R_{ij} with $R_{ij} := \{(ik+1,j),\ldots,((i+1)k,j)\}$ $(i = 0,\ldots,\pi_j - 1)$ and $R_{\pi_j j} := \{(\pi_j k + 1, j),\ldots,(r_j,j)\}$. Thus, each set R_{ij} contains for $i \leq \pi_j - 1$ exactly k elements.

Let v_{ij} denote the maximum of the weights of items from R_{ij}, i.e.

$$v_{ij} := w_{(i+1)k,j} \qquad i = 0,\ldots,\pi_j - 1$$

and $v_{\pi_j j} := w_{r_j,j}$. Define new sets X_{ij} such that X_{ij} consists of k elements of profit $\tilde{p}_j$ and weight v_{ij} $(i = 0,\ldots,\pi_j - 1, j = 0,\ldots,f(\tfrac{1}{\varepsilon^2}))$ and $X_{\pi_j j}$ consists of q_j elements of weight $v_{\pi_j j}$ and profit $\tilde{p}_j$. Set $X_j := \bigcup_i X_{ij}$.

A similar partition is done with the set I_4. Set $R_{-1} := I_4$ with $r_{-1} := |R_{-1}| = |I_4|$. Denote the items from R_{-1} by $(i,-1)$ $(i = 1,\ldots,r_{-1})$ and assume them to be sorted in non-decreasing weight order, i.e. $w_{1,-1} \leq w_{2,-1} \leq \ldots \leq w_{r_{-1},-1}$.

As above, let π_{-1} and q_{-1} be nonnegative integers such that $r_{-1} = \pi_{-1}k + q_{-1}$ and $0 \leq q_{-1} \leq k - 1$. We obtain with Lemma 1

$$\pi_{-1} \leq \frac{r_{-1}}{k} < \frac{m/2}{\lfloor m\varepsilon^3 \rfloor} \leq \frac{1}{\varepsilon^5}.$$

Partition R_{-1} into $\pi_{-1} + 1$ subsets $R_{i,-1}$ $(i = 0,\ldots,\pi_{-1} - 1)$ with $R_{i,-1} := \{(ik+1,-1),\ldots,((i+1)k,-1)\}$ and $R_{\pi_{-1},-1} := \{(\pi_{-1}k+1,-1),\ldots,(r_{-1},-1)\}$. Set

$$v_{i,-1} := w_{(i+1)k,-1} \qquad i = 0,\ldots,\pi_{-1} - 1$$

and $v_{\pi_{-1},-1} := w_{r_{-1},-1}$. Let $X_{i,-1}$ consist of k elements of weight $v_{i,-1}$ $(i = 0,\ldots,\pi_{-1} - 1)$ and $X_{\pi_{-1},-1}$ consist of q_{-1} elements of weight $v_{\pi_{-1},-1}$. Finally, set $R := \bigcup_j R_j$, $X_{-1} := \bigcup_i X_{i,-1}$ and $X := \bigcup_{i,j} X_{ij}$. All elements of X have profit $\tilde{p}_{-1} := 0$.

The next lemma shows that we can find for each feasible set in B a corresponding feasible set in $R \cup I_2$ which differs not too much in profit from the original set and has weight not larger as before.

Lemma 5. *Let $B' \subseteq B$ be feasible with $I_2 \cup I_4 \subseteq B'$. Then there is a set $\tilde{R} \subseteq R$ with $R_{-1} \subseteq \tilde{R}$ and $I_2 \cup \tilde{R}$ feasible such that $p(\tilde{R} \cup I_2) \geq p(B') - \varepsilon LB$ and $w(\tilde{R} \cup I_2) \leq w(B')$.*

We call *feasible bin type* a list of arrays t_{ij} $(i = 0, \ldots, \pi_j, j = -1, \ldots, f(\frac{1}{\varepsilon^2}))$ with nonnegative integer entries such that $t_{ij} \leq k$ for $i = 0, \ldots, \pi_j - 1$ and $t_{\pi_j j} \leq q_j \leq k$ for $j = -1, \ldots, f(\frac{1}{\varepsilon^2})$ and that $\sum_{i,j} t_{ij} v_{ij} \leq c$ holds. Thus, a feasible bin type represents a subset of elements of X which can be put into a knapsack without exceeding the capacity c. Let $t^{(1)}, t^{(2)}, \ldots, t^{(f)}$ be an enumeration of all feasible bin types.

By definition, each set X_{ij} consists of at most $m\varepsilon^3$ identical items. We say that each set X_{ij} contains one *item type*. Thus, there are at most $O(\sum_{j=-1}^{f(1/\varepsilon^2)} \pi_j) = O(\frac{1}{\varepsilon^7})$ different item types corresponding to the different elements of X. Because items in X have all weight greater than εc, we get $\sum_{i,j} t_{ij} \leq \frac{1}{\varepsilon}$ and since the number of different item types is finite, the number of feasible bin types f is finite too.

Let us denote with $w(t) := \sum_{i,j \geq 0} t_{ij} v_{ij}$ the *weight of feasible bin type* t and with $p(t) := \sum_{i,j} t_{ij} \tilde{p}_j$ the *profit of feasible bin type* t. Note that for the weight of feasible bin type (and in principle also for the profit) we do not count the items of X_{-1}. We are ready to formulate the integer program $IP(\ell)$. This program tests whether there is a feasible set of items of X with minimal weight such that some given lower and upper bounds on the profit and the weight, respectively, are fulfilled. $IP(\ell)$ is defined as follows:

$$\text{minimize} \quad \sum_{d=1}^{f} w(t^{(d)}) x_d$$

$$\text{subject to} \quad \sum_{d=1}^{f} x_d \leq m, \tag{1}$$

$$\sum_{d=1}^{f} t_{ij}^{(d)} x_d \leq k, \quad i = 0, \ldots, \pi_j - 1, j = 0, \ldots, f(\frac{1}{\varepsilon^2}) \tag{2}$$

$$\sum_{d=1}^{f} t_{\pi_j j}^{(d)} x_d \leq q_j, \quad j = 0, \ldots, f(\frac{1}{\varepsilon^2}) \tag{3}$$

$$\sum_{d=1}^{f} t_{i,-1}^{(d)} x_d = k, \quad i = 0, \ldots, \pi_{-1} - 1 \tag{4}$$

$$\sum_{d=1}^{f} t_{\pi_{-1},-1}^{(d)} x_d = q_{-1}, \tag{5}$$

$$\sum_{d=1}^{f} p(t^{(d)}) x_d \geq \ell \varepsilon LB, \tag{6}$$

$$\sum_{d=1}^{f} w(t^{(d)}) x_d \leq m(1 - \varepsilon)c - w(I_2) - w(I_4), \tag{7}$$

$$x_d \geq 0 \text{ integer} , d = 1, \ldots, f.$$

Constraint (1) ensures that at most m feasible bin types are selected. Constraints (2) to (3) force that not more items than possible are taken from X_{ij} ($i = 0, \ldots, \pi_j, j = 0, \ldots, f(\frac{1}{\varepsilon^2})$). By (4) and (5) it is guaranteed that all elements from X_{-1}, representing the items of I_4, are put into the knapsacks. Finally, (6) and (7) ensure that the feasible set determined by $IP(\ell)$ has profit at least $\ell \varepsilon LB + p(I_4)$ and weight at most $m(1 - \varepsilon)c - w(I_2)$. As the number f of variables is constant if ε is fixed, this integer linear program can be solved in time polynomial in the number of constraints by applying Lenstra's algorithm [3]. Moreover, as the number of constraints is a constant as well, the overall running time of this algorithm is constant. (Note that we could skip also constraint (7) and consider only programs $IP(\ell)$ which have solutions not exceeding $m(1 - \varepsilon)c - w(I_2) - w(I_4)$. Note also that a solution for $IP(\ell)$ can be found in time $O(m^f)$ without using the results of Lenstra.)

4 The Algorithm and Proof of the Main Result

In this section we will present the promised PTAS for (MKP) and it will be proven that the given procedure is really a PTAS. Our algorithm works in principle as follows: First, the sets $I_1, \ldots, I_4, R_{ij}, X_{ij}$ are determined. For all "possible" integer values of ℓ the program $IP(\ell)$ is solved, then for each ℓ the solution obtained is converted into a feasible set for items of $I_3 \cup I_4$. Then, all items of I_2 are filled in the knapsacks by First Fit. Then, small items are assigned by the generalized Greedy algorithm as defined in the introduction. Finally, among all the produced solutions the one with maximal profit is choosen. Note that X_{-1} (with at most $m/2$ elements by Lemma 1) is a feasible solution for $IP(0)$. Therefore, it is guaranted that the PTAS generates at least one feasible assignment. A formal description of the PTAS will be given below:

1. Compute ε. Partition the item set N into sets $I_1, \ldots, I_4$.

2. By reducing the number of items of I_3 and rounding down the profits generate sets R_{ij} $(i = 0, \ldots, \pi_j, j = 0, \ldots, f(\frac{1}{\varepsilon^2}))$. Generate sets $R_{i,-1}$ $(i = 0, \ldots, \pi_{-1})$.

3. Convert sets R_{ij} into sets X_{ij}, thus producing a finite number of sets with at most $\lfloor m\varepsilon^3 \rfloor$ identical items per each set.

4. **For** $\ell = 0, \ldots, \lfloor \frac{2}{\varepsilon} \rfloor$ **do begin**

   4.1 If the integer program $IP(\ell)$ has a solution, generate a feasible subset of X with $s_{ij}(\ell)$ elements from X_{ij}, otherwise goto 5.

   4.2 Convert the solution obtained by $IP(\ell)$ into an assignment with items of $I_3 \cup I_4$. This is done by replacing X_{-1} by I_4 and by replacing the $s_{ij}(\ell)$ items from X_{ij} $(i = 0, \ldots \pi_j, j = 0, \ldots, f(\frac{1}{\varepsilon^2}))$ by the items from I_{3j} with weights $w_{(i+1)k,j}, w_{(i+1)k-1,j}, \ldots, w_{(i+1)k-s_{ij}(\ell)+1,j}$.

   4.3 Add all items from I_2 by First Fit to the knapsacks.

   4.4 Assign small items by the generalized Greedy algorithm generating a feasible set $z(\ell) = z_B(\ell) \cup z_S(\ell)$ with $z_B(\ell) \subseteq B$ and $z_S(\ell) \subseteq S$.

   end

5. Choose the solution $z(\ell)$ with maximal profit.

Let $I^* = B^* \cup S^*$ denote the optimal solution set with $B^* \subseteq B$ and $S^* \subseteq S$. Define $\tilde{I}^* = \tilde{B}^* \cup \tilde{S}^*$ with $\tilde{B}^* \subseteq B$ and $\tilde{S}^* \subseteq S$ as the optimal solution for (MKP) with the additional constraint that $I_2 \subseteq \tilde{B}^*$. The next lemma shows that there is an ℓ such that $z_B(\ell)$ is "very close" to $\tilde{B}^*$ in profits and weights.

Lemma 6. *There is an integer* $\ell \in \{0, \ldots, \lceil \frac{2}{\varepsilon} \rceil\}$ *such that the integer program* $IP(\ell)$ *has a solution and such that for the corresponding feasible solution for* (MKP), $z(\ell)$ *holds that* $p(z_B(\ell)) \geq p(\tilde{B}^*) - 34LB\varepsilon$ *and* $w(z_B(\ell)) \leq w(\tilde{B}^*)$.

Proof. By Lemma 5 with $B' = \tilde{B}^*$ there is a set $\tilde{R}^* \subseteq R$ with $R_{-1} \subseteq \tilde{R}^*$ and $I_2 \cup \tilde{R}^*$ feasible such that $p(\tilde{R}^* \cup I_2) \geq p(\tilde{B}^*) - \varepsilon LB$ and $w(\tilde{R}^* \cup I_2) \leq w(\tilde{B}^*)$. Set $\tilde{R}_j^* := \tilde{R}^* \cap R_j$ and $\tilde{R}_{ij}^* := \tilde{R}^* \cap R_{ij}$ $(i = 0, \ldots, \pi_j, j = -1, \ldots, f(\frac{1}{\varepsilon^2}))$ with $\tilde{r}_{ij}^* := |\tilde{R}_{ij}^*|$. Note that $\tilde{R}^* = \bigcup_{j \geq -1} \tilde{R}_j^*$. By definition of I_{3j} and R_j we have

$$p(\tilde{R}^*) = \hat{\ell} \frac{LB}{m} \varepsilon^2 + p(I_4) \qquad (8)$$

for some integer $\hat{\ell} \in \{0, \ldots, \lfloor \frac{2m}{\varepsilon^2} \rfloor\}$ and

$$w(\tilde{R})^* \leq mc - w(I_2) \qquad (9)$$

since $\tilde{R}^* \cup I_2$ is feasible . Set

$$\ell := \max \left\{ \left\lfloor \hat{\ell} \frac{\varepsilon}{m} \right\rfloor - 32, 0 \right\}.$$

We will show that $IP(\ell)$ has a solution:

For this reason define $X_{-1,j}$ for $j = 0, \ldots, f(\frac{1}{\varepsilon^2})$ as sets of k dummy items with both profits and weights zero. Consider the sets Y_{ij} $(i = 0, \ldots, \pi_j, j =$

$0, \ldots, f(\frac{1}{\varepsilon^2}))$ which consist of $\tilde{r}_{ij}^*$ items of the sets $X_{i-1,j}$ ($i = 0, \ldots, \pi_j$, $j = 0, \ldots, f(\frac{1}{\varepsilon^2})$). Because $|X_{ij}| = k$ for $i \leq \pi_j - 1$, the sets Y_{ij} are well-defined. Set $Y_j := \bigcup_{i \geq 0} Y_{ij}$ and $Y := \bigcup_{j \geq 0} Y_j$. Because of $|Y_{ij}| = |\tilde{R}_{ij}^*|$ one can construct a one-to-one correspondence between items of Y_{ij} and the items of $\tilde{R}_{ij}^*$ ($j \geq 0$). Furthermore, the non-decreasing weight ordering of sets R_j implies that items of R_{ij} have larger weights than items of $X_{i-1,j}$. Consequently, for each item $y \in Y_{ij}$ and its corresponding item $\tilde{y} \in \tilde{R}_{ij}^*$ the inequality $w(y) \leq w(\tilde{y})$ holds. Thus, Y is feasible with

$$w(Y) \leq w(\tilde{R}^*) - w(I_4) \tag{10}$$

and the feasible assignment for Y, which shall be denoted by $\mathcal{Y}$, can be found by assigning the items of Y to the same knapsacks as the corresponding items from $\tilde{R}^*$.

Items from R_j and X_j have the same profit $\tilde{p}_j$. Since Y_{0j} consists of k items of profit zero, we can conclude

$$p(Y_j) \geq p(\tilde{R}_j^*) - k\tilde{p}_j \qquad j = 0, \ldots, f(\frac{1}{\varepsilon^2}).$$

Using the fact that items in $R \setminus R_{-1}$ have maximum profit $4\frac{LB}{m}$, summation of the above inequalities and including I_4 yields

$$\begin{aligned} p(Y) + p(I_4) &\geq p(\tilde{R}^*) - k\left(\frac{4}{\varepsilon^2} + 1\right)\frac{4LB}{m} > \\ &> p(\tilde{R}^*) - m\varepsilon^3 \frac{5}{\varepsilon^2} \frac{4LB}{m} = p(\tilde{R}^*) - 20LB\varepsilon. \end{aligned} \tag{11}$$

We still did not consider the items of $R_{-1} = I_4$ which are also a part of $\tilde{R}^*$. The feasible assignment $\mathcal{Y}$ is extended to a new assignment just by assigning the items of $X_{i,-1}$ to the same knapsacks as the items of $R_{i+1,-1}$ for $i = 0, \ldots, \pi_{-1} - 2$, by assigning q_{-1} items of $X_{\pi_{-1}-1,-1}$ to the same knapsacks as the items of $R_{\pi_{-1},-1}$ and finally by assigning the remaining k items of X_{-1} to the same knapsacks as the items of $R_{0,-1}$. Note that only these remaining k items can have larger weights than the corresponding items of R_{-1}. There is no reason that this assignment is feasible, but there are at most $k \leq m\varepsilon^3$ knapsacks with weight exceeding the capacity c. By Lemma 1 we have $|X_{-1}| = |I_4| < m/2$. So we can find k knapsacks without items from X_{-1}. We remove the items in these knapsacks and replace them by the mentioned k items from X_{-1}. All removed items have profit $\leq 4\frac{LB}{m}$ and profit-weight ratio $\leq \frac{4LB}{m\varepsilon c}$. This creates a feasible set $Y^1 \subseteq X$ with corresponding assignment $\mathcal{Y}^1$ and $X_{-1} \subseteq Y^1$. The difference in profit between Y and Y^1 is

$$\begin{aligned} p(Y^1) &\geq p(Y) + p(I_4) - \frac{4LB}{m\varepsilon c}kc \geq \\ &\geq p(Y) + p(I_4) - m\varepsilon^3 \frac{4LB}{m\varepsilon} = p(Y) + p(I_4) - 4LB\varepsilon^2, \end{aligned} \tag{12}$$

and the weight of Y^1 differs from Y by

$$w(Y^1) \leq w(Y) + w(I_4). \tag{13}$$

Combining (9), (10) and (13) we obtain

$$w(Y^1) \leq mc - w(I_2).\qquad(14)$$

By converting the items of Y^1 into a feasible set I consisting of items of $I_3 \cup I_4$, the profits of the items are increased and the weights are decreased. Thus, the total weight of good ratio items in Y^1 is not larger than in I and it can be concluded with Lemma 3 that

$$w((Y^1 \setminus \bar{G}) \cup I_2) \leq w((I \setminus \bar{G}) \cup I_2) < \frac{mc}{2}.\qquad(15)$$

By (15) we can remove arbitrarily bad ratio items from Y^1 (corresponding to items from I_3) and stop when we get a set Y^2 with

$$w(Y^2) \leq m(1-\varepsilon)c - w(I_2).\qquad(16)$$

By (14) the bad ratio items to be removed consist of a set of items with total weight not exceeding $m\varepsilon c$ plus some "critical" item i. With $m > 1/\varepsilon$ we get

$$p(Y^2) \geq p(Y^1) - m\varepsilon c\frac{4LB}{mc} - p_i \geq p(Y^1) - 4LB\varepsilon - \frac{4LB}{m} > p(Y^1) - 8LB\varepsilon.\quad(17)$$

The combination of (11), (12) and (17) shows

$$p(Y^2) \geq p(\tilde{R}^*) - 28LB\varepsilon - 4LB\varepsilon^2 > p(\tilde{R}^*) - 32LB\varepsilon.$$

From (8) follows that

$$p(Y^2) > \hat{\ell}\frac{LB}{m}\varepsilon^2 + p(I_4) - 32LB\varepsilon = \left(\hat{\ell}\frac{\varepsilon}{m} - 32\right)LB\varepsilon + p(I_4)$$

and because of $p(Y^2) \geq p(I_4)$ we have finally

$$p(Y^2) \geq \ell LB\varepsilon + p(I_4).\qquad(18)$$

So, it can be seen that Y^2 corresponds to a feasible solution of $IP(\ell)$ and we conclude $p(IP(\ell)) \geq \ell LB\varepsilon + p(I_4)$ and $w(IP(\ell)) \leq w(Y^2)$.

By recalculating the items of I_3 and I_4 from $IP(\ell)$ in Step 4.2 of the algorithm all weights are decreasing and by (16) and Lemma 2 all items of I_2 can be assigned by First Fit such that the obtained set is still feasible. We have

$$w(z_B(\ell)) \leq w(IP(\ell)) + w(I_2) \leq w(Y^2) + w(I_2) \leq w(\tilde{R}^*) + w(I_2) \leq w(\tilde{B}^*).$$

Step 4.2 increases the profits of the items. Hence with (8) and (18),

$$p(z_B(\ell)) \geq \ell LB\varepsilon + p(I_4) + p(I_2) \geq \left(\hat{\ell}\frac{\varepsilon}{m} - 33\right)LB\varepsilon + p(I_4) + p(I_2) =$$

$$= p(\tilde{R}^*) + p(I_2) - 33LB\varepsilon \geq p(\tilde{B}^*) - 34LB\varepsilon.$$

□

Theorem 1. *The proposed algorithm is a PTAS.*

Proof. Consider integer program $IP(\ell)$ with ℓ as defined in Lemma 6. From this lemma we know that the total remaining capacity of the knapsacks before assigning the small items in loop ℓ of Step 4.4 is at least $mc - w(\tilde{B}^*) \geq w(\tilde{S}^*)$. Recall that using the generalized Greedy algorithm an upper bound for the difference between $p(\tilde{S}^*)$ and $p(z_S(\ell))$ is the sum of (at most) m critical items. Hence,

$$p(z_S(\ell)) \geq p(\tilde{S}^*) - m\frac{4LB\varepsilon}{m} = p(\tilde{S}^*) - 4LB\varepsilon.$$

This gives together with Lemma 6

$$z^H \geq p(z_B(\ell)) + p(z_S(\ell)) \geq p(\tilde{B}^*) + p(\tilde{S}^*) - 38LB\varepsilon = p(\tilde{I}^*) - 38LB\varepsilon.$$

Lemma 4 with $I = I^*$ yields $p(\tilde{I}^*) \geq p(I^*) - 4LB(\frac{1}{m} + \varepsilon)$. Using $m > 1/\varepsilon$ and $z^* \geq LB$,

$$z^H \geq z^* \left(1 - 38\varepsilon - 4(\frac{1}{m} + \varepsilon)\right) > z^*(1 - 46\varepsilon).$$

Setting $\tilde{\varepsilon} := 46\varepsilon$, we get a $(1 - \tilde{\varepsilon})$-approximation algorithm and thus a PTAS.

It remains to consider the special case $m \leq 1/\varepsilon^4$. Its solution can be roughly described as follows. Form the sets R_j $(j = 0, 1, \ldots)$ from I_3 as in the general case. There are $O(\frac{1}{\varepsilon^2})$ sets R_j with at most m/ε elements, moreover $|I_4| < m/2$ and $|I_2| < \frac{1}{\varepsilon}\frac{m}{2}$. Consequently, set R containing items from all R_j and from I_2 and I_4 has only a finite number of elements. Analogously to Lemma 5 it can be shown that there is a feasible set $R^* \subseteq R$ such that $p(R^*) \geq p(B^*) - \varepsilon LB$ and $w(R^*) \leq w(B^*)$. This set R^* can be found in constant time by complete enumeration of all assignments of R. Then the generalized Greedy algorithm is applied for the small items with a additional loss in profit of at most $4LB\varepsilon$. This completes the proof of our theorem. $\square$

References

1. Caprara A., Kellerer H, Pferschy U.:, The multiple subset sum problem. Technical Report (1998) Faculty of Economics, University of Graz (submitted)
2. Kellerer H., Pferschy U.: A new fully polynomial approximation scheme for the knapsack problem. Proceedings of APPROX 98, Springer Lecture Notes in Computer Science **1444** (1998) 123-134
3. Lenstra H.W.: Integer programming with a fixed number of variables. Mathematics of Operations Research **8** (1983) 538-548
4. Martello S. Toth P.: Knapsack problems: Algorithms and computer implementations. J. Wiley & Sons, Chichester (1990).

Appendix

Proof (of Lemma 1). Since $z^* \leq 2LB$, there are at most $m/2$ knapsacks with knapsack profit exceeding $4LB/m$. Assume, there is an item $i \in I_4$ with $i \notin I^*$. Then, take a knapsack K with $w(K) \leq 4LB/m$ remove all items of K and put i into K. This increases the total profit, a contradiction to the optimality of I^*. The second assertion follows immediately. □

Proof (of Lemma 2). Consider an feasible assignment of I to the m knapsacks and fill in the items of I_2 by First Fit. Since the weights of items in I_2 are not greater than εc, each knapsack can be filled with items of I_2 such it has weight greater than $(1 - \varepsilon)c$ until we run out of items of I_2. Therefore, $I \cup I_2$ is feasible. □

Proof (of Lemma 3). Assume $G' \cup I_2$ is not feasible. Analogously to the proof of Lemma 1 we can assign the items of I_2 to a feasible assignment of G' by First Fit until we get a feasible set $G' \cup I_2'$ with $w(G' \cup I_2') \geq \frac{mc}{2}$ and $I_2' \subseteq I_2$. All elements of this set are good ratio items which implies

$$p(G' \cup I_2') > \frac{4LB}{mc} w(G' \cup I_2') \geq \frac{4LB}{mc} \frac{mc}{2} = 2LB,$$

a contradiction to $z^* \leq 2LB$. □

Proof (of Lemma 4). Consider an feasible assignment of I to the m knapsacks. We change this assignment now by removing successively items from $\bar{G}$ and replacing them by items from I_2 as long as the new knapsack weight does not exceed the capacity c. In this exchange procedure we start with an arbitrary knapsack and turn to the next one when all items from $\bar{G}$ in this knapsack are exchanged. All items of I_2 can be filled into the knapsacks, because otherwise the total weight of items of G in the new feasible assignment would be greater than $m(1 - \varepsilon)c$ in contradiction to Lemma 3.

Let the original assignment for I consist of knapsacks $K^1, \ldots, K^m$ and the new assignment (after inserting the items of I_2) consist of knapsacks $\tilde{K}^1, \ldots, \tilde{K}^m$. The corresponding feasible set shall be called $\tilde{I}$. W.l.o.g., we may assume that

a) knapsacks $\tilde{K}^1, \ldots \tilde{K}^{m_1}$ $(0 \leq m_1 \leq m)$ contain no items from $\bar{G}$ and have weight exceeding $(1 - \varepsilon)c$,

b) we run out of items of I_2 after removing an item $i \in \bar{G}$ from knapsack $\tilde{K}^{m_1+1}$, i.e. $w(K^{m_1+1}) \leq w(\tilde{K}^{m_1+1}) + w_i$,

c) all other knapsacks remain unchanged.

Since we exchanged only bad ratio items by good ratio items the profit $p(\tilde{I})$ can only decrease compared to $p(I)$ if the weights of the knapsacks decrease too.

Consequently, we get as an upper bound for the total loss in profit from the first m_1 knapsacks and from item i

$$p(I) - p(\tilde{I}) \le \frac{4LB}{mc} m_1 \varepsilon c + \frac{4LB}{m} < 4LB \left(\frac{1}{m} + \varepsilon \right).$$

□

Proof (of Lemma 5). The set $\tilde{R}$ is constructed by exchanging the elements from $B' \cap I_3$ with corresponding elements from R. (Note that $I_4 = R_{-1}$.) Let b'_j denote the number of elements of I_{3j} in set B'. Since items of I_3 have weight exceeding εc, there can be at most $\lceil \frac{1}{\varepsilon} - 1 \rceil$ items of I_3 in one knapsack and we have $b'_j \le \min\{m\lceil \frac{1}{\varepsilon} - 1\rceil, |I_{3j}|\}$ which is the cardinality of R_j ($j \ge 0$). Therefore, we can exchange the b'_j elements of I_{3j} in B' by the b'_j elements of R_j with minimal weight.

By definition each item in $\tilde{R}$ has weight not greater than the corresponding item in B'. Thus, $I_2 \cup \tilde{R}$ is feasible. The maximal difference in profit between item from B' and $\tilde{R}$ is $\frac{LB}{m} \varepsilon^2$. Since B' consists of at most $m\lceil \frac{1}{\varepsilon} - 1 \rceil$ items from I_3, the total change in profit is bounded from above by $\frac{m}{\varepsilon} \frac{LB}{m} \varepsilon^2 = \varepsilon LB$. □

Set Cover with Requirements and Costs Evolving over Time

Milena Mihail

College of Computing and Department of Industrial and Systems Engineering
Georgia Institute of Technology, Atlanta GA 30332
mihail@cc.gatech.edu

Abstract. We model certain issues of future planning by introducing time parameters to the set cover problem. For example, this model captures the scenario of optimization under projections of increasing covering demand and decreasing set cost. We obtain an efficient approximation algorithm with performance guarantee independent of time, thus achieving planning for the future with the same accuracy as optimizing in the standard static model.

From a technical point of view, the difficulty in scheduling the evolution of a (set cover) solution that is "good over time" is in quantifying the intuition that "a solution which is suboptimal for time t may be chosen, if this solution reduces substantially the additional cost required to obtain a solution for $t' > t$". We use the greedy set picking approach, however, we introduce a new criterion for evaluating the potential benefit of sets that addresses precisely the above difficulty.

The above extension of the set cover problem arose in a toolkit for automated design and architecture evolution of high speed networks. Further optimization problems that arise in the same context include survivable network design, facility location with demands and natural extensions of these problems under projections of increasing demands and decreasing costs; obtaining efficient approximation algorithms for the latter questions are interesting open problems.

1   Introduction: Set Cover with Time Parameters

Fluctuations in cost and demand is a natural phenomenon of free markets and the relevance of cost effective schedules under such fluctuations is fundamental for both clients and service providers. When cost and demand fluctuations are totally unpredictable the models are necessarily on-line and, by now, we have a rich theory of on-line algorithms with heuristics that have found concrete practical applications [2] [4] [14]. However, in many cases, fairly accurate projections for the evolution of cost and demand over time are known in advance and the performance measures of on-line models no longer apply. This paper focuses on the latter context.

The concern that motivated this paper involves the evolution of networks like SONET, ATM, FRAME RELAY, WDM, e.t.c. which are experiencing a sharp increase in service demand (e.g. bandwidth) together with a substantial decrease in cost for the upgrade of the infrastructure that will cover this demand. Tools for automated design of such networks input service demand and equipment cost projected over several points in the future, and output a network solution that evolves over these points in the future. At the algorithmic core of such tools it is typical to find problems reminiscent of set cover and its variants. Very roughly, elements represent bandwidth demand and sets represent systems that can cover a collection of demands. For example, this approach was explicitly taken in Bellcore's SONET planning tool [6]; see Figure 1. We thus need adaptations of the set cover problem that capture the following intuition: If equipment cost (cost of sets) is expected to drop as bandwith demand (requirements of elements) is expected to rise, we wish to explore the option of bying equipment at a later point for a smaller cost. On the other hand, it may also be beneficial to buy slightly more equipment than what is necessary at present, if this additional equipment will cover a large amount of future demand at very low extra cost. The set cover problem with parameters evolving over time of Section 2 captures these considerations. Note that this is not an on-line model. Here the future is known in advance, but it introduces an additional dimension of complexity to the problem. We measure our performance by *comparing the approximation factor of the time variant of the problem to the best known approximation factor of the static problem.*

Each pair of nodes A and B of the network is represented by an element i.

A set S_j represents a specific SONET architecture that can be embedded in the network. It contains element i if and only if embedding the specific architecture can satisfy one unit of demand from A to B.

The requirement of element i, r(i), represents the point–to–point bandwidth demand between nodes A and B.

The cost c(j) of set S_j is the cost of the SONET architecture represented by S_j.

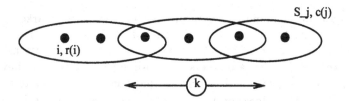

Figure 1 [6] describes a commercial SONET planning tool implementing an adaptation of the classical greedy algorithm for set cover. At a suitable level of abstraction, the heuristic of the planning tool can be viewed as follows: while there exists point-to-point bandwidth demand not covered by the SONET architectures selected so far, select a new architecture that is most efficient for the current iteration.

In Section 2 we give a reduction of set cover with time parameters to the standard set cover problem. This reduction carries over approximations and suggests a factor $\mathcal{H}(kT)$ algorithm for set cover with time parameters (e.g. following Chvatal [5]), where k is the cardinality of the largest set in the original problem, T is the number of points in time and kT is the cardinality of the largest set in the set cover instance that arises in the reduction.

In Section 3 we improve this factor to an optimal $\mathcal{H}(k)$ by introducing a new criterion for picking sets and adapting the standard duality based performance guarantee accordingly. The new criterion captures the following intuition: (a) cost suboptimal solutions should be considered for time t, if such solutions reduce substantially the additional cost required to obtain a solution for $t' > t$, (b) there is benefit in postponing picking sets that are not necessary for time t, if the cost at time $t' > t$ drops substantially and (c) while there is potential benefit in sets whose cost drops over time, this benefit should be counter-measured against the potential redundancy of such sets, if their effectiveness will be eventually covered by other sets that are necessary at earlier times.

The survivable network design problem and versions of the facility location problem are further examples of combinatorial optimization problems that arise repeatedly in automated network design [1] [9] [12] [13] [15]; major progress for these problems has been reported recently [3] [8] [10] [11] [16]. The survivable network design problem and the facility location problem have natural extensions with time parameters which remain open. We give an outline in Section 4.

2 Reduction to Set Cover and a factor $\mathcal{H}(kT)$ Approximation

The formal definition of SET COVER WITH REQUIREMENTS AND COSTS EVOLVING OVER TIME is as follows. There is a universe of n elements and a set system of m sets denoted by S_j, $1 \leq j \leq m$. Let k denote the maximum cardinality of a set $k = \max_{1 \leq j \leq m} |S_j|$. As in the standard set cover problem, elements have covering requirements and sets have costs. In this extended model however, requirements and costs evolve over T discrete points in time. In particular, for each time t : $1 \leq t \leq T$, each element i needs to be covered by $r(i,t)$ sets that have been picked on or before time t, while picking one copy of set S_j at time t has cost $c(j,t)$. We assume that the future evolution of requirements and costs are known in advance and we consider a "buying" scenario where, if a set is picked at some point in time, it is never removed — for example, the purchase and installation of SONET architectures incurs tens or hundreds of millions of cost in equipment and management; once installed, such architectures are not removed. We wish to pick sets that satisfy the requirements at every point in time and are of minimal total cost. Formally, where $x(j,t)$ denotes the number of copies of S_j picked at time t, we have to solve:

$$\begin{aligned}
\min \quad & \sum_{t=1}^{T} \sum_{j=1}^{m} c(j,t)x(j,t) \\
\text{subject to} \quad & \sum_{t'=1}^{t} \sum_{j:i \in S_j} x(j,t') \geq r(i,t) \quad 1 \leq i \leq n\,,\ 1 \leq t \leq T \\
& x(j,t) \in \aleph_0 \quad 1 \leq j \leq m\,,\ 1 \leq t \leq T
\end{aligned}$$

We first give a reduction to the standard SET COVER problem. In particular, for a universe of n elements, a set system of m sets denoted by S_j, $1 \leq j \leq m$ and where $k = \max_{1 \leq j \leq m} |S_j|$, in the SET COVER problem each element i has a covering requirement $r(i)$, picking one copy of set S_j has cost $c(j)$ and we wish to find a minimum cost collection of sets that satisfy the covering requirements:

$$\begin{aligned}
\min \quad & \sum_{j=1}^{m} c(j)x(j) \\
\text{subject to} \quad & \sum_{j:i \in S_j} x(j) \geq r(i) \quad 1 \leq i \leq n \\
& x(j) \in \aleph_0 \quad 1 \leq j \leq m
\end{aligned}$$

Recall also the classical GREEDY ALGORITHM FOR SET COVER which repeatedly picks sets that reduce the total number of requirements at minimum average cost per unit of covered requirement. More specifically, for each element i with requirement $r(i)$ consider a stack of $r(i)$ chips labeled p_{ir}, $1 \leq r \leq r(i)$, and consider a further labeling of each chip as either *covered* or *uncovered*. Define the *potential* of a set S_j with respect to such a labeling as the average cost at which the set covers uncovered chips: $P(S_j) = c(j)/|\{i : \exists p_{ir} \in S_j \text{ and } p_{ir} \text{ is uncovered}\}|$. The algorithm then is:

GREEDY ALGORITHM FOR SET COVER
$x(j) = 0$, $\forall j$;
label chip p_{ir} "uncovered", $\forall i, r$;
while there exist uncovered chips do
        set $P(S_j) = |\{i : \exists p_{ir} \in S_j \text{ and } p_{ir} \text{ is uncovered}\}|$, $\forall j$;
        for some S_{j_0} that minimizes $c(j)/P(S_j)$ set $x(j_0) = x(j_0) + 1$;
        for all i,
            if some uncovered chip $p_{ir} \in S_{j_0}$ then
                label chip p_{ir} "covered" for exactly one such r;
        set $cost(p_{ir}) = c(j_0)/P(S_{j_0})$;

Now the following performance guarantee is well known and follows by duality considerations [5]:

Theorem 1. [Chvatal]. *The cost of the* GREEDY ALGORITHM FOR SET COVER *is within a* $\mathcal{H}(k)$ *multiplicative factor of the cost of any optimal solution:* $\sum_{j=1}^{m} c(j)x(j) \leq \mathcal{H}(k) \cdot \text{OPT}$.

We may now give the reduction from SET COVER WITH PARAMETERS EVOLVING OVER TIME to SET COVER. See Figure 2. For each element i and each time t of SET COVER WITH PARAMETERS EVOLVING OVER TIME we introduce an element I_{it} with requirement $r(i,t)$ for SET COVER, and for each set S_j and each time t of SET COVER WITH TIME PARAMETERS we introduce a new set $S_{jt} = \{I_{it'} : i \in S_j , t' \geq t\}$ of cost $c(j,t)$ for SET COVER. Realize that the maximum set cardinality is kT, thus Chvatal's Theorem suggests that the greedy algorithm for set cover achieves a $\ln kT$ approximation factor. In the next Section we will modify the criterion for picking sets and achieve approximation factor $\ln k$. This is optimal in view of Feige's bound [7].

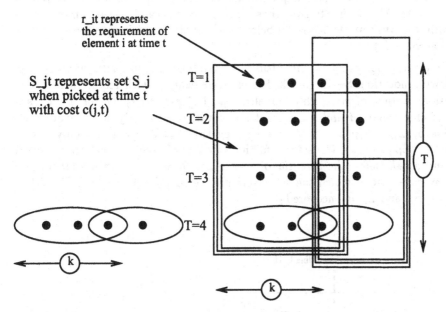

Figure 2 Indicating the the reduction of SET COVER WITH PARAMETERS EVOLVING OVER TIME to SET COVER. This reduction increases the size of the problem by a factor T.

3 A factor $\mathcal{H}(k)$ Modified Greedy Algorithm

How can we improve the GREEDY ALGORITHM of Section 2 when applied to SET COVER instances that arise from the reduction from SET COVER WITH RE-QUIREMENTS AND COSTS EVOLVING OVER TIME? Realize that a good heuristic for the latter set cover problem should capture the following: (a) cost suboptimal solutions must be considered for time t, if such solutions reduce substantially the additional cost required to obtain a solution for $t' > t$, (b) there is benefit in postponing picking sets that are not necessary for time t, if the cost at time $t' > t$ drops substantially and (c) while there is potential benefit in sets whose cost drops over time, this benefit should be counter-measured against the potential redundancy of such sets, if their effectiveness will be eventually covered by other sets that are necessary at earlier times. Realize further that the potential of sets S_{jt} arising in the reduction indeed capture (a) and (b). In particular, for (a), note that a set S_{jt} includes elements representing requirements for times $t' \geq t$ which may increase the potential of S_{jt}, while for (b), note that a substantial drop of the cost of a set S_j at time t' is represented by the cost of the set $S_{jt'}$ which must consequently become of high potential. However, the reduction does not capture (c). In particular, a set $S_{jt'}$ of very low cost could be chosen at first to satisfy the requirement of an element at time t'. On the other hand, this element may also have requirements at time $t < t'$ which will eventually result in the choice of sets $S_{j't}$, thus making the choice of $S_{jt'}$ redundant. The MODIFIED GREEDY ALGORITHM below modifies the set picking criterion to take into account (c).

For the description of the MODIFIED GREEDY ALGORITHM we need the following notation. See Figure 3. For each element I_{it} with requirement $r(i, t)$ consider a stack of $r(i, t)$ chips labeled p_{itr}, $1 \leq r \leq r(i, t)$. Define a *line* as a set of chips where i and r are fixed and t varies arbitrarily, and denote such a line by $L_{ir} = \{p_{itr} : 1 \leq t \leq T\}$. Say that $L_{ir} \in S_{jt}$ if and only if $t \leq \min\{t' : p_{it'r} \in L_{ir}\}$. Consider a further labeling of each line as either *covered* or *uncovered*. Define the *potential* of a set S_{jt} with respect to such a labeling as the average cost at which the set covers uncovered lines: $P(S_{jt}) = C(j, t)/|\{i : \exists L_{ir} \in S_{jt} \text{ and } L_{ir} \text{ is uncovered}\}|$.

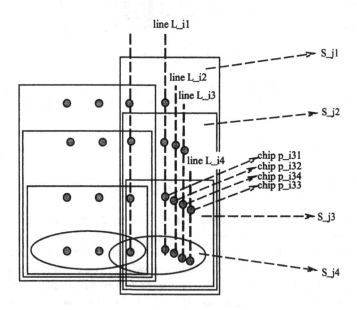

Figure 3 Indicating sets covering entire lines. For example, S_{j3} covers line L_{i4}, but S_{j3} does not cover lines L_{i3}, L_{i2} and L_{i1}.

Now we modify the set picking criterion as follows:

MODIFIED GREEDY ALGORITHM FOR SET COVER WITH PARAMETERS EVOLV-
ING OVER TIME

$x(j,t) = 0, \ \forall j,t;$

label line L_{ir} "uncovered", $\forall i, r$;

while there exist uncovered lines do

        set $P(S_{jt}) = |\{i : \exists L_{ir} \in S_{jt} \text{ and } L_{ir} \text{ is uncovered}\}|, \ \forall j, t$;

        for some $S_{j_0 t_0}$ that minimizes $c(j,t)/P(S_{jt})$ set $x(j_0, t_0) = x(j_0, t_0) + 1$;

        for all i,

            if some uncovered line $L_{ir} \in S_{j_0 t_0}$ then

                label line L_{ir} "covered" for exactly one such r;

        set $cost(L_{ir}) = c(j_0, t_0)/P(S_{j_0 t_0})$;

For the performance guarantee observe:

Lemma 2. *For all sets S_{jt}, $\sum_{i \in S_j} \max_{L_{ir} \in S_{jt}} cost(L_{ir}) \leq \mathcal{H}(k) \cdot c(j,t)$.*

Proof. Assume without loss of generality that $S_j = \{1, \ldots, k'\}$. Also assume without loss of generality that for fixed t, among all lines $L_{ir} \in S_{jt}$ (as r varies), L_{ir_i} was the last line to be covered by the MODIFIED GREEDY ALGORITHM. Finally assume without loss of generality that, for all $1 \leq i' \leq i \leq k'$, L_{ir_i} was covered at a previous or at the same iteration of the MODIFIED GREEDY ALGORITHM as line $L_{i'r_{i'}}$. Then, since S_{jt} could have covered L_{ir_i} at cost no more than $c(j,t)/i$ we have:

$$cost(L_{ir}) \leq \frac{c(j,t)}{i} \quad , \quad 1 \leq i \leq k'$$

Thus,

$$\sum_{i \in S_j} \max_{L_{ir} \in S_{jt}} cost(L_{ir}) = \sum_{i \in S_j} cost(L_{ir_i})$$

$$\leq \left(\frac{1}{k'} + \frac{1}{k'-1} + \ldots + 1 \right) c(j,t)$$

$$\leq \mathcal{H}(k) \cdot c(j,t) .$$

Theorem 3. Performance Guarantee. *The cost of the solution of MODIFIED GREEDY ALGORITHM is within a $\ln k$ multiplicative factor of the cost of any optimal solution: $\sum_{t=1}^{T} \sum_{j=1}^{m} c(j,t) x(j,t) \leq \mathcal{H}(k) \cdot \text{OPT}$*

Proof. Follows by applying the Lemma to the sets of some optimal solution. In particular, let us fix an optimal solution, and suppose that it contains $x^*(j,t)$ copies of set S_{jt}. Then

$$\mathcal{H}(k) \cdot \text{OPT} = \mathcal{H}(k) \sum_{t=1}^{T} \sum_{j=1}^{m} c(j,t) x^*(j,t)$$

$$\geq \sum_{t=1}^{T} \sum_{j=1}^{m} x^*(j,t) \sum_{i \in S_j} \max_{L_{ir} \in S_{jt}} cost(L_{ir}) \ , \ \text{by the Lemma}$$

$$\geq \sum_{i=1}^{n} \sum_{r=1}^{\max_t r(i,t)} cost(L_{ir}) \text{ , by counting}$$

$$= \sum_{t=1}^{T} \sum_{j=1}^{m} c(j,t)x(j,t) \text{ , also by counting}$$

4 Survivabe Network Design and Facility Location

In the survivable network design problem we are given a weighted undirected graph and a requirement function over the cuts of the graph. We wish to pick a minimum cost subgraph such that each cut is crossed by at least as many edges as its requirement. Formally, for an undirected graph $G(V, E)$, $n = |V|$, cost function c on its edges: $E \xrightarrow{c} Q_+$, cut requirement function $f\colon 2^V \xrightarrow{f} \aleph_0$ and where $\delta(S)$ is the set of edges in E with exactly one endpoint in S, the survivable network design problem is expressed by the integer program below.

$$\min \sum_{e \in E} c(e)x(e)$$
$$\sum_{e \in \delta(S)} x(e) \geq f(S) \quad \forall S \subseteq V$$
$$x(e) \in \{0, 1\} \quad \forall e \in E$$

The survivable network design problem models "survivability/reliability" considerations and has a long history in practice and in theory [1] [4] [8] [10] [12] [13] [16]. The extension of this problem with time parameters involves a cut requirement function which increases with time and costs of edges which decrease with time. Formally, for an undirected graph $G(V, E)$, $n = |V|$, $[T] = \{1, 2, \ldots, T\}$ points in time, cost function c on the edges: $E \times [T] \xrightarrow{c} Q_+$ and cut requirement function $f\colon 2^V \times [T] \xrightarrow{f} \aleph_0$, the network design problem with requirements and costs evolving over time is expressed by the integer program below.

$$\min \sum_{t=1}^{T} \sum_{e \in E} c(e,t)x(e,t)$$
$$\sum_{t'=1}^{t} \sum_{e \in \delta(S)} x(e,t') \geq f(S,t) \quad \forall S \subseteq V, \forall t \in [T]$$
$$\sum_{t=1}^{T} x(e,t) \in \{0, 1\} \quad \forall e \in E$$

Obtaining an approximation for the above problem is open. Such an approximation would be of concrete practical importance, for example, in upgrading the architectures of survivable CCSN, SONET, and WDM telecommunications networks.

The facility location problem has many variants; here we outline a representative one. There is a collection of facilities: each facility i can be opened at cost $f(i)$ for each unit of capacity $u(i)$. There is also a collection of cities: each city j has demand $d(j)$ which must be routed to open facilities. One unit of demand from city j can be routed to facility i at cost $c(i, j)$. We wish choose facilities that minimize the total cost:

$$\min \sum_{i \in F, j \in C} c(i,j)x(i,j) + \sum_{i \in F} y(i)f(i)$$
$$\sum_{i \in F} x(i,j) \geq d(j) \qquad \forall j \in C$$
$$u(i)y(i) \geq \sum_{j \in C} x(i,j) \qquad \forall i \in F , \forall j \in C$$
$$x(i,j), y(i) \in \aleph_0 \qquad \forall i \in F , \forall j \in C$$

Now in the facility location problem with parameters evolving over time we have the demands increasing with time, and the costs of opening facilities and routing demands decreasing with time. We may write:

$$\min \sum_{t=1}^{T} \sum_{i \in F, j \in C} c(i,j,t)x(i,j,t) + \sum_{i \in F} y(i,t)f(i,t)$$
$$\sum_{t'=1}^{t} \sum_{i \in F} x(i,j,t') \geq d(j,t) \qquad \forall j \in C , \forall t \in [T]$$
$$u(i) \sum_{t'=1}^{t} y(i,t') \geq \sum_{t'=1}^{t} \sum_{j \in C} x(i,j,t') \qquad \forall i \in F , \forall j \in C , \forall t \in [T]$$
$$x(i,j,t), y(i,t) \in \aleph_0 \qquad \forall i \in F , \forall j \in C , \forall t \in [T]$$

Obtaining an approximation for the above problem (or some suitable variant) is open. Such an approximation would be of importance, for example, in upgrading the architectures of ATM and Frame Relay telecommunications networks [15].

References

1. M. Ball, T. Magnati, C. Monma, and G Hemhauser, *Handbook in Operations Research and Management Science*, Vol 8, North-Holland (1992).
2. A. Borodin and R. El-Yaniv, Online Computation and Competitive Analysis, Cambridge University Press, 1998.
3. M. Charikar, S. Guha, E. Tardos, and D. Shmoys, "A constant-factor Approximation Algorithm for the k-median Problem", to appear in STOC Proc., 1999.
4. D. Hochbaum, Approximation Algorithms for NP-Hard Problems, PSW Publishing Company, Boston MA, 1997.
5. V. Chvatal, *A Greedy Heuristic for the Set Covering Probl em*, Mathematics of Operations Research, 4 (1979), pp. 233-235.
6. S. Cosares, D.Deutch, I. Saniee, and O. Wasem, "SONET Toolkit: A Decision Support System for the Design of Cost-Effective Fiber Optic Networks, *Interfaces* Vol25, Jan-Feb 1995, pp.20-40.
7. U. Feige, "A Threshold of ln n for Approximating Set Cover", in Proceedings of *STOC* 1996.
8. M. Goemans, A. Goldberg, S. Plotkin, D. Schmoys, E. Tardos, and D. Williamson, *Improved Approximation Algorithms for Network Design Problems*, Proc. SODA 94.
9. 3rd *INFORMS Telecommunications Conference*, Special Sessions on "Network Design Aspects about ATM" and "Design and Routing for Telecommunications Networks", May 1997.
10. K. Jain, "A Factor 2 Approximation Algorithm for the Generalized Steiner Network Problem", FOCS Proc., 1998.
11. K. Jain and V.V. Vazirani, "Primal-Dual Approximation Algorithms for Metric Facility Location and k-Median Problems", submitted, also in http://www.cc.gatech.edu/fac/Vijay.Vazirani.
12. T. Magnati and R.T. Wong, "Network Design and Transportation Planning": Models and Algorithms, *Transportation Science* 18, pp. 1-55, 1984.

13. M. Mihail, D. Shallcross, N. Dean, and M. Mostrel, *A Commercial Application of Survivable Network Design*, Proc. SODA 96.

14. R.Motwani and P. Raghavan, Randomized Algorithms, Cambridge University Press, 1995.

15. Iraj Saniee and Dan Bienstock, "ATM Network Design: Traffic Models and Optimization Based Heuristics", 4th *INFORMS Telecommunications Conference*, March 1998.

16. D. Williamson, M. Goemans, M. Mihail, and V. Vazirani, *A Primal-Dual Approximation Algorithm for Generalized Steiner Network Problems, Combinatorica* 15:435-454, December 1995.

Multicoloring Planar Graphs and Partial k-Trees

Magnús M. Halldórsson[1] and Guy Kortsarz[2]

[1] Science Institute, University of Iceland, Reykjavík, Iceland. mmh@hi.is.
[2] Dept. of Computer Science, Open University, Ramat Aviv, Israel.
guyk@tavor.openu.ac.il.

Abstract. We study the multicoloring problem with two objective functions: minimizing the *makespan* and minimizing the *multisum*. We focus on partial k-trees and planar graphs. In particular, we give polynomial time approximation schemes (PTAS) for both classes, for both preemptive and non-preemptive multisum colorings.

1 Introduction

Scheduling dependent jobs on multiple machines is modeled as a graph *coloring* problem, when all jobs have the same (unit) execution times, and as graph *multicoloring* for arbitrary execution times. The vertices of the graph represent the jobs and an edge in the graph between two vertices represents a dependency between the two corresponding jobs that forbids scheduling these jobs at the same time.

An *instance* to multicoloring problems is a pair (G, x), where $G = (V, E)$ is a graph, and x is a vector of *color requirements* (or *lengths*) of the vertices. For a given instance, we denote by n the number of vertices, by $p = \max_{v \in V} x(v)$ the maximum color requirement, and by $S(G) = \sum_v x(v)$ the sum of the color requirements of the vertices. A *multicoloring* of G is an assignment $\Psi : V \mapsto 2^N$, such that each vertex $v \in V$ is assigned a set of $x(v)$ distinct colors and adjacent vertices receive non-intersecting sets of colors.

A multicoloring Ψ is called *non-preemptive* if the colors assigned to v are contiguous, i.e. if for any $v \in V$, $(\max_{i \in \Psi(v)} i) - (\min_{i \in \Psi(v)} i) + 1 = x(v)$. If arbitrary sets of colors are allowed, the coloring is *preemptive*. The preemptive version corresponds to the scheduling approach commonly used in modern operating systems [SG98], where jobs may be interrupted during their execution and resumed at a later time. The non-preemptive version captures the execution model adopted in real-time systems where scheduled jobs must run to completion.

One of the traditional optimization goals is to minimize the total number of colors assigned to G. In the setting of a job system, this is equivalent to finding a schedule minimizing the time within which *all* the jobs have been completed. Such an optimization goal favors the system. However, from the point of view of the jobs themselves, another important goal is to minimize the average completion time of the jobs.

We study multicoloring graphs in both the preemptive and non-preemptive models, under both the makespan and sum-of-completion times measures defined as follows. Denote by $f_\Psi(v) = \max_{i \in \Psi(v)} i$ the largest color assigned to v by multicoloring Ψ. The *multisum* of Ψ on G is

$$\text{SMC}(G, \Psi) = \sum_{v \in V} f_\Psi(v) .$$

Minimizing the makespan is simply minimizing $\max_v \{f_\Psi(v)\}$. The problem of finding a preemptive (non-preemptive) multicoloring with minimum sum (makespan) is denoted p-sum (p-makespan), while the non-preemptive version is np-sum (np-makespan, respectively). When all the color requirements are equal to 1, the makespan problem is simply the usual coloring problem, while the sum versions reduce to the well-known *sum coloring* (SC) problem.

The Sum Multicoloring (SMC) problem has numerous applications, including traffic intersection control [B92,BH94], session scheduling in local-area networks [CCO93], compiler design and VLSI routing [NSS99].

Related work: The p-makespan problem is NP-hard even when restricted to hexagon graphs [MR97], while a 4/3-approximation is known [NS97]. The problem is polynomial solvable on outerplanar graphs [NS97], and trivial on bipartite graphs (cf. [NS97]).

The sum coloring problem was introduced in [KS89]. A summary of recent work can be found in [BBH$^+$98], along with approximation results on various classes of graphs. For partial k-trees, Jansen [J97] gave a polynomial algorithm for the Optimal Chromatic Cost Problem (OCCP) that generalizes the sum coloring problem.

The paper most relevant to our study is [BHK$^+$98], where the np-sum and p-sum problems are thoroughly studied. We shall make use of the following result of [BHK$^+$98]:

Fact 1 *Let (G, x) be a graph with lengths x, and with a k-coloring given. Then, we can find a non-preemptive multicoloring of G with a sum of at most $(1.55k + 1)\mathcal{S}(G)$ and at most $4kp$ colors.*

In [HK$^+$99], efficient exact algorithms for np-sum are given for trees and paths, while a polynomial time approximation schema (PTAS) is given for the preemptive case.

The sum coloring problem is known to be hard to approximate within $n^{1-\epsilon}$, for any $\epsilon > 0$, unless $NP = ZPP$ [BBH$^+$98,FK98]. Also, it is NP-hard to approximate on bipartite graphs within $1 + \epsilon$, for some $\epsilon > 0$ [BK98]. Clearly, these limitations carry over to the sum multicoloring problems.

Our results:

Partial k-trees: In Section 3 we deal with multicoloring problems on partial k-trees, graphs whose treewidth is bounded by k. We design a general algorithm CompSum for that goal, that outputs an optimum solution for np-makespan and np-sum on partial k-trees in time $O(n \cdot (p \cdot \log n)^{k+1})$. It also solves p-sum and p-makespan on partial k-trees when p is small ($O(\log n / \log \log n)$). The algorithm is converted to a $1 + O(1/n)$-approximation of np-makespan and np-sum on partial k-trees.

In Section 4, we give PTASs for both **p-makespan** and **p-sum**, that hold for any p. These schemes satisfy a strong additional property: only $O(\log n)$ preemptions are used. These results are applied in the approximation of planar graphs.

Planar graphs: In Section 5, we give PTASs for sum multicoloring planar graphs, separately for the preemptive and non-preemptive cases. These imply the first PTAS for the sum coloring problem. These algorithms are complemented with a matching NP-hardness result for sum coloring.

For both of these classes in both models, the best bounds previously known were fixed constant factors [BHK+98].

The preemptive and non-preemptive cases turn out to require very different treatments. Non-preemptiveness restricts the form of valid solutions, which helps dramatically in designing efficient exact algorithm. On the other hand, approximation also becomes more difficult due to these restrictions. The added dimension of lengths of jobs, whose distribution can be arbitrary, introduces new difficulties into the already hard problems of coloring graphs. Perhaps our main contribution lies in the techniques of partitioning the multicoloring instance, both "horizontally" into segments of the color requirements (Section 4) and "vertically" into subgraphs similar length vertices (Section 5).

The full version of the paper [HK99] covers additional related topics including bicriteria approximation and generalized cost functions.

2 Preliminaries

Notation: The *minimum multisum* of a graph G, denoted by pSMC(G), is the minimum SMC(G, Ψ) over all multicolorings Ψ. We denote the minimum contiguous multisum of G by npSMC(G). We denote by $\mu(G, x)$ the minimum makespan, i.e. the number of colors required to preemptively multicolor (G, x). We let $OPT(G)$ denote the cost of the optimal solution, for the respective problem at hand.

Tools for approximation: We can bound the number of colors used by any optimal multicoloring. Note that there are trees whose optimum sum coloring requires $\Omega(\log n)$ colors [KS89]. Hence, the bound is tight even for the sum coloring version.

Lemma 1 (Color count). *In any k-colorable graph, both preemptive and non-preemptive optimum sum multicolorings use at most $O(p \cdot k \cdot \log n)$ colors. More generally, for any positive i, at least half of the vertices yet to be completed by step $ip(k + 1)$ are completed by step $(i + 1)p(k + 1)$.*

Proof. Consider the set of n' vertices not completed by step $ip(k+1)$, for some i. A coloring of these n' vertices that completes less than half of them by step $(i+1)(k+1)p$ incurs a delay of more than $n'/2 \cdot ((i+1)(k+1)p) + n'/2 \cdot (i(k+1)p) = n'(i+1/2)(k+1)p$. We can construct the following alternative coloring. Any job initiated before time $i(k+1)p$ is run to completion. Then, use any k-coloring and order the color classes in non-decreasing order of cardinality. The delay incurred by this approach is at most $n'p$ for the first step, and $(k-1)p/2$ amortized per vertex by the k-coloring, for a total delay of at most $n'(ip(k+1)+p+(k-1)p/2) = n'(i+1/2)(k+1)p$. □

Vertices with very small color requirements can be disposed of easily.

Assumption 2 (Polynomial lengths) *When seeking a PTAS, one may assume that for any vertex v, $x(v) \geq p/n^3$.*

Proof. Let *Small* be the set of vertices v with $x(v) < 2p/n^3$. Color first, one by one, non-preemptively, all the vertices of *Small*. The maximum color used for a vertex in *Small* is at most $n \cdot 2p/n^3 = 2p/n^2$. Hence, the total delay of the vertices of $V \setminus Small$ (or the number of colors that the coloring of $V \setminus Small$ must be offset by due to *Small*) is bounded by $2p/n$. Further, the sum of the coloring of *Small* is bounded by $n^2 \cdot 2p/n^3 = 2p/n$. As $OPT(G) \geq p$, we pay an additive $O(p/n) = O(OPT(G)/n)$ factor. This holds both for the preemptive and the non-preemptive case.

The following well-known decomposition lemma of Baker [B94] will be used repeatedly in what follows. For our purposes, the essential property of the class of k-outerplanar graphs, defined in [B94], is that it has treewidth at most $3k-1$ [B88].

Lemma 2 (Planar decomposition). *Let G be a planar graph, and k be a positive integer. Then G can be decomposed into two vertex-disjoint graphs: G_b, which is k-outerplanar, and G_s, which is outerplanar with $O(n/k)$ vertices and $O(S(G)/k)$ weight.*

We now prove a basic scaling lemma.

Lemma 3 (Preemptive scaling). *Let $\epsilon > 0$ and $c = c_\epsilon$ be large enough. Let $I = (G, x)$ be a multicoloring instance where for each v, $x(v)$ is divisible by q and $x(v)/q \geq c \cdot \log n$. Let I/q denote the instance resulting by dividing each $x(v)$ by q. Then, the makespans are related by*

$$q \cdot \mu_{I/q} \leq (1+\epsilon) \cdot \mu_I. \tag{1}$$

Proof. We prove this using a probabilistic argument. Consider an optimum makespan solution $OPT(I)$. We shall form a solution ψ for I/q. Include each independent set of $OPT(I)$ into ψ with probability $(1 + \epsilon_2)/q$ (with ϵ_2 to be determined later). The expected makespan is $1/q \cdot (1 + \epsilon_2) \cdot \mu_I$. So, by Markov's inequality, with probability at least $1 - 1/(1 + \epsilon_3)$, the makespan is at most $1/q \cdot (1 + \epsilon_3) \cdot (1 + \epsilon_2) \cdot \mu_I$. For this solution to be legal for I/q, we need to show that each vertex v gets at least $x(v)/q$ colors. We now show that this holds with non-zero probability. The number of colors each v gets is a binomial variable with mean $(1 + \epsilon_2) \cdot x(v)/q \geq (1 + \epsilon_2) \cdot c \cdot \log n$. For a binomial variable X with mean μ, Chernoff's bound gives that

$$\Pr(X < (1 - \delta)\mu) \leq exp(-\delta^2 \cdot \mu/2).$$

By choosing $c = (2 + 2/\epsilon_2)^2$, we bound the probability that v gets less than $x(v)/q$ colors by $1/n^2$. Hence, with probability at least $1 - 1/(1 + \epsilon_3) - 1/n$ all vertices get their required number of colors, and simultaneously, the makespan is at most $(1 + \epsilon_3) \cdot (1 + \epsilon_2) \cdot \mu_I/q$. By choosing ϵ_2 and ϵ_3 appropriately and for large enough n, Inequality (1) immediately follows, because we can take the formed solution ψ for I/q and repeat each subset q times. $\qquad\square$

It is now possible to validate the following assumption.

Assumption 3 (Polynomial lengths) *When seeking a PTAS, one may assume that $p < n^2$.*

3 Multicoloring partial k-trees

In this section, we study multicoloring problems on partial k-trees. We note that the results here hold for a fairly general type of a cost function or measure that includes makespan and multisum functions. Its description is omitted.

The scenario is as follows. We are given a (assumed to be large) family $\mathcal{F}$ of colorings, and we look for the best coloring in this family. The family $\mathcal{F}$ can contain for example all possible colorings (in which case we simply look for an optimal solution). However, with the point of view of approximations $\mathcal{F}$ may contain colorings that are only *close* to the optimum coloring. We denote by $\mathcal{D}(\mathcal{F}, v)$ the number of different colorings v has among the family $\mathcal{F}$. Let $\mathcal{D}(\mathcal{F})$ denote $\mathcal{D}(\mathcal{F}) = \max_v\{\mathcal{D}(\mathcal{F}, v)\}$. One of our main goals will be finding "good" families $\mathcal{F}$ with small $\mathcal{D}$ value.

We give a general algorithm CompSum for finding the minimum cost function in $\mathcal{F}$. The algorithm follows a direction similar to the unit-length case in [J97], thus its description and analysis is omitted.

Theorem 4. CompSum *runs in time bounded by $O(poly(n) \cdot \mathcal{D}^{k+1})$.*

Our first result is for **np-sum** on partial k-trees. In this case, by Lemma 1, the number of possible colorings of a vertex v, $\mathcal{D}(v)$, is bounded by $\mathcal{D} \leq O(p \cdot \log n)$, since we only need to specify the first color.

Corollary 1. *The **np-sum** and **np-makespan** problems admit an exact algorithm for partial k-trees that runs in $O(n \cdot (p \cdot \log n)^{k+1})$ time.* $\qquad\square$

Our second example deals with the preemptive case for **p-sum** and **p-makespan**, when the lengths are "small", or $O(\log n/\log\log n)$. In this case $\mathcal{D}$ is polynomial in n, which is explained as follows. We know by Lemma 1 that the number of colors used by an optimum solution for **p-sum** on partial k-trees is at most $O(p \cdot \log n)$. Thus, for each vertex we need to choose up to p colors, in the range $1, \ldots, O(p \cdot \log n)$. The number of different possible preemptive assignment of colors to a vertex v is $\binom{O(p \cdot \log n)}{p}$, which is polynomial in n due to the bound on p. Hence, the following corollary.

Corollary 2. *The **p-sum** and **p-makespan** problems on partial k-trees admit polynomial solutions in the case $p = O(\log n/\log\log n)$.* $\qquad\square$

The algorithm CompSum is polynomial only when p is polynomially bounded. We now show np-sum can in general be approximated within a very small ratio. The same can be shown to hold for np-makespan.

Theorem 5. np-sum *admits a* $1 + O(1/n)$-*ratio,* $n^{O(k)}$-*time algorithm on partial* k-*trees.*

Proof. We may assume that $p \geq n^5$, as otherwise algorithm CompSum runs in polynomial time. Define $q = \lfloor p/n^5 \rfloor \cdot n$. Given an instance I with color requirements $x(v)$, consider the instance I' whose color requirements $x'(v)$ are given by $x'(v) = \lfloor x(v)/q \rfloor \cdot q$. Clearly, npSMC($I'$) $\leq$ npSMC(I).

Consider the instance $I'' = (G, x'')$, where $x''(v) = x'(v) \cdot (1 + 1/n)$, which we note is integral. We now argue that npSMC(I'') $=$ npSMC(I') $\cdot (1 + 1/n)$. Since the color requirements in I' are all divisible by n, in any optimum the least color given to a vertex is congruent to 1 modulo n. Consider the optimum OPT' of I' with independent sets $S_1, S_2, \ldots$ Now, plug another copy of each independent set S_j in OPT', for j congruent to 1 modulo n. Namely, replace OPT' by $S_1, S_1, S_2, S_3 \ldots, S_n, S_{n+1}, S_{n+1}, S_{n-2} \ldots$, etc. The resulting multisum is bounded by npSMC(I')$(1 + 1/n)$. The inequality in the other direction is proved similarly. Now observe that

$$x''(v) \geq (x(v) - q) \cdot (1 + 1/n) = x(v) - q - q/n + x(v)/n \geq x(v),$$

where the last inequality follows since $x(v)$ is large by Assumption 2 and since $q \leq p/n^4$. Hence, npSMC(I'') $\geq$ npSMC(I).

Now, we apply algorithm CompSum on the instance I''. This is done by first scaling $x(v)$ by a factor of q, and afterwards taking q consecutive copies of each resulting independent set. We delete vertices from their last independent sets, once they get their required number of colors. Since $x''(v) \geq x(v)$, we get a valid coloring ψ for I, of cost

$$\text{SMC}(I, \psi) \leq \text{SMC}(I'', \psi) = \text{npSMC}(I'') = \text{npSMC}(I') \cdot (1 + 1/n) \leq \text{npSMC}(I) \cdot (1 + 1/n).$$

This gives the required result. $\qquad\square$

4 Preemptive multicoloring of partial k-trees

We show in this section that one can obtain near-optimal solutions to p-makespan on partial k-trees with the additional property of using few preemptions. This leads to a good approximation of p-sum, that also uses few preemptions.

Theorem 6. *The* p-makespan *problem on partial* k-*trees admits a PTAS that uses* $O(\log n)$ *preemptions.*

The theorem follows immediately from Lemma 4 below along with Theorem 4. A family of colorings is said to be *universal* if it depends only on the number of vertices n, not on the graph.

Lemma 4. *There is a universal family of multicolorings $\mathcal{F}$ with $\mathcal{D}(\mathcal{F})$ polynomial in n, such that for any k-colorable graph G, $\mathcal{F}$ includes a coloring that approximates the makespan of G by a $1 + \epsilon$ factor. Additionally, for any vertex, the number of preemptions used by any coloring in $\mathcal{F}$ is $O(\log n)$.*

Proof. Let c be a constant to be determined later. Let $\omega = \lfloor \lg p \rfloor + 1$ denote the number of bits needed to represent p, and let $\alpha = \lceil \lg c/\epsilon + \log \log n + 1 \rceil$.

We shall partition the color requirements x into two parts x' and x'', yielding two instances such that $x(v) = x'(v) + x''(v)$. Representing color requirements as a bit-string of length w, let $x'(v)$ be the value given by the α most-significant bits of $x(v)$, and $x''(v)$ the value of the remaining bits. We schedule the instance $I'' = (G, x'')$ followed by the instance $I' = (G, x')$, which combined yields a schedule of the original instance $I = (G, x)$. I'' is scheduled non-preemptively by any graph k-coloring of G. To schedule I', first form the instance I'/q obtained by dividing $x'(v)$ by $q = 2^{\omega - \alpha}$ (which causes no remainder). Take the optimal schedule of I'/q, and repeat each of its independent sets q times in sequence, to obtain a proper schedule of I'.

By choosing c above large enough, we get by the Scaling Inequality (1) that the makespan of the resulting schedule of I' is bounded by $q \cdot \mu_{I'/q} \leq (1 + \epsilon/2)\mu_{I'}$. Also, if $c \geq 2k$, the makespan of the k-coloring of I'' is at most $kp\epsilon_1/c < (\epsilon/2)p < (\epsilon/2)\mu_I$. Thus, the makespan of the combined schedule I is at most a $1 + \epsilon$ factor from optimal. The length p' of the longest task in I' is at most $2^\alpha = O(c/\epsilon \cdot \log n)$. Since the graph is k-colorable, $\mu_{I'/q} \leq p'k$. Hence, the number of preemptions used in total per vertex is at most $p'k + k = O(c/\epsilon \cdot \log n)$. Let $\mathcal{F}$ denote the family of all possible multicolorings in the above restricted form. For each vertex v,

$$\mathcal{D}(\mathcal{F}, v) \leq \binom{p' \cdot k}{p'} \leq 2^{p'k} = n^{O(kc/\epsilon_1)} = n^{O(k/\epsilon^3)}.$$

Taking q consecutive copies of each color does not affect this bound. $\qquad\square$

It should be clear from the proof that the different colorings of a given vertex can be computed efficiently.

We now give a PTAS for the sum measure, building on the makespan result.

Theorem 7. *The p-sum problem on partial k-trees admits a PTAS using $O(\log n)$ preemptions per vertex.*

This theorem follows immediately from the following lemma.

Lemma 5. *There is a universal family of multicolorings $\mathcal{F}$ with $\mathcal{D}$ polynomial in n, such that for any k-colorable graph G, $\mathcal{F}$ includes a schedule that approximates p-sum(G) within $1+\epsilon$. Additionally, each coloring in $\mathcal{F}$ has $O(\log n)$ preemptions per vertex.*

We show this by transforming an exact p-sum solution to an approximate solution with the desired restricted structure of the colorings in $\mathcal{F}$. The exact schedule is divided into layers, and each layer considered as a makespan instance, for which we use the restricted approximate solutions of Lemma 4. In addition, we use a small fraction of the colors to schedule the tasks round-robin, in order to ensure that each color set be limited to a compact region of the color space.

Proof. Let ψ^* be an optimal p-sum schedule of G. Partition the colors $\{1, 2, \ldots\}$ into a sequence of geometrically increasing segments. The length d_i of the i-th segment is $(1 + \epsilon_4)^i$ (ignoring round-off). The segments partition the solution and the instance into a collection of solutions ψ_i^* and instances (G, x_i), where $x_i(v)$ equals the number of colors assigned by ψ^* to v within segment i.

The approximate schedule ψ that we construct from ψ^* uses the following partition of the color space. It consists of a sequence of *levels*, each of which consists of a *main segment* s_i of length $(1 + \epsilon_5)d_i$ and k *round-robin segments* $r_{i,j}$, $j = 1, \ldots, k$, of length $(\epsilon_5/k)d_i$ each. The round-robin segments are used in accordance with a graph k-coloring $f : V \mapsto N$ of G. Thus, all vertices in a given color class j can simultaneously use the segment $r_{i,j}$, in each layer i.

We get ψ from ψ^* as follows. Values in $\psi^*(v)$ that are smaller than $(\epsilon_5/2k)x(v)$ or larger than $(k/\epsilon_5)x(v)$ are stored in round-robin segments $r_{i,f(v)}$, beginning with the segment that contains color $\epsilon_5/k \cdot x(v)$. The segments $\psi_i^*(v)$ of $\psi^*(v)$ with values in the intermediate range $[\epsilon_5/2k, k/\epsilon_5]x(v)$ are mapped via Lemma 4 to the main segments s_i. Namely, we treat each such segment as separate instance (G, x_i) and make a transformation as in Lemma 4. These form a valid schedule. Let us now verify the cost and preemptiveness properties of the schedule, as well as the number of possible colorings of a vertex.

An ϵ_5/k-fraction of the color space is available in each vertex's round-robin segments. The at most $(\epsilon_5/2k)x(v)$ small values are therefore completed before color $x(v)$. The large values are at most $x(v)$, and are completed before color $(k/\epsilon_5)x(v)$. This follows since when the d_i length of the segments sum to at least $k/\epsilon_5 \cdot x(v)$, the round-robin segments cover $x(v)$ colors. It follows that the only main segments used are those with colors in the range $[\epsilon_5/2k, k/\epsilon_5]x(v)$. The same holds for the round-robin segments.

Thus, there are only $\log_{1+\epsilon_4} 2 \cdot (k/\epsilon_5)^2$ levels involved, which is a constant. In each of these levels a vertex v has $O(\log n)$ preemptions. Hence, the total number of preemptions for each vertex is $O(\log n)$. Let $\mathcal{F}$ denote the family of all possible legal colorings in the above restricted form. Thus, in a way similar to Lemma 4, $D(\mathcal{F}, v)$ is polynomially bounded in n.

Each level i contributes $(1 + 2\epsilon_5)d_i$ to the finishing time of v. One ϵ_5 term comes from the round-robin, and the other, from adding $\epsilon_5 \cdot d_i$ to each level. Note that a vertex v is also affected by the possible change done to its highest color (in the highest level for v), for an additional factor of $1 + \epsilon_4$. The maximum color assigned in ψ to each v is then at most $(1 + 2\epsilon_5)(1 + \epsilon_4)$ times its optimal finish time. We can select ϵ_4 and ϵ_5 so as to make this $1 + \epsilon$, which gives the desired result. $\qquad\square$

5   Sum multicoloring planar graphs

It is clear that p-makespan and np-makespan are NP-hard on planar graphs, as they extend the NP-hard minimum coloring problem on planar graphs (cf.

[GJ79]). We prove that already the sum coloring problem, SC, is NP-complete for planar graphs. The proof is omitted in this abstract.

Theorem 8. *The* SC *problem and* SMC *problems are NP-complete on planar graphs.*

We match the hardness result with approximation schemas, starting with the easier preemptive case.

Theorem 9. *The* p-sum *problem on planar graphs admits a $(1+\epsilon)$-approximation algorithm that runs in $n^{O(1/\epsilon^3)}$ time.*

The following lemma relates approximations of planar graphs to those of partial k-trees. The theorem follows then from a combination with Theorem 7.

Lemma 6. *A ρ-approximation for* p-sum *on partial k-trees for any fixed k, implies a $\rho(1+\epsilon)$-approximation for planar graphs, for any $\epsilon > 0$.*

Proof. Given a constant k, decompose G into G_1 and G_2, with G_1 k^2-outerplanar, and G_2 outerplanar, following Lemma 2. Thus, we may assume that $S(G_2) \leq S(G)/k^2$. Use the assumed approximation to get solutions ψ_1 and ψ_2 whose sums are bounded by $\rho \cdot OPT(G_1)$ and $\rho \cdot OPT(G_2)$. Now, use a biased round-robin, as follows: after each group of $k - 1$ independent sets in ψ_1, insert the next independent set of ψ_2. Clearly, the finish times of each of the vertices in G_1 is multiplied by at most $1 + 1/k$, and that of a vertex in G_2, by at most k. Now, use the fact that $OPT(G_2) = O(S(G_2)) = O(S(G)/k^2)$. The resulting sum is bounded by:

$$\rho((1+1/k)OPT(G_1) + k \cdot OPT(G_2)) \leq \rho((1+1/k)OPT(G_1) + O(S(G)/k))$$
$$\leq \rho(1 + O(1/k)) \cdot OPT(G). \qquad \square$$

We now turn to the non-preemptive case, starting with a sequence of lemmas that lead to the derivation of a PTAS.

Lemma 7. *Let (G, x) be a k-colorable instance. Then, for any positive $c > 1$, at most $S(G)/(c \cdot p)$ vertices remain to be completed in an optimal multicoloring (preemptive or non-preemptive) of (G, x) by step $p(1.55k + 1 + (k + 1)\lg c)$.*

Proof. From Fact 1, the optimal multisum is at most $(1.55k+1)S(G)$. Thus, by step $(1.55k+1)p$, at most $S(G)/p$ vertices remain to be completed. By Lemma 1, the number of remaining vertices is halved every $p(k + 1)$ steps. Thus, after additional $p(k + 1)\lg c$ steps, the number of remaining vertices is down to at most $S(G)/(cp)$. $\qquad \square$

Lemma 8 (Compact lengths). *Let (G, x) be a planar instance with all color requirements in the range $[t, O(t \log n / \log \log n)]$, for some t. Then, for any $\epsilon > 0$,* np-sum(G) *admits a $1 + \epsilon$-approximation using $O(p \cdot \log \epsilon^{-1})$ colors.*

Proof. Let c be a constant to be determined later, and let $d = cp/(S(G)/n)$ and $b = 7.2 + 5\lg c$. We apply the following approach.

1. Partition V via Lemma 2 into V_1 and V_2, where V_1 induces a d-outerplanar graph G_1 while $|V_2| \leq n/d$.
2. Sum multicolor G_1 nearly-optimally, by rounding and scaling so that maximum length is $O(\log n/\log\log n)$ before applying CompSum and repeating the colors accordingly. Use the first $b \cdot p$ colors, and let $\hat{V}$ be the set of vertices not fully colored by these colors.
3. Color $V_2 \cup \hat{V}$ using a graph 4-coloring algorithm, yielding a multicoloring with at most $4p$ colors.

The cost of coloring V_1, and thus that of coloring $V_1 - \hat{V}$, is at most OPT. By Lemma 7, $\hat{V}$ contains at most $S(G)/(cp)$ vertices. Also, V_2 contains at most $n/d = S(G)/(cp)$ vertices. Hence, the cost of coloring $V_2 \cup \hat{V}$ is at most

$$(b+4)p \cdot 2S(G)/(cp) = \frac{22.4 + 10\lg c}{c}S(G).$$

Now set c to make the above expression at most $\epsilon S(G) \leq \epsilon OPT$. Thus $c = O(1/\epsilon \cdot \log 1/\epsilon)$. The additional cost of rounding is also easily bounded by ϵOPT. Then, the total cost of the coloring is at most $(1 + 2\epsilon)OPT$.

The complexity of our algorithm depends primarily on CompSum. The scaled instance is solved in time

$$((p/t)\log n)^{O(d)}n = (\log n)^{O(cp/t)} = n^{O(c)}.$$

Also, a $1 + 1/\lg\lg n$-ratio holds when $p = O(\log n/(\log\log n)^4)$.

The number of colors used is $(b+4)p = (11.2 + 5\lg c)p = O(p\log\epsilon^{-1})$. $\quad\square$

Markov's inequality shows that at most $1/t$ fraction of the elements of a set of non-negative numbers are greater than t times the average. It is easy to show it to be tight for any fixed value of t, while it cannot be tight for multiple values of t simultaneously. We show that if we are free to choose t from a range of values, the resulting bound on the tail is improved by a logarithmic factor.

Lemma 9 (Breakpoint lemma). *Let $x_1, x_2, \ldots, x_n$ be non-negative real numbers with average μ. Define $g(t)$ to be the number of x_i greater than or equal to t, i.e. $g(t) = |\{x_i : x_i \geq t\}|$. Then, for any r and s, $r \leq s$, there exists an integer t, $s \leq t \leq r$, such that*

$$g(t) \leq \frac{1}{\log(r/s)} \cdot \frac{\mu n}{t}.$$

Proof. Define the indicator functions $I_i(x)$ as 1 where $x \leq x_i$ and 0 elsewhere. Thus, $g(x) = \sum_i I_i(x)$. We have that

$$\int_0^\infty g(x)dx = \sum_i \int_0^\infty I_i(x)dx = \sum_i x_i = \mu n. \tag{2}$$

Let $\tau = \min_{x \in [s,r]} x \cdot g(x)$. Then,

$$\int_s^r g(x)dx \geq \tau \int_s^r \frac{1}{x}dx = \tau \left[\ln x\right]_s^r = \tau \ln(r/s). \tag{3}$$

Observing that the left-hand side of (3) is at most that of (2), we have from the right-hand sides that $\tau \leq \mu n / \ln(r/s)$, which yields the claim. $\quad\square$

The proof of the lemma yields a stronger property. It shows that there exists a t between s and r such that

$$g(t) \leq \frac{1}{\ln r/s} \cdot f(s,r) \cdot \mu n / t \quad \text{where} \quad f(s,r) = \frac{\int_s^r g(x)dx}{\int_0^\infty g(x)dx} = \frac{\int_s^r g(x)dx}{\mu n}. \tag{4}$$

Proposition 1. *Let (G, x) be a multicoloring instance and q be a natural number. We can generate a collection $b_1, b_2, \ldots, b_m$ of breakpoints, with $m \leq 2 \log_q p$, where adjacent breakpoints are at most a factor q apart, such that*

$$\sum_{i=1}^m g(b_i) \cdot b_i \leq \frac{1}{\ln \sqrt{q}} S(G).$$

Proof: Let b_0 be the smallest x_i value, and inductively let b_i be the breakpoint t obtained by applying the breakpoint Lemma 9 on the set of color requirements $x(v)$ of all the vertices, with $s = b_{i-1} \cdot \sqrt{q}$ and $r = b_{i-1} \cdot q$. Terminate the sequence once b_i exceeds the maximum length, p.

Since $b_i \geq b_{i-1}\sqrt{q}$, we have that $b_i \geq q^{i/2}$, and the loop terminates within $2 \log_q p$ iterations. In each iteration, the ratio r/s is at least $\sqrt{q}$ and the mean stays unchanged at $S(G)/n$. By Inequality 4,

$$g(b_i) \leq \frac{1}{\ln \sqrt{q}} f(b_{i-1}\sqrt{q}, b_{i-1}q) \cdot S(G)/b_i.$$

Thus,

$$\sum_i b_i g(b_i) \leq \frac{S(G)}{\ln \sqrt{q}} \cdot \sum_i f(b_{i-1}\sqrt{q}, b_{i-1}q).$$

Note that $b_i \geq b_{i-1}\sqrt{q}$ and thus the intervals $[b_{i-1}\sqrt{q}, b_{i-1}q)$ are disjoint. Since f is modular, it follows that $\sum_i f(b_{i-1}\sqrt{q}, b_{i-1}q) \leq f(0, \infty) = 1$. $\quad\square$

We are now ready to present our approximation schema:

Find breakpoints $b_1, b_2, \ldots$ by Proposition 1, with $q = \sqrt{\log n}$.
Partition G into subgraphs G_i, induced by $V_i = \{v : b_{i-1} \leq x(v) < b_i\}$.
Solve instances (G_i, x) independently, using Lemma 8, and schedule them in that order.

The cost of the multicoloring is derived from two parts: the sum of the costs of the subproblems, and the delay costs incurred by the colorings of the subproblems. The latter equals the sum over the subproblems, of the number of colors used times the number of vertices delayed (i.e. number of vertices of higher lengths). The number of colors used was shown to be $\theta(b_i \cdot \log 1/\epsilon)$ by Lemma 8, while $g(b_i)$ represents the number of vertices delayed. By Proposition 1, this cost is thus at most $\theta(\frac{1}{\log \log n} \cdot \log \frac{1}{\epsilon} \cdot \mathcal{S}(G))$.

The cost of subproblem i was bounded by Lemma 8 to be at most $(1 + \epsilon)OPT(G_i)$. Since $\sum_i OPT(G_i) \le OPT(G)$, the total cost of the coloring is at most $(1 + \epsilon + o(1))OPT(G)$. In fact, we can derive a $1 + \lg\lg\lg n / \lg\lg n$-approximation. Hence, we have derived the following:

Theorem 10. *There is a PTAS for* np-sum *on planar graphs.*

References

[B88]   H. L. Bodlaender. Planar Graphs with Bounded Treewidth. Tech. Report RUU-CS-88-14, Dept. of Comp. Sci., Univ. of Utrecht, Mar. 1988.

[B92]   M. Bell. Future directions in traffic signal control. *Transportation Research Part A*, 26:303–313, 1992.

[B94]   B. S. Baker. Approximation algorithms for NP-complete problems on planar graphs. *J. ACM*, 41:153–180, Jan. 1994.

[BBH+98]   A. Bar-Noy, M. Bellare, M. M. Halldórsson, H. Shachnai, and T. Tamir. On chromatic sums and distributed resource allocation. *Information and Computation*, 140:183–202, 1998.

[BH94]   D. Bullock and C. Hendrickson. Roadway traffic control software. *IEEE Transactions on Control Systems Technology*, 2:255–264, 1994.

[BK98]   A. Bar-Noy and G. Kortsarz. The minimum color-sum of bipartite graphs. *Journal of Algorithms*, 28:339–365, 1998.

[BHK+98]   A. Bar-Noy, M. M. Halldórsson, G. Kortsarz, H. Shachnai, and R. Salman. Sum Multi-Coloring of Graphs. To appear in *ESA '99*.

[CCO93]   J. Chen, I. Cidon and Y. Ofek. A local fairness algorithm for gigabit LANs/MANs with spatial reuse. *IEEE Journal on Selected Areas in Communications*, 11:1183–1192, 1993.

[FK98]   U. Feige and J. Kilian. Zero Knowledge and the Chromatic number. *Journal of Computer and System Sciences*, 57(2):187-199, October 1998.

[GJ79]   M. R. Garey and D. S. Johnson. Computers and Intractability: A Guide to the Theory of NP-completeness. W. H. Freeman, 1979.

[HK99]   M. M. Halldórsson, G. Kortsarz. Multicoloring Planar Graphs and Partial k-Trees. At www.raunvis.hi.is/~mmh/publications.html.

[HK+99]   M. M. Halldórsson, G. Kortsarz, A. Proskurowski, H. Shachnai, R. Salman, and J. A. Telle. Multi-Coloring Trees. To appear in *COCOON '99*.

[J97]   K. Jansen. The Optimum Cost Chromatic Partition Problem. *Proc. CIAC '97*. LNCS 1203, 1997.

[KS89]   E. Kubicka and A. J Schwenk. An Introduction to Chromatic Sums. *Proceedings of the ACM Computer Science Conf.*, pp. 39-45, 1989.

[MR97]   C. McDiarmid and B. Reed. Channel assignment and weighted coloring. Manuscript, 1997.

[NS97]   L. Narayanan and S. Shende. Static Frequency Assignment in Cellular Networks. Manuscript, 1997.

[NSS99]   S. Nicoloso, M. Sarrafzadeh, and X. Song. On the Sum Coloring Problem on Interval Graphs. *Algorithmica*, 23:109–126, 1999.

[SG98]   A. Silberschatz and P. Galvin. Operating System Concepts. Addison-Wesley, 5th Edition, 1998.

Testing the Diameter of Graphs

Michal Parnas[1] and Dana Ron[2]

[1] The Academic College of Tel-Aviv-Yaffo, Tel-Aviv, ISRAEL.
michalp@server.mta.ac.il
[2] Department of EE – Systems, Tel-Aviv University Ramat Aviv, ISRAEL.
danar@eng.tau.ac.il

Abstract. We propose a general model for testing graph properties, which extends and simplifies the bounded degree model of [GR97]. In this model we present a family of algorithms that test whether the diameter of a graph is bounded by a given parameter D, or is ϵ-far from any graph with diameter at most $\beta(D)$. The function $\beta(D)$ ranges between $D + 4$ and $4D + 2$, depending on the algorithm. All our algorithms run in time polynomial in $1/\epsilon$.

1 Introduction

Testing Graph Properties [GGR98] is the study of the following family of tasks. Let P be a predetermined graph property (such as connectivity or 3-colorability). The goal of the testing algorithm is to decide whether a given graph $G = (V, E)$ has property P, or whether it differs significantly from any graph having the property. In other words, the algorithm should accept every graph that has the property, and reject every graph for which many edge modifications should be performed so that the graph has the property. To this end the algorithm is given *access* to the graph in the form of being able to *query* on the incidence relationship between vertices. The testing algorithm is allowed a constant probability of error, and should perform its task by observing as few vertices and edges in the graph as possible.

In order to formulate the above family of problems more precisely, we need to answer two questions: (1) How are graphs represented, and what form of access to the graph is the testing algorithm given? (2) How do we measure the difference or distance between graphs (and what do we mean by "many" edge modifications)?

Adjacency-Matrix Model. Goldreich et. al. [GGR98] considered the adjacency matrix representation of graphs, where the testing algorithm is allowed to probe into the matrix. That is, the algorithm can query whether there is an edge between any two vertices of its choice. They define the distance between graphs as the fraction of entries in the adjacency matrix on which the two graphs differ. By this definition, for a given distance parameter ϵ, the algorithm should reject every graph that requires more than $\epsilon \cdot |V|^2$ edge modifications in order to acquire the tested property. Hence, the notion of distance is directly determined by the representation of the graphs.

This representation is most appropriate for dense graphs, and the results for testing in this model are most meaningful for such graphs. However, if we are interested in testing properties of sparse graphs, then the above model might not be suitable. For example, if a graph has $o(|V|^2)$ edges then it is trivially close (for any constant distance) to having every property that the empty graph has.

Bounded-length Incidence-Lists Model. This prompted the consideration of a different representation of graphs which can be used when testing properties of sparse graphs. A natural alternative is that based on *incidence lists*, which is studied in [GR97]. In [GR97], graphs are represented by incidence lists of *length* d, where d is a bound on the degree of the graph. Here the testing algorithm can query, for every vertex v and index $i \in \{1, \ldots, d\}$, who is the i'th neighbor of v. Analogously to the adjacency matrix model, the distance between graphs is defined to be the fraction of entries on which the graphs differ according to this representation. Since the total number of incidence-list entries is $d \cdot |V|$, a graph should be rejected if the number of edges modifications required in order to obtain the property is greater than $\epsilon \cdot d|V|$.

This model is most appropriate for testing graphs that not only have an upper bound d on their degree, but for which the number of edges is $\Omega(d \cdot |V|)$. In particular this model is suitable for testing graphs that have a constant bound on their degree. However, if we want to study non-constant degree bounds then we may run into a difficulty similar to the one that arose in the adjacency matrix model. Namely, a graph may be considered close to having a certain property simply because it has relatively few edges with respect to $d \cdot |V|$, and so it is close to the empty graph. In particular, this problem arises when there is a large variance between the degrees of vertices in the graph.

Functional Representations. The above two models were defined to fit into the framework of *Property Testing of functions* [RS96, GGR98]. That is, the two graph representations described above correspond to *functional* representation of graphs. In the first case these are functions from pairs of vertices to $\{0, 1\}$. In the second case these are functions from a vertex and an index in $\{1, \ldots, d\}$ to another vertex (or 0 if the neighbor does not exist). In both cases, the distance between graphs is determined by the representation. Namely, it is the distance between the corresponding functions, which is simply the fraction of domain-elements on which the two functions differ. The results of testing are most meaningful, if the number of edges in the tested graph is of the order of the size of the domain of the function. However, for the general case, we may need to seek a different representation, and an alternative distance measure.

A Non-Functional Model for Testing Graph Properties

In this paper we propose a general model which is appropriate for testing properties of all (non-dense) graphs. To this end we have chosen to divorce ourselves from the functional representation of graphs. Furthermore, the question of distance measure is treated separately from the question of representation. In this model graphs are represented by incidence lists of *varying lengths*. For each vertex v, the testing algorithm can obtain the degree of v, denoted $d(v)$, and can query, for every $i \leq d(v)$, who is the i'th neighbor of v. Thus, there is no fixed functional relationship (and in particular, no fixed domain). The distance between graphs is defined independently of the representation. It is measured by the fraction of edge modifications that should be performed in order to obtain the property, where this fraction is defined *with respect to the number of graph edges*. More precisely, since we do not want to assume that the algorithm is given the exact number of edges in the graph, we assume it is given an upper bound m on this number. For any given ϵ, a graph is said to be ϵ-*far* from having the property (and should be rejected), if the number of edge modifications performed so that the graph obtains the property is greater than $\epsilon \cdot m$. The implicit assumption is that m is in fact not much larger than the actual number of edges in the graph (but does not necessarily equal to this number).

By considering this generalization, we "open the door" to studying properties of graphs whose vertices vary significantly in their degrees. Furthermore, some problems are more interesting in this model, in the sense that removing the degree bound makes them less restricted. For example, the Diameter Problem presented in this paper, is less interesting in the bounded degree model, since a bound d on the degree implies a lower bound on the diameter of a graph. Another example is testing whether a graph has a clique of a given size k (or is far from any graph having a clique of size k). This question is meaningless in the bounded degree model for any $k > d + 1$.

In our testing model we also allow for the following relaxation of the testing task (which was previously studied in [KR98]). Consider a parameterized property $P = \{P_s\}$, where P_s may be for example: "Having diameter at most s". Then for any given parameter s, the algorithm is required: (1) to accept every graph with property P_s; (2) to reject every graph that is ϵ-far from having property $P_{\beta(s)}$, where $\beta(\cdot)$ is a certain *boundary function*. Thus, while in the previous definition of testing, $\beta(\cdot)$ was always the identity function, here we allow for different functions. However, we strive to obtain results where $\beta(s)$ is as close as possible to s. This relaxation is suitable whenever it is sufficient to know that a graph is close to having property $P_{\beta(s)}$. This seems to be true of many parameterized properties.

Testing the Diameter of Graphs We consider the problem of testing whether the diameter of a graph is within a certain upper bound. Our main result is a family of algorithms which determine (with probability at least 2/3), whether the diameter of a graph is bounded by a given parameter D, or is ϵ-far from any graph with diameter at most $\beta(D)$. The function $\beta(D)$ is a small linear function, which ranges between $D + O(1)$ and $O(D)$, depending on the algorithm. In particular:

- For $\beta(D) = 2D + 2$, we have an algorithm that works for every ϵ and has one-sided error (that is, it always accepts a graph that has diameter at most D).
- For every $\beta(D) = (1 + \frac{1}{2^i - 1}) \cdot D + 2$ (where $i > 1$), we have a testing algorithm that has two-sided error, and works for ϵ that is lower bounded as a function of i, $|V|$, and m. For example, when $i = 2$, we obtain $\beta(D) = \frac{4}{3}D + 2$, and the

algorithm works for $\epsilon = \Omega\left(\frac{n^{3/4}\log n}{m}\right)$, which in the worst case of $m = O(n)$ implies $\epsilon = \tilde{\Omega}(n^{-1/4})$.

All algorithms have query complexity and running time $\tilde{O}(\epsilon^{-3})$. In fact, the complexity improves as the number of edges m grows compared to $|V|$. Specifically, the complexity is bounded by $\tilde{O}\left(\left(\frac{m}{|V|}\cdot\epsilon\right)^{-3}\right)$.

In view of our positive results it is interesting to note that the problem of finding the minimum number of edges that should be added to a graph in order to transform it into a graph with diameter at most D is NP-hard [LMSL92]. Furthermore, in [LMSL92] evidence is given to the difficulty of approximating this minimum number.

Transforming results from the bounded-degree model Many of the results proved in [GR97] (for the bounded degree model) can be transformed smoothly to our general model. Due to space limitations, discussion of these transformations is deferred to the long version of this paper [PR99].

Organization The remainder of the paper is structured as follows. In Section 2 we present the general model. In Section 3 we define the diameter problem and present a family of testing algorithms for this problem. In Section 4 we prove a series of lemmas in which we show how to reduce the diameter of a graph by adding a small number of edges. Building on these lemmas, we prove in Section 5, the main theorem which states the performance and correctness of the testing algorithms.

2 Model Definition

Let $G = (V, E)$ be an undirected graph where $|V| = n$. We represent graphs by *incidence lists* of possibly varying lengths, where the length of each list (i.e., the degree of the vertex the list corresponds to) is provided at the head of the list. ordered. For any vertex v, let $d(v)$ denote the degree of v in the graph G.

Let $P = \{P_s\}$ be a *parameterized* graph property (e.g., P_s may be the property of having diameter of size at most s).

Definition 2.1 *Let P_s be a fixed parameterized property, $0 < \epsilon < 1$, and m a positive integer. A graph G having at most m edges is ϵ-far from property P_s (with respect to the bound m), if the number of edges that need to be added and/or removed from G in order to obtain a graph having property P_s, is greater than $\epsilon \cdot m$. Otherwise, G is ϵ-close to P_s.*

A *testing algorithm* for (parameterized) property P_s, with boundary function $\beta(\cdot)$, is given a (size) parameter $s > 0$, a distance parameter $0 < \epsilon < 1$, a bound $m > 0$, and query access to an unknown graph G having at most m edges. Namely, in accordance with the above representation of graphs, the algorithm can query, for any vertex v and index $1 \leq i \leq d(v)$, what is the i'th vertex incident to v.[3] The output of the algorithm is either **accept** or **reject**. We require that:

1. If G has property P_s, then the algorithm should output **accept** with probability at least $2/3$;
2. If G is ϵ-far from property $P_{\beta(s)}$, then the algorithm should output **reject** with probability at least $2/3$.

We shall be interested in bounding the query complexity and running time of testing algorithms as a function of the distance parameter ϵ, and possibly the size parameter s.

3 The Diameter Testing Problem

We present a family of algorithms that test whether the diameter of a graph is bounded by a given parameter D. The algorithms differ in the following aspects: (1) The boundary function $\beta(\cdot)$; (2) The query and time complexities; (3) The values of ϵ for which they can be applied. For brevity of the presentation we shall think of m as being the actual number of edges in the graph.

We first establish that if ϵ is above some threshold (dependent on D, n and m), then every *connected* graph with n vertices and m edges is ϵ-close to having diameter D. Therefore, for these values of ϵ (or more precisely, if ϵ is at least twice this lower bound) the testing algorithm can perform a connectivity test [GR97] (with distance parameter $\frac{\epsilon}{2}$), and accept if this test passes. Hence our interest lies mainly in smaller values of ϵ. Note that we may also assume that $m \geq n - 1$ or otherwise we know that the graph is disconnected and may reject it without testing. The following theorem, whose proof is omitted, was proved independently by Alon, Gyárfás and Ruszinkó [AGR99].

Theorem 3.1 *Every connected graph on n vertices can be transformed into a graph of diameter at most D by adding at most $\frac{2}{D-1} \cdot n$ edges.*

As an immediate corollary we get:

Corollary 3.2 *Every connected graph with n vertices and m edges is ϵ-close to having diameter D for every $\epsilon \geq \frac{2}{D} \cdot \frac{n}{m}$.*

Having established what is the "interesting" range of ϵ, we next state our main theorem. For any given n, m, and ϵ, define $\epsilon_{n,m} \stackrel{\text{def}}{=} \frac{m}{n} \cdot \epsilon$. Recall that we assume that $m \geq n - 1$, and so, with the exception of the case $m = n - 1$, we have $\epsilon_{n,m} \geq \epsilon$. Notice also, that for any fixed n, as m grows and the graph is denser, $\epsilon_{n,m}$ increases.

---

[3] Thus it is implicitly assumed that $n = |V(G)|$ is known (at least to within a constant factor of 2) just so the testing algorithm can refer to each vertex by its $\lceil \log n \rceil$-bits name.

Theorem 3.3 *1. There exists a testing algorithm for diameter D with boundary function $\beta(D) = 4D+2$, whose query and time complexities are $O(1/\epsilon_{n,m}^3)$. Furthermore, the testing algorithm has 1-sided error, that is, it always accepts graphs with diameter at most D.*

2. There exists a testing algorithm for diameter D with boundary function $\beta(D) = 2D+2$, whose query and time complexities are $O(\frac{1}{\epsilon_{n,m}^3} \cdot \log^2(1/\epsilon_{n,m}))$, and which has 1-sided error.

3. For any integer $2 \le i \le \log(D/2+1)$, there exists a testing algorithm for diameter D with boundary function $\beta(D) = D(1 + \frac{1}{2^i-1}) + 2$, whose query and time complexities are $O(\frac{1}{\epsilon_{n,m}^3} \cdot \log^2(1/\epsilon_{n,m}))$, and which works for every

$$\epsilon = \Omega\left(\frac{n^{1-\frac{1}{i+2}} \cdot \log n}{(i+2) \cdot m}\right).$$

We first consider the application of Item 3 for different settings of i. Note that the lower bound on ϵ translates to $\epsilon_{n,m} = \Omega\left(n^{-\frac{1}{2^{i+1}}} \cdot \frac{\log n}{i+2}\right)$, and we shall find it more convenient to refer to the latter bound. If $i = 2$ then Item 3 implies that we can distinguish between graphs with diameter D and graphs which are ϵ-far from diameter $\frac{4}{3}D + 2$ for every $\epsilon_{n,m} = \Omega(n^{-1/4} \cdot \log n)$. As we increase i, our boundary function $\beta(D)$ gets closer and closer to D, while the lower bound on $\epsilon_{n,m}$ becomes a larger inverse root of n. At the extreme setting of $i = \log(D/2+1)$, we get $\beta(D) = D + 4$, and $\epsilon_{n,m} = \Omega\left(n^{-\frac{1}{\log(D/2)+2}} \cdot \frac{\log n}{\log D}\right)$. Hence for very large D (say $D = N^\alpha$ for some $\alpha \le 1$), this result is applicable for constant $\epsilon_{n,m}$, while for $D = \text{poly}(\log n)$, it is applicable for $\epsilon_{n,m} = \Omega(n)^{\frac{1}{\log\log n}}$ (and in particular for $\epsilon_{n,m} = 1/\text{poly}(\log n)$).

We now describe the testing algorithm on which the theorem is based. Define the *C-neighborhood* of vertex v to be the set of all vertices at distance at most C from v (including v itself), and denote it by $\Gamma_C(v) = \{u \mid \text{dist}(u,v) \le C\}$, where $\text{dist}(u,v)$ denotes the distance between u and v in G. The basic testing algorithm, uniformly selects a sample of $O(1/\epsilon_{n,m})$ vertices, and checks whether the C-neighborhood of all but an α-fraction of these vertices contains at least k vertices. Since in the worst case, each such search may traverse $\Theta(k^2)$ edges, the query and time complexities of this algorithm are $O(k^2/\epsilon_{n,m})$. The testing algorithms referred to by Theorem 3.3, are essentially variants of this algorithm, and they differ only in the settings of the parameters C, k and α.

Algorithm Test-Diameter

- Given D, n, m and ϵ as input, let C, k, and α be set as functions of these parameters (where the particular setting depends on the variant of the algorithm).
- If $k = \Omega(n)$ then determine whether the graph has diameter at most D or not by observing the whole graph. Otherwise:
  1. Uniformly select $S = \Theta(\frac{1}{\epsilon_{n,m}})$ starting vertices.

2. For each vertex selected, perform a Breadth First Search until k vertices are reached, or all vertices at distance at most C have been reached (whichever occurs first).

3. If the fraction of starting vertices that reach distance C before observing k vertices (i.e., their C-neighborhood is of size less than k) is at most α, then accept, otherwise reject.

4 Reducing the Diameter of a Graph

Our goal is to reduce the diameter of graphs for which the C-neighborhoods of all but possibly a small fraction of the vertices, contain k vertices. The main technique we use to obtain a diameter of some bounded size is to select a set of R vertices as "representatives", such that each vertex in the graph is at distance at most ℓ from some representative. We then connect all representatives to one designated representative v_0. Thus, the number of edges added to the graph is $R - 1$. The distance between any two vertices u, w in the new graph is at most $2\ell + 2$, because u is at distance at most ℓ from some representative v_u, w is at distance at most ℓ from some representative v_w, and the distance between v_u and v_w is at most 2 since both are connected to v_0. Our technique is reminiscent of those applied in [DF85, Awe85, AP90, LS93].

The following results differ in the way the representatives are selected, and thus exhibit a tradeoff between the number R of representatives and the distance ℓ of each vertex to its nearest representative.

Lemma 4.1 *If the C-neighborhood of each vertex contains at least k vertices, then the graph can be transformed into a graph with diameter at most $4C + 2$ by adding at most $\frac{1}{k} \cdot n$ edges.*

Proof: For any vertex v, we shall view the set of vertices in the C-neighborhood of v as a *ball of radius C*, centered at v. We shall say that a vertex u is in the *C-boundary* of the ball if it is at distance at most C from some vertex in the ball (i.e., it is at distance at most $2C$ from the center). We partially cover the graph with (disjoint) balls of radius C in the following iterative manner: The center of the first ball can be any vertex. At any subsequent step, the next center selected can be any vertex not contained in any of the previous balls nor in their C-boundaries. Thus, every new ball is disjoint from all previous balls. When no such new center can be found we know that every vertex is at distance at most $2C$ from some previously selected center. Furthermore, the number of centers is at most $\frac{1}{k} \cdot n$.

Now connect the centers of all balls to the center of the first ball. The number of edges added is at most $\frac{1}{k} \cdot n - 1$, and the diameter of the resulting graph is at most $4C + 2$, since every vertex is at distance at most $2C$ from some center. $\square$

In the next lemma we slightly modify the premise of Lemma 4.1 so as to allow a small fraction of vertices whose C-neighborhoods contain less than k vertices.

The proof follows the same lines as the proof of Lemma 4.1, except here we need to treat separately the (relatively few) vertices that have less than k vertices in their C-neighborhood. These vertices are not selected as centers of balls and are directly connected to the first center selected.

Lemma 4.2 *If the C-neighborhood of at least $(1 - \frac{1}{k})n$ of the vertices contains at least k vertices, then the graph can be transformed into a graph with diameter at most $4C + 2$ by adding at most $\frac{2}{k} \cdot n$ edges.*

In the following Lemma we improve (by a factor of 2) the size of the diameter we can achieve, at a cost of a small increase in the required size of the C-neighborhood of most vertices. This is done by ensuring that the representatives selected be at distance at most C from every vertex (as opposed to $2C$ in the previous lemma), while allowing their C-neighborhoods to overlap.

Lemma 4.3 *If the C-neighborhood of at least $(1 - \frac{\delta}{2})n$ of the vertices contains at least $k = \frac{4}{\delta}\ln(4/\delta)$ vertices for some $\delta < 1$, then the graph can be transformed into a graph with diameter at most $2C + 2$ by adding at most δn edges.*

In order to prove the lemma we shall need the following Claim. The proof of the claim is very similar to that given in [AS92] for bounding the size of a dominating set in a graph whose minimum degree is bounded below.

Claim 4.4 *Let $S_1, \ldots, S_Y$ be sets over $\{1, \ldots, n\}$, such that $Y \leq n$, and each set contains at least $X \geq \frac{2}{\delta}\ln(2/\delta)$ elements for some $\delta < 1$. Then there exists a blocking set $T \subset \{1, \ldots, n\}$ of $\{S_1, \ldots, S_Y\}$, with at most δn elements. Namely, T is such that for every $i \in \{1, \ldots, Y\}$, we have $S_i \cap T \neq \emptyset$.*

Proof of Lemma 4.3 Let B be the set of vertices whose C-neighborhood contains less than k vertices. By the Theorem's premise, $|B| \leq \frac{\delta}{2}n$. By definition, for every v and u, if $u \in \Gamma_C(v)$ then $v \in \Gamma_C(u)$. Suppose there exists a set T of vertices of size at most $\frac{\delta}{2}n$, such that for every $v \in V \setminus B$, $T \cap \Gamma_C(v) \neq \emptyset$. Namely, the set T is a set of centers of balls of radius C whose union covers all vertices in $V \setminus B$. Then, we can arbitrarily select one vertex in T, denoted v_0, and add an edge between every other vertex in T and v_0 and between every vertex in B and v_0. In this way we obtain a graph of diameter at most $2C + 2$ by adding $|T - 1| + |B| \leq \delta n$ edges.

The existence of a set T as desired follows directly from Claim 4.4 and the premise of the Theorem concerning the size of the C-neighborhoods of all vertices in $V \setminus B$. Simply let $S_1, \ldots, S_Y$ be the C-neighborhood sets of vertices in $V \setminus B$. $\square$

5 Proof of Main Theorem

The proofs of Items 1 and 2 in Theorem 3.3 will be shown to follow from the results in the previous section. The proof of the third item is slightly more involved and will require to establish additional claims concerning properties of graphs that have diameter at most D. As stated before, all algorithms referred to by the theorem are based on the basic algorithm **Test-Diameter**. Recall that the algorithm selects S vertices, and accepts if the C-neighborhood of at least $(1 - \alpha)S$ of the selected vertices contains k vertices.

Definition 5.1 *A vertex whose C-neighborhood contains less than k vertices will be called* bad.

The basic proof line will be to show that if the graph has diameter at most D, then the fraction of bad vertices is small (0 in some cases), and therefore the algorithm will accept (with probability at least 2/3). If the graph is ϵ-far from diameter $\beta(D)$ for the various boundary functions $\beta(\cdot)$, then the sample selected by the algorithm will (with probability at least 2/3), include a fraction greater than α of bad vertices and hence the graph will be rejected.

5.1 Proofs of Items 1 and 2 in Theorem 3.3

Item 1 corresponds to the variant of the algorithm **Test-Diameter** where $C = D$, $k = \frac{2}{\epsilon_{n,m}}$, and $\alpha = 0$. In this case the sample size will be $S = \frac{4}{\epsilon_{n,m}}$. Namely, (for the non-trivial case of $\frac{1}{\epsilon_{n,m}} = \Omega(\frac{1}{n})$), the algorithm requires that the D-neighborhood of *every* starting vertex selected be of size at least $k = \frac{2}{\epsilon_{n,m}}$. Clearly, every graph with diameter at most D will pass the test, and so it remains to show that graphs that are ϵ-far from diameter $\beta(D) = 4D + 2$ will be rejected with probability at least 2/3.

By Lemma 4.2 if the D-neighborhood of at most $\frac{1}{k}n = \frac{\epsilon_{n,m}}{2}n$ of the vertices contains less than $k = \frac{2}{\epsilon_{n,m}}$ vertices, then the graph can be transformed into a graph with diameter at most $4D + 2$ by adding at most $\frac{1}{k} \cdot n = \epsilon_{n,m} \cdot n = \epsilon \cdot m$ edges. This implies that every graph which is ϵ-far from diameter $4D + 2$ must have more than $\frac{\epsilon_{n,m}}{2} \cdot n$ bad vertices. For any such graph, the probability that a sample of size at least $S = \frac{4}{\epsilon_{n,m}}$ will not contain such an "incriminating" bad vertex is at most $\left(1 - \frac{\epsilon_{n,m}}{2}\right)^S < e^{-2} < 1/3$ and the correctness of the algorithm follows.

Item 2 corresponds to the variant of the algorithm **Test-Diameter** where $C = D$, $k = \frac{4}{\epsilon_{n,m}} \cdot \ln \frac{4}{\epsilon_{n,m}}$, $\alpha = 0$, and the sample size is $S = \frac{4}{\epsilon_{n,m}}$. Its correctness follows from Lemma 4.3, similarly to the above.

5.2 Proof of Item 3 in Theorem 3.3

For this item we set $C = \frac{D}{2} + \frac{D}{2^{i+1}-2}$, $k = \frac{4}{\epsilon_{n,m}} \ln \frac{4}{\epsilon_{n,m}}$, $\alpha = \frac{\epsilon_{n,m}}{4}$, and $S = \frac{48}{\epsilon_{n,m}}$. Note that $C < D$, and this will be the source of our 2-sided error. Let χ_i be a random variable, such that $\chi_i = 1$ if the i'th vertex sampled is bad, and $\chi_i = 0$ otherwise.

We first show that every graph that is ϵ-far from diameter $\beta(D) = (1 + \frac{1}{2^i-1}) \cdot D + 2$ is rejected with probability at least 2/3. By Lemma 4.3, every graph that is ϵ-far from having diameter $2C + 2 = D(1 + \frac{1}{2^i-1}) + 2$ has at least $\frac{\epsilon_{n,m}}{2} \cdot n$ bad vertices. Thus, $\Pr[\chi_i = 1] \geq \frac{\epsilon_{n,m}}{2}$. By a multiplicative Chernoff bound, if we uniformly select $S \geq \frac{48}{\epsilon_{n,m}}$ vertices, then the probability that we obtain less than a fraction of $\alpha = \frac{\epsilon_{n,m}}{4}$ bad vertices (i.e., less than half the expected fraction), is bounded by $e^{-3} < 1/3$. Therefore, the graph is rejected with probability at least $2/3$.

We now turn to the case in which the graph has diameter at most D. The algorithm walks to distance smaller than D from each selected vertex and rejects the graph in case the fraction of selected vertices that reach less than k vertices is greater than $\alpha = \frac{\epsilon_{n,m}}{4}$. Thus, we cannot claim anymore that a graph with diameter D will always be accepted. Instead, we shall want to bound the number (fraction) of vertices whose C-neighborhoods are small. We show:

Lemma 5.2 *For every graph with diameter at most D, for every C of the form $C = \frac{D}{2} + \frac{D}{2^{i+1}-2}$ where $i \geq 2$, and for every k, the number of vertices whose C-neighborhood is of size less than k is at most k^{i+1}.*

We prove the lemma momentarily, but first show how the completeness of Item 3 in Theorem 3.3 follows (i.e., that every graph with diameter at most D is accepted with probability at least $2/3$). By the lemma, the number of bad vertices of every graph with diameter at most D is at most $k^{i+1} = \left(\frac{4}{\epsilon_{n,m}} \cdot \log \frac{4}{\epsilon_{n,m}} \right)^{i+1}$.

By our assumption on ϵ as a function of n and m, we have that $\epsilon_{n,m} = \Omega(n^{-\frac{1}{i+2}} \cdot \log n/(i+2))$. Therefore, for the appropriate setting of the constants in the $\Omega(\cdot)$ notation, we get that the number of bad vertices is at most $\left(\frac{4}{\epsilon_{n,m}} \cdot \log \frac{4}{\epsilon_{n,m}} \right)^{i+1} < \frac{\epsilon_{n,m}}{8} \cdot n$. Thus, in this case: $\Pr[\chi_i = 1] < \frac{\epsilon_{n,m}}{8}$.

Therefore, by a multiplicative Chernoff bound, the probability that the algorithm rejects is at most: $e^{-2} < 1/3$, and this completes the proof of Item 3.

Proof of Lemma 5.2: Assume, contrary to the claim that there are more than k^{i+1} bad vertices whose C-neighborhood contains less than k vertices. Since $C > \frac{D}{2}$, it follows that the $\frac{D}{2}$-neighborhood of all bad vertices contains less than k vertices as well. Since the graph has diameter at most D, every two vertices must have at least one common vertex in their respective $\frac{D}{2}$-neighborhoods. Let us fix some bad vertex u. By the above, u has less than k vertices in its $\frac{D}{2}$-neighborhood, and it shares at least one vertex with every one of the other k^{i+1} bad vertices. Therefore, there exists at least one vertex v in u's $\frac{D}{2}$-neighborhood that belongs to the $\frac{D}{2}$-neighborhood of at least k^i additional bad vertices.

Let these vertices be denoted $u_1, \ldots, u_t$, and consider a tree T_v of height at most $\frac{D}{2}$ rooted at v and containing all these vertices. Namely, we construct the tree in at most t steps. In each step we add a new vertex u_j (among $u_1, \ldots, u_t$) that does not yet belong to the tree, and a path of length at most $\frac{D}{2}$ from u_j to v. (If the new path forms a cycle we may always remove the new edges that create the cycle with the edges already in the tree.) The resulting tree has height $h \leq \frac{D}{2}$, contains at least $t \geq k^i$ vertices, and all its leaves belong to $\{u_1, \ldots, u_t\}$. We make the following claim.

Claim 5.3 *For any tree of height h and size at least t, and for any $a < h$, there exists a leaf in the tree whose $(h+a)$-neighborhood contains at least $t^{\frac{1}{\log((h+a)/a)+1}}$ vertices.*

We prove the claim momentarily, but first show how it can be applied to complete the proof of Lemma 5.2. Let $a = \frac{D}{2^{i+1}-2}$, so that $C = \frac{D}{2} + a$. Clearly, the C-neighborhood (in G) of every bad vertex u_j, contains at least as many vertices as the C-neighborhood of u_j restricted to the tree T_v. But according to Claim 5.3, there exists at least one bad vertex u_j, that is a leaf in T_v, whose C-neighborhood in the tree contains at least $t^{\frac{1}{\log(2^i)+1}} = t^{\frac{1}{i+1}} \geq k$ contradicting the fact that u_j is bad. Lemma 5.2 follows.

Proof of Claim 5.3: We define a function $f(t, h, a)$ which is the minimum over all trees T of height h having t vertices, of the maximum size of the $(h + a)$-neighborhood of a leaf in the tree (where the neighborhood is restricted to the tree). That is:

$$f(t, h, a) = \min_T \max_{u \ a \ leaf} |\Gamma_{h+a}(u)|.$$

Let $s(h, a) \overset{\text{def}}{=} \lceil \log((h + a)/a) \rceil$. We claim that for every $a \geq 1$, $f(t, h, a) \geq t^{1/s(h,a)}$, and prove the claim by induction.

For every h, if $a \geq h$ or $t \leq h$ or $t = 1$, then the claim clearly holds, as in these cases $f(t, h, a) = t$. Consider any $h \geq 1$, $t > h$ and $a < h$ and assume the claim holds for every $t' < t$, $h' < h$, and $a' > a$. Let us look at any leaf w in the tree. The $(h + 1)$-neighborhood of w contains all vertices at distance at most a from the root. Furthermore, if the number of these vertices is b, then there exists at least one sub-tree of height $h - a$, rooted at one of the vertices at distance a from the root that contains at least $(t - b)/(b - 1) + 1 = (t - 1)/(b - 1)$ vertices. To see why this is true, observe that there are $t - b$ vertices at distance greater than a from the root, and they belong to at most $(b - 1)$ sub-trees. Thus, there exists at least one sub-tree with at least $(t-b)/(b-1)$ vertices at distance greater than a fro the root. Adding the root of this sub-tree (at distance exactly a from the root), we get the above expression. Thus, there exists at least one leaf that can reach $f((t - 1)/(b - 1), h - a, 2a)$ vertices in its sub-tree. Taking care not to count the root of this sub-tree twice, we get

$$f(t, h, a) \geq \min_{b > 1} \left(b - 1 + f\left(\frac{t - 1}{b - 1}, h - a, 2a\right) \right)$$

By the induction hypothesis we have

$$f(t, h, a) \geq \min_{b > 1} \left(b - 1 + \left(\frac{t - 1}{b - 1}\right)^{1/s(h-a,2a)} \right)$$

We next show that $f(t, h, a) \geq t^{1/(s(h-a,2a)+1)}$. If $b - 1 \geq t^{1/(s(h-a,2a)+1)}$, then we are done. Otherwise $(b - 1 < t^{1/(s(h-a,2a)+1)})$, and we have

$$\left(\frac{t - 1}{b - 1}\right)^{1/s(h-a,2a)} > (t - 1)^{1/(s(h-a,2a)+1)} > t^{1/(s(h-a,2a)+1)} - 1$$

Since $b > 1$, the bound on $f(t, h, a)$ follows.

It remains to show that $t^{1/(s(h-a,2a)+1)} \geq t^{1/s(h,a)}$. If $(h + a)/(2a)$ is an exponent of 2, that is $(h + a)/(2a) = 2^i$ for some $i \geq 0$, then

$$s(h - a, 2a) + 1 = \lceil \log((h + a)/(2a)) \rceil + 1 = i + 1 = \lceil \log((h + a)/(a)) \rceil = s(h, a)$$

and the induction step is completed. Otherwise, $(h + a)/(2a) = 2^{i-1} + X$, where $i \geq 1$, and $0 < X < 2^{i-1}$ (recall that $h > a$ so $h + a > 2a$). In this case it is still true that $\lceil \log((h + a)/(2a)) \rceil + 1 = i + 1$, and $\lceil \log((h + a)/(a)) \rceil = \lceil \log(2^i + 2X) \rceil = i + 1$, and the Claim follows. $\square$

References

[AGR99] N. Alon, A. Gyárfás, and M. Ruszinkó. Decreasing the diameter of bounded degree graphs. To appear in Journal of Graph Theory, 1999.

[AP90] B. Awerbuch and D Peleg. Sparse partitions. In *Proceedings of the Thirty-First Annual Symposium on Foundations of Computer Science*, pages 503–513, 1990.

[AS92] N. Alon and J. H. Spencer. *The Probabilistic Method*. John Wiley & Sons, Inc., 1992.

[Awe85] B. Awerbuch. Complexity of network synchronization. *Journal of the Association for Computing Machinery*, 32:804–823, 1985.

[DF85] M. E. Dyer and A. M. Frieze. A simple heuristic for the p-centre problem. *Operations Research Letters*, 3(6):285–288, 1985.

[GGR98] O. Goldreich, S. Goldwasser, and D. Ron. Property testing and its connection to learning and approximation. *Journal of the Association for Computing Machinery*, 45(4):653–750, 1998. An extended abstract appeared in FOCS96.

[GR97] O. Goldreich and D. Ron. Property testing in bounded degree graphs. In *Proceedings of the Thirty-First Annual ACM Symposium on the Theory of Computing*, pages 406–415, 1997.

[KR98] M. Kearns and D. Ron. Testing problems with sub-learning sample complexity. In *Proceedings of the Eleventh Annual ACM Conference on Computational Learning Theory*, pages 268–279, 1998.

[LMSL92] C. L. Li, S. T. McCormick, and D. Simchi-Levi. On the minimum cardinality bounded diameter and the bounded cardinality minimum diameter edge addition problems. *Operations Research Letters*, 11(5):303–308, 1992.

[LS93] N. Linial and M. Saks. Low diameter graph decompositions. *Combinatorica*, 13:441–454, 1993.

[PR99] M. Parnas and D. Ron. Testing the diameter of graphs. Available from http://www.eng.tau.ac.il/~danar, 1999.

[RS96] R. Rubinfeld and M. Sudan. Robust characterization of polynomials with applications to program testing. *SIAM Journal on Computing*, 25(2):252–271, 1996.

Improved Testing Algorithms for Monotonicity

Yevgeniy Dodis[1] and Oded Goldreich[2] and Eric Lehman[1] and Sofya
Raskhodnikova[1] and Dana Ron[3] and Alex Samorodnitsky[4]

[1] Lab for Computer Science, MIT, 545 Technology Sq. Cambridge, MA 02139. email:
{yevgen,e_lehman,sofya}@theory.lcs.mit.edu.
[2] Dept. of Computer Science and Applied Mathematics, Weizmann Institute of
Science, Rehovot, ISRAEL. e-mail: oded@wisdom.weizmann.ac.il.
[3] Dept. of EE – Systems, Tel Aviv University, Ramat Aviv, ISRAEL. e-mail:
danar@eng.tau.ac.il.
[4] DIMACS Center, Rutgers University, Piscataway, NJ 08854. email:
salex@av.rutgers.edu.

Abstract. We present improved algorithms for testing monotonicity of
functions. Namely, given the ability to query an unknown function f :
$\Sigma^n \mapsto \Xi$, where Σ and Ξ are finite ordered sets, the test always accepts
a monotone f, and rejects f with high probability if it is ϵ-far from being
monotone (i.e., every monotone function differs from f on more than an
ϵ fraction of the domain). For any $\epsilon > 0$, the query complexity of the
test is $O((n/\epsilon) \cdot \log |\Sigma| \cdot \log |\Xi|)$. The previous best known bound was
$\tilde{O}((n^2/\epsilon) \cdot |\Sigma|^2 \cdot |\Xi|)$.
We also present an alternative test for the boolean range $\Xi = \{0, 1\}$
whose query complexity $O(n^2/\epsilon^2)$ is independent of alphabet size $|\Sigma|$.

1 Introduction

Property Testing (cf., [14, 10]) is a general formulation of computational tasks in
which one is to determine whether a given object has a predetermined property
or is "far" from any object having the property. Thus, property testing captures
a natural notion of approximation, where the measure approximated is the object's "distance" to having the property. Typically one aims at performing this
task within complexity smaller than the size of the object, while employing a
randomized algorithm and given oracle access to a natural encoding of the object (as a function). Thus, we are talking of determining with high probability
whether a function, to which we have oracle access, belongs to some class or is
"far" from this class (i.e., one needs to modify the function value at many places
so to obtain a function in the class).

Much work in this area was devoted to testing algebraic properties of functions such as linearity (e.g., [5, 1, 4, 3]) and low-degree properties (e.g., [5, 8,
14, 13, 2]). Recently, some attention was given to testing combinatorial properties of functions; firstly, for functions representing graphs [10, 11, 12], and
more recently for functions per se [7, 9]. The most natural combinatorial property of functions is monotonicity, and indeed [9] focuses on testing monotonicity.
The basic problem studied there is the following. Given a distance parameter

ϵ and oracle access to a function $f : \{0,1\}^n \mapsto \{0,1\}$, determine whether f is monotone or is "ϵ-far" from being monotone. Monotonicity is defined in the natural manner: One considers the standard partial order $\prec$ on binary strings (i.e., $x_1 x_2 \cdots x_n \prec y_1 y_2 \cdots y_n$ iff $x_i \leq y_i$ for every i and $x_i < y_i$ for some i), and f is said to be monotone if $f(x) \leq f(y)$ for every $x \prec y$. The definition extends naturally to functions defined on the standard partial order of strings over an arbitrary alphabet, Σ, and having an arbitrary range Ξ. That is,

Definition 1 (monotone functions and testing): *Let Σ and Ξ be sets with total order $\leq_\Sigma$ and $\leq_\Xi$, respectively. We consider the partial order, $\prec$, defined on equal-length strings over Σ by $x_1 x_2 \cdots x_n \prec y_1 y_2 \cdots y_n$ iff $x_i \leq_\Sigma y_i$ for every i and $x_i \neq y_i$ for some i.*

- *A function $f : \Sigma^n \mapsto \Xi$ is monotone if $f(x) \leq_\Xi f(y)$ holds for every $x \prec y$.*
- *A relative distance of $f : \Sigma^n \mapsto \Xi$ from the class of monotone functions, $\epsilon_M(f)$, is the minimum over all monotone functions $g : \Sigma^n \mapsto \Xi$ of $\mathrm{dist}(f, g) \stackrel{\text{def}}{=} |\{x \in \Sigma^n : f(x) \neq g(x)\}| \,/\, |\Sigma|^n$.*
- *A function $f : \Sigma^n \mapsto \Xi$ is ϵ-far from monotone if $\epsilon_M(f) \geq \epsilon$.*
- *A probabilistic oracle machine M is said to be a tester of monotonicity if*

$$\mathrm{Prob}[M^f(\epsilon, n) = 1] \geq \frac{2}{3} \quad \text{for any monotone function } f, \qquad (1)$$

$$\mathrm{Prob}[M^f(\epsilon, n) = 0] \geq \frac{2}{3} \quad \text{for } f \text{ which is } \epsilon\text{-far from monotone.} \qquad (2)$$

Note that all notions are defined w.r.t. Σ and Ξ, and so at times we prefer terms which explicitly mention this dependence.

The main result of [9] is a tester of monotonicity for the case $\Sigma = \Xi = \{0, 1\}$ having query and time complexities of the form $\mathrm{poly}(n)/\epsilon$. Specifically, the analysis of the query complexity in [9] yields a bound of $\tilde{O}(n^2/\epsilon)$, and it was also shown that $\Omega(n/\epsilon)$ is a lower bound on the query complexity of their algorithm. For general Σ and Ξ, the bounds obtained in [9] were proportional to $|\Sigma|^2 \cdot |\Xi|$. Here we improve both the algorithm and the analysis in [9] to obtain the following.

Theorem 1 (main result): *There exists a tester of monotonicity with query complexity*

$$q(\epsilon, n) \stackrel{\text{def}}{=} O\left(\frac{n \cdot (\log |\Sigma|) \cdot (\log |\Xi|)}{\epsilon} \right).$$

The tester works by selecting independently $q(\epsilon, n)/2$ pairs of n-long strings over Σ, and comparing the two f-values obtained for the elements of each pair.[5]

---

[5] Since the algorithm is comparison-based, its complexity depends only on the size of the image of the function. Thus, one may replace Ξ in the above bound by $\Xi_f = \{f(x) : x \in \Sigma^n\}$. In particular, $\log |\Xi_f| \leq n \cdot \log |\Sigma|$, so our bound is never worse than $O(n^2 \cdot (\log |\Sigma|)^2/\epsilon)$.

Thus, the global feature of being monotone or far from it, is determined by a sequence of many independent random local checks. Each local check consists of selecting a pair, (x, y), so that (w.l.o.g) $x \prec y$, according to some fixed distribution and checking whether $f(x) \leq_\Xi f(y)$. If we ever find a pair for which this does not hold (i.e., local violation of monotonicity), then we reject. Otherwise we accept. Thus, we never reject a monotone function, and the challenge is to analyze the dependence of rejection probability on the distance of the given function from being monotone.

The only thing left unspecified in the above description of the testing algorithm is the distribution by which the pairs are selected. In case $\Sigma = \{0, 1\}$ there seems to be a very natural choice. Uniformly select $i \in [n] \stackrel{\text{def}}{=} \{1, ..., n\}$, independently and uniformly select $z_1, \ldots, z_{i-1}, z_{i+1}, \ldots, z_n \in \{0, 1\}$, and set $x = z_1 \cdots z_{i-1} 0 z_{i+1} \cdots z_n$ and $y = z_1 \cdots z_{i-1} 1 z_{i+1} \cdots z_n$. Our improvement over [9], in this case (where $\Sigma = \Xi = \{0, 1\}$), comes from a better (and in fact tight for $\Xi = \{0, 1\}$) analysis of this test: Let $\delta_M(f)$ denote the fraction of pairs (x, y) as above for which $f(x) > f(y)$. We show that $\delta_M(f) \geq \epsilon_M(f)/n$ improving on the bound $\delta_M(f) \geq \epsilon_M(f)/n^2 \log(1/\epsilon_M(f))$ in [9] (whereas by [9] there exist functions f for which $\delta_M(f) = 2\epsilon_M(f)/n$).

In case of non-binary $\Sigma = \{1, \ldots, d\}$ there seem to be several natural possibilities: Even if we restrict ourselves (as above) to select only pairs of strings which differ on a single coordinate i, there is still the question of how to select the corresponding pair of symbols. We study two distributions on pairs $(k, \ell) \in \Sigma \times \Sigma$. The first distribution p_1 is uniform over a carefully chosen subset of pairs (k, ℓ) with $k < \ell$. The second distribution p_2 is uniform over all pairs (k, ℓ) with $k < \ell$.

A key result of this work is the reduction of the analysis of testing algorithms as above for any n and Σ, and for $\Xi = \{0, 1\}$, to their behavior in the special case of $n = 1$ (where we simply select pairs (k, ℓ) according to one of the above distributions and check the order between $f(k)$ and $f(\ell)$). Using this reduction we derive the following theorem.

Theorem 2 (Monotonicity Testing of Boolean functions): *There exist efficiently samplable distributions on pairs $(x, y) \in \Sigma^n \times \Sigma^n$ with $x \prec y$ so that for every function $f : \Sigma^n \mapsto \{0, 1\}$ the following holds:*

1. *If (x, y) is drawn according to one distribution (derived from p_1) then*

$$\text{Prob}[f(x) > f(y)] = \Omega\left(\frac{\epsilon_M(f)}{n \cdot (\log |\Sigma|)}\right).$$

2. *If (x, y) is drawn according to another distribution (derived from p_2) then*

$$\text{Prob}[f(x) > f(y)] = \Omega\left(\frac{\epsilon_M(f)^2}{n^2}\right).$$

We note that the first item of the theorem can also be derived by applying our reduction and using an alternative distribution on pairs in Σ^2 which was previously suggested in [7], and analyzed for the case $n = 1$. The second item leads to an algorithm having query complexity $O(n^2/\epsilon^2)$. It is possible to obtain an algorithm having complexity $O((n/\epsilon)\log^2(n/\epsilon))$ if the algorithm is not required to select independent pairs of strings. The alternative algorithm, suggested by Noga Alon, works by picking $i \in \{1,\ldots,n\}$ uniformly, and then querying f on $O(1/\epsilon)$ strings that differ only on the ith coordinate. The analysis of this algorithm can also be shown to reduce to the $n = 1$ case (using analogous claims to those presented here).

The reader may be tempted to say that since our algorithm is comparison-based, the analysis should also hold for non-boolean functions. However, this is false. For example, by Item (2) above, boolean functions over Σ may be tested for monotonicity within complexity independent of $|\Sigma|$. In contrast, a lower bound in [7] asserts that arbitrary functions over Σ (e.g., with $\Xi = \Sigma$) cannot be tested for monotonicity within complexity independent of $|\Sigma|$ (but rather require complexity $\Omega(\log|\Sigma|)$ for some fixed distance parameter $\epsilon > 0$). Thus, a natural question arises: *Under what conditions and at what cost can results regarding testing of monotonicity of boolean functions be transformed to results for testing monotonicity of arbitrary functions?* Our most general result is the following.

Theorem 3 (Monotonicity Testing – Range Reduction): *Consider the task of testing monotonicity of functions defined over any partially ordered set S (with p.o. $\prec_S$). Suppose that for some distribution on pairs $(x,y) \in S \times S$ with $x \prec_S y$ and for every function $f : S \mapsto \{0,1\}$*

$$\text{Prob}[f(x) > f(y)] \geq \frac{\epsilon_M(f)}{C} \; ,$$

where C depends on S only. Then, for every Ξ and every function $f : S \mapsto \Xi$ for pairs selected according to the same distribution

$$\text{Prob}[f(x) > f(y)] \geq \frac{\epsilon_M(f)}{C \cdot \log_2 |\Xi|} \; .$$

Theorem 1 follows by combining Part 1 of Theorem 2 and Theorem 3 with $C = O(n \cdot \log|\Sigma|)$.

Organization:

We start with some preliminaries in Section 2. In Section 3 we show how the analysis of our algorithm in the boolean-range case for arbitrary n and Σ, reduces to the case $n = 1$. The algorithms for the case $n = 1$ (each corresponding to a different distribution on pairs in $\Sigma \times \Sigma$), are provided in Subsection 3.3, and the proof of Theorem 2, in Subsection 3.4. Finally, in Section 4 we prove a general reduction from an arbitrary range to the boolean range, and derive Theorem 3. Due to space limitations, many details of proofs are deferred to the technical-report version of this paper [6]

2 Preliminaries

Let Σ and Ξ be sets with total order $\leq_\Sigma$ and $\leq_\Xi$, respectively. We consider the partial order, $\prec$, defined on equal-length strings over Σ as in the introduction, and shorthand $\leq_\Xi$ by $\leq$.

For any pair of functions $f, g : \Sigma^n \mapsto \Xi$, we define the *distance* between f and g, denoted $\mathrm{dist}(f,g)$, to be the fraction of instances $x \in \Sigma^n$ on which $f(x) \neq g(x)$. As in the introduction, we let $\epsilon_\mathrm{M}(f)$ denote the minimum distance between f and any monotone function $g : \Sigma^n \mapsto \Xi$. Let us formally define the algorithmic schema studied in this paper. The schema uses an arbitrary probability distribution $p : \Sigma \times \Sigma \mapsto [0,1]$. Without loss of generality, we assume that the support of p is restricted to pairs (k,ℓ) with $k < \ell$. The function t referred to below, depends on p.

ALGORITHMIC SCHEMA: Given parameters ϵ, n, Σ, Ξ, and oracle access to an arbitrary function $f : \Sigma^n \mapsto \Xi$, repeat the following steps up to $t(\epsilon, n, |\Sigma|, |\Xi|)$ times:

1. Uniformly select dimension $i \in [n]$, prefix $\alpha \in \Sigma^{i-1}$, and suffix $\beta \in \Sigma^{n-i}$.
2. Select (k, ℓ) according to p. Let $x = \alpha k \beta$, $y = \alpha \ell \beta$.
3. If $f(x) > f(y)$ (i.e., (x,y) witnesses that f is not monotone), then **reject**.

If all iterations were completed without rejecting then **accept**.

We focus on the analysis of a single iteration of the above test. Such an iteration is fully specified by the distribution, denoted $D_p^n : \Sigma^n \times \Sigma^n \mapsto [0,1]$, by which pairs (x,y) are selected. That is, $D_p^n(x,y) = \frac{p(k,\ell)}{n \cdot |\Sigma|^{n-1}}$ if $x = \alpha k \beta$ and $y = \alpha \ell \beta$, for some α, β, and $D_p^n(x,y) = 0$ otherwise. Observe that $D_p^n(x,y) > 0$ only if $x \prec y$. Let $\mathrm{DETECT}(f, D_p^n)$ be the probability that a pair (x,y) selected according to D_p^n witnesses that f is not monotone; that is,

$$\mathrm{DETECT}(f, D_p^n) \stackrel{\mathrm{def}}{=} \mathrm{Prob}_{(x,y) \sim D_p^n}[f(x) > f(y)] \tag{3}$$

(where the above definition can of course be applied to any distribution D on pairs $x \prec y$). Our goal is to find distributions D_p^n (determined by the distributions p) for which $\mathrm{DETECT}(f, D_p^n)$ is "well" lower-bounded as a function of $\epsilon_\mathrm{M}(f)$. If D_p^n is such that $\mathrm{DETECT}(f, D_p^n) \geq \delta(\epsilon, n, |\Sigma|, |\Xi|)$ *for any* $f : \Sigma^n \mapsto \Xi$ with $\epsilon_\mathrm{M}(f) \geq \epsilon$, then setting $t(\epsilon, n, |\Sigma|, |\Xi|) = \Theta(1/\delta(\epsilon, n, |\Sigma|, |\Xi|))$ yields a tester for monotonicity.

THE PARTIAL ORDER GRAPH: It will be convenient to view the partial order over Σ^n as a directed (acyclic) graph, denoted G_Σ^n. The vertices of G_Σ^n are the strings in Σ^n and directed edges correspond to comparable pairs (i.e. (x,y) is an edge iff $x \prec y$). An edge (x,y) is said to be **violated** by f if $f(x) > f(y)$. We denote by $\mathrm{VIOL}(f)$ the set of violated edges of f. We remark that most of the definitions in this section naturally extend to any partially ordered set S in place of Σ^n.

3 Dimension Reduction for Boolean Functions

In this section we restrict our attention to boolean functions $f : \Sigma^n \mapsto \{0,1\}$. Without loss of generality assume $\Sigma = \{1 \ldots d\}$, so $|\Sigma| = d$. In what follows we reduce the analysis of the performance of our algorithmic schema for any n and Σ (and $\Xi = \{0,1\}$) to its performance for the case $n = 1$ (the "line"). In Subsection 3.3 we describe and analyze several algorithms for the line. Recall that by our algorithmic schema any such algorithm is determined by a probability distribution p on pairs $(k, \ell) \in \Sigma \times \Sigma$. We conclude this section by combining the reduction with these algorithms to derive Theorem 2.

3.1 A Sorting Operator

First, we describe *sort* operators which can transform any boolean function over Σ^n into a monotone function (as we prove below).

Definition 2 *For every $i \in [n]$, the function $S_i[f] : \Sigma^n \mapsto \{0,1\}$ is defined as follows: For every $\alpha \in \Sigma^{i-1}$ and every $\beta \in \Sigma^{n-i}$, we let $S_i[f](\alpha 1 \beta), \ldots, S_i[f](\alpha d \beta)$ be assigned the values of $f(\alpha 1 \beta), \ldots, f(\alpha d \beta)$, in sorted order.*

For every $i \in [n]$ and every pair $(k, \ell) \in \Sigma^2$ so that $k < \ell$, let

$$\Delta_{i,(k,\ell)}(f) \overset{\text{def}}{=} \{(x,y) \in \text{VIOL}(f) : \exists \alpha \in \Sigma^{i-1}, \beta \in \Sigma^{n-i} \text{ s.t. } x = \alpha k \beta, y = \alpha \ell \beta\}.$$

Thus, $\bigcup_{i,(k,\ell)} \Delta_{i,(k,\ell)}(f)$ is the set of all violated edges of f that differ in a single coordinate. These are the only violated edges of f that can be potentially detected by our algorithmic schema. In the next lemma we show that by sorting in one dimension we do not increase the number of violations in any other dimension.

Lemma 4 *For every $f : \Sigma^n \mapsto \{0,1\}$, dimensions $i, j \in [n]$, and $1 \le k < \ell \le d$,*

$$|\Delta_{i,(k,\ell)}(S_j[f])| \le |\Delta_{i,(k,\ell)}(f)|.$$

Proof sketch. The important observation is that in order to prove the lemma we may consider the function f restricted at all dimensions but the two in question. Furthermore, proving the lemma boils down to asserting a claim about sorting zero-one matrices. Specifically, assume without loss of generality that $i < j$. We fix any $\alpha \in \Sigma^{i-1}$, $\beta \in \Sigma^{j-i-1}$, and $\gamma \in \Sigma^{n-j}$, and define the function $f' : \Sigma^2 \mapsto \{0,1\}$ by $f'(\sigma \tau) \overset{\text{def}}{=} f(\alpha \sigma \beta \tau \gamma)$. We now consider the 2-by-d zero-one matrix whose first row corresponds to the values of $f'(k, \cdot)$, and whose second row corresponds to the values of $f'(\ell, \cdot)$. The lemma amounts to showing that sorting the two rows of such a matrix does not increase the number of unsorted columns. (For details see [6]). $\square$

3.2 Dimension Reduction

With Lemma 4 at our disposal, we are ready to state and prove that the analysis of the algorithmic schema (for any n) reduces to its analysis for the special case $n = 1$. Let A denote one iteration of the algorithmic schema, p be any distribution on pairs $(k, \ell) \in \Sigma \times \Sigma$ such that $k < \ell$, and D_p^n be the corresponding distribution induced on edges of G_Σ^n. The dimension reduction lemma upper bounds $\epsilon_M(f)$ and lower bounds $\text{DETECT}(f, D_p^n)$ by the corresponding quantities for $n = 1$.

Lemma 5 (Dimension Reduction for Boolean Range)
Let $f : \Sigma^n \mapsto \{0, 1\}$. Then there exist functions $f_{i,\alpha,\beta} : \Sigma \mapsto \{0, 1\}$, for $i \in [n]$, $\alpha \in \{0, 1\}^{i-1}$ and $\beta \in \{0, 1\}^{n-i}$, so that the following holds (all expectations below are taken uniformly over $i \in [n]$, $\alpha \in \{0, 1\}^{i-1}$ and $\beta \in \{0, 1\}^{n-i}$):

1. $\epsilon_M(f) \leq 2n \cdot \mathbf{E}_{i,\alpha,\beta}(\epsilon_M(f_{i,\alpha,\beta}))$.
2. $\text{DETECT}(f, D_p^n) \geq \mathbf{E}_{i,\alpha,\beta}(\text{DETECT}(f_{i,\alpha,\beta}, p))$.

Proof sketch. For $i = 1, \ldots, n + 1$, we define $f_i \overset{\text{def}}{=} S_{i-1} \cdots S_1[f]$. Thus, $f_1 \equiv f$, and since $|\Delta_{i,(k,\ell)}(f_{i+1})| = 0$ for every k, ℓ and i, by Lemma 4 we have that f_{n+1} is monotone. It follows that

$$\epsilon_M(f) \leq \text{dist}(f, f_{n+1}) \leq \sum_{i=1}^{n} \text{dist}(f_i, f_{i+1}). \tag{4}$$

Next, for $i = 1 \ldots n$, $\alpha \in \{0, 1\}^{i-1}$ and $\beta \in \{0, 1\}^{n-i}$, define the function $f_{i,\alpha,\beta} : \Sigma \mapsto \{0, 1\}$, by $f_{i,\alpha,\beta}(x) = f_i(\alpha x \beta)$, for $x \in \Sigma$. We claim that

$$\text{dist}(f_i, f_{i+1}) \leq 2 \cdot \mathbf{E}_{\alpha,\beta}(\epsilon_M(f_{i,\alpha,\beta})). \tag{5}$$

This claim can be proved by observing that f_{i+1} is obtained from f_i by sorting, separately, the elements in each $f_{i,\alpha,\beta}$. (The factor of 2 is due to the relationship between the distance of a vector to its sorted form and its distance to monotone.)
Combining Eq. (4) and (5), the first item of the lemma follows:

$$\epsilon_M(f) \leq \sum_{i=1}^{n} \text{dist}(f_i, f_{i+1}) \leq 2 \cdot \sum_{i=1}^{n} \mathbf{E}_{\alpha,\beta}(\epsilon_M(f_{i,\alpha,\beta})) = 2n \cdot \mathbf{E}_{i,\alpha,\beta}(\epsilon_M(f_{i,\alpha,\beta})).$$

From the definition of algorithm A, it can be shown that

$$\text{DETECT}(f, D_p^n) = \frac{1}{n \cdot d^{n-1}} \sum_{i=1}^{n} \sum_{(k,\ell)} p(k, \ell) \cdot |\Delta_{i,(k,\ell)}(f)| . \tag{6}$$

Using Lemma 4, we have

$$|\Delta_{i,(k,\ell)}(f)| \geq |\Delta_{i,(k,\ell)}(S_1[f])| \geq \cdots \geq |\Delta_{i,(k,\ell)}(S_{i-1} \cdots S_1[f])| = |\Delta_{i,(k,\ell)}(f_i)|.$$

By combining Eq. (6) with the above and the definition of $f_{i,\alpha,\beta}$, the second item of the lemma follows. $\square$

3.3 Testing Monotonicity on a Line (the $n = 1$ case)

This section describes algorithms for the case $n = 1$, for any Σ and Ξ. In accordance with our algorithmic schema, such algorithms are defined by a probability distribution $p : \Sigma^2 \mapsto [0, 1]$ (with support only on pairs (k, ℓ) with $k < \ell$).

Note that for $n = 1$, we have $D_p^n \equiv p$. We present two such distributions, denoted p_1 and p_2, and provide bounds on $\text{DETECT}(f, p_j)$, for each j. Due to space limitations, we only state our results.

DISTRIBUTION p_1: This distribution is uniform on a set $P \subset \Sigma \times \Sigma$ which is defined as follows. The set P consists of pairs (k, ℓ), where $0 < \ell - k \leq 2^t$ and 2^t is the largest power of 2 which divides either k or ℓ. That is, let $\text{power}_2(i) \in \{0, 1 ..., \log_2 i\}$ denote the largest power of 2 which divides i. Then,

$$P \stackrel{\text{def}}{=} \{(k, \ell) \in \Sigma \times \Sigma : 0 < \ell - k \leq 2^{\max(\text{power}_2(k), \text{power}_2(\ell))}\} \qquad (7)$$

and $p_1(k, \ell) = \frac{1}{|P|}$ for every $(k, \ell) \in P$, and is 0 otherwise.

Proposition 6 *For any* Ξ *and* $f : \Sigma \mapsto \Xi$, $\text{DETECT}(f, p_1) \geq \frac{1}{O(\log d)} \cdot \epsilon_{\text{M}}(f)$.

The second distribution works well for the boolean range only.

DISTRIBUTION p_2: This distribution is uniform over all pairs (k, ℓ) such that $k < \ell$. That is, $p_2(k, \ell) = 2/((d - 1)d)$ for $1 \leq k < \ell \leq d$.

Proposition 7 *For any* $f : \Sigma \mapsto \{0, 1\}$, $\text{DETECT}(f, p_2) \geq \epsilon_{\text{M}}(f)^2$.

3.4 Proof of Theorem 2

In this subsection we combine Lemma 5 with the results for the case $n = 1$ provided in Subsection 3.3, and derive Theorem 2

Combining Lemma 5 and Proposition 6 (applied only to $\Xi = \{0, 1\}$), we have

$$
\begin{aligned}
\text{DETECT}(f, D_{p_1}^n)) &\geq \mathbf{E}_{i,\alpha,\beta}(\text{DETECT}(f_{i,\alpha,\beta}, \; p_1)) &&\text{[By Part 2 of the lemma]}\\
&\geq \mathbf{E}_{i,\alpha,\beta}(\epsilon_{\text{M}}(f_{i,\alpha,\beta})/O(\log d)) &&\text{[By the proposition]}\\
&\geq \frac{\epsilon_{\text{M}}(f)}{2n \cdot O(\log d)} = \Omega(\frac{\epsilon_{\text{M}}(f)}{n \log d}) &&\text{[By Part 1 of the lemma]}
\end{aligned}
$$

which establishes the the first item in the theorem.

Combining Lemma 5 and Proposition 7, we have

$$
\begin{aligned}
\text{DETECT}(f, D_{p_2}^n)) &\geq \mathbf{E}_{i,\alpha,\beta}(\text{DETECT}(f_{i,\alpha,\beta}, \; p_2)) &&\text{[By Part 2 of the lemma]}\\
&\geq \mathbf{E}_{i,\alpha,\beta}(\epsilon_{\text{M}}(f_{i,\alpha,\beta})^2/2) &&\text{[By the proposition]}\\
&\geq [\mathbf{E}_{i,\alpha,\beta}(\epsilon_{\text{M}}(f_{i,\alpha,\beta}))]^2/2 &&\text{[as } \mathbf{E}(X^2) \geq [\mathbf{E}(X)]^2]\\
&\geq (\epsilon_{\text{M}}(f)/2n)^2/2 = \Omega(\epsilon_{\text{M}}(f)^2/n^2) &&\text{[By Part 1 of the lemma]}
\end{aligned}
$$

which establishes the second item in the theorem.

4 Testing Monotonicity over General Ranges

We now reduce the problem of testing arbitrary-range functions to the simpler problem of testing boolean functions, which was considered in the preceding section. This reduction works not only for functions with domain Σ^n, but more generally when the domain is any partially ordered set S. The reduction is characterized by Theorem 3, which states that a certain type of monotonicity test for functions of the form $f : S \mapsto \{0, 1\}$ also works well for functions of the form $f : S \mapsto \Xi$. Here Ξ is a finite totally ordered set of size r, which we can regard as the integers in the interval $[0, r - 1]$. Furthermore, for simplicity, we assume that $r = 2^s$ for some integer s. All references to "edges" are references to edges of the partial order graph, whose vertices are strings in the domain S and directed edges correspond to ordered comparable pairs (i.e. (x, y) is an edge iff $x \prec y$).

To ensure that a function far from monotone can be readily detected by our test, we lower bound $\text{DETECT}(f, D)$ in terms of $\epsilon_M(f)$. Equivalently, we are looking for a good upper bound on $\epsilon_M(f)$ in terms of $\text{DETECT}(f, D)$. We reduce the task of obtaining an upper bound for functions with an arbitrary range to that of obtaining such an upper bound for functions with binary range.

The general idea of the reduction is to incrementally transform a function f into a monotone function, while ensuring that for each repaired violated edge, the value of the function is changed at only a few points. This transformation allows us to find a monotone function close to f and to upper bound $\epsilon_M(f)$ by the distance from f to that function. The transformation produces the following chain of functions: $f \mapsto f_1 \mapsto f_2 \mapsto f_3$, where f_3 is monotone. The distance between any two consecutive functions in the chain is equal to the distance to monotone of some auxiliary function with a smaller range. Thus, we obtain an upper bound on $\epsilon_M(f)$ in terms of the distance to monotone of smaller-range functions. In addition, edges violated by the auxiliary functions are also violated by f, and we can obtain a lower bound on $\text{DETECT}(f, D)$ in terms of the corresponding probability for the smaller-range auxiliary functions. Using the inductive assumption for smaller-range functions and the two claims above, we finally obtain the needed upper bound on $\epsilon_M(f)$ in terms of $\text{DETECT}(f, D)$.

Subsection 4.1 describes and analyzes operators SQUASH, MONO, and CLEAR later used to define functions f_1, f_2, and f_3 described above. Subsection 4.2 proves the range reduction lemma which upper bounds $\epsilon_M(f)$ and lower bounds $\text{DETECT}(f, D)$ by the corresponding quantities for smaller range functions. This section is concluded by the proof of Theorem 3 in Subsection 4.3.

4.1 Operators SQUASH, MONO, and CLEAR

First, we introduce operators, later used for obtaining functions f_1, f_2, and f_3.

Definition 3 *The operators* SQUASH, MONO, *and* CLEAR *each map a function* $f : S \mapsto [0, r - 1]$ *to a related function with the same domain and the same or smaller range. In particular,* MONO[f] *is some arbitrary monotone function* *at distance $\epsilon_M(f)$ from the function f. The operators* SQUASH *and* CLEAR *are defined below; in these definitions a and b are elements of $[0, r-1]$ and $a < b$.*

$$\text{SQUASH}[f, a, b](x) = \begin{cases} a & \text{if } f(x) \le a \\ b & \text{if } f(x) \ge b \\ f(x) & \text{otherwise} \end{cases}$$

$$\text{CLEAR}[f, a, b](x) = \begin{cases} \text{MONO}[\text{SQUASH}[f, a, b]](x) \\ \qquad \text{if } \text{MONO}[\text{SQUASH}[f, a, b]](x) \ne \text{SQUASH}[f, a, b](x) \\ f(x) \qquad \text{otherwise} \end{cases}$$

Lemma 8 states the main properties of these operators. Define the *interval of a violated edge* (x, y) *with respect to function* f to be the interval $[f(y), f(x)]$ (since the edge is violated by f, $f(x) > f(y)$). We say that two intervals *cross* if they intersect in more than one point.

Lemma 8 *The functions* SQUASH$[f, a, b]$ *and* CLEAR$[f, a, b]$ *have the following properties, for all $f : S \mapsto [0, r-1]$ and all $a, b \in [0, r-1]$ such that $a < b$:*

1. VIOL(SQUASH$[f, a, b]$) $\subseteq$ VIOL(f), *i.e.* SQUASH *does not introduce any new violated edges.*
2. VIOL(CLEAR$[f, a, b]$) $\subseteq$ VIOL(f), *i.e.* CLEAR *does not introduce any new violated edges.*
3. CLEAR$[f, a, b]$ *has no violated edges whose intervals cross* $[a, b]$.
4. *The interval of a violated edge with respect to* CLEAR$[f, a, b]$ *is contained in the interval of this edge with respect to* f.
5. dist(f, CLEAR$[f, a, b]$) = ϵ_M(SQUASH$[f, a, b]$).

Proof sketch. Part 1 immediately follows from the definition of SQUASH Parts 2–4 can be proved by looking at any $(x, y) \in$ VIOL(CLEAR$[f, a, b]$) and considering four cases depending on whether each of CLEAR$[f, a, b](x)$ and CLEAR$[f, a, b](y)$ lie inside or outside of the interval $[a, b]$ (see [6] for details). Part 5 follows from the definition of the CLEAR and MONO operators:

$$\text{dist}(f, \text{CLEAR}[f, a, b]) = \text{dist}(\text{MONO}[\text{SQUASH}[f, a, b]], \text{SQUASH}[f, a, b])$$
$$= \epsilon_M(\text{SQUASH}[f, a, b]). \quad \square$$

4.2 Range Reduction

We are now ready to define functions in the chain $f \mapsto f_1 \mapsto f_2 \mapsto f_3$, as well as auxiliary smaller-range functions f_1', f_2', and f_3'. Lemma 9 defines these functions and summarizes their properties. The transition from f to f_1 transforms violated edges with one endpoint in the lower half of the range and the other endpoint in the upper half into edges with both endpoints in the same half of the range. Then we repair violated edges with both endpoints in the lower half of the range to obtain f_2 and finally, upper half of the range to obtain f_3.

Lemma 9 (Range Reduction) *Given* $f : S \mapsto [0, r-1]$, *define*

$$f' = \text{SQUASH}[f, \tfrac{r}{2} - 1, \tfrac{r}{2}], \quad f_1' = \text{SQUASH}[f_1, 0, \tfrac{r}{2} - 1], \quad f_2' = \text{SQUASH}[f_2, \tfrac{r}{2}, r - 1],$$
$$f_1 = \text{CLEAR}[f, \tfrac{r}{2} - 1, \tfrac{r}{2}], \quad f_2 = \text{CLEAR}[f_1, 0, \tfrac{r}{2} - 1], \quad f_3 = \text{CLEAR}[f_2, \tfrac{r}{2}, r - 1].$$

These functions have the following properties, for any probability distribution D.

1. $\text{DETECT}(f, D) \geq \text{DETECT}(f', D)$
2. $\text{DETECT}(f, D) \geq \text{DETECT}(f_1', D) + \text{DETECT}(f_2', D)$
3. $\epsilon_M(f) \leq \epsilon_M(f') + \epsilon_M(f_1') + \epsilon_M(f_2')$

Proof sketch. All references to "parts" are references to parts of Lemma 8.

(*1*) $\text{VIOL}(f') \subseteq \text{VIOL}(f)$, since, by part 1, SQUASH never adds new violated edges.

(*2*) It is enough to show that $\text{VIOL}(f_1')$ and $\text{VIOL}(f_2')$ are *disjoint* subsets of $\text{VIOL}(f)$. By parts 1 and 2, $\text{VIOL}(f_1')$ and $\text{VIOL}(f_2')$ are subsets of $\text{VIOL}(f)$. It remains to prove that $\text{VIOL}(f_1')$ and $\text{VIOL}(f_2')$ are disjoint. By part 3, there is no edge violated by f_1 whose interval crosses $[\tfrac{r}{2} - 1, \tfrac{r}{2}]$. Hence, the edges violated by f_1 are partitioned into two disjoint subsets: "low" edges with intervals contained in $[0, \tfrac{r}{2} - 1]$ and "high" edges with intervals contained in $[\tfrac{r}{2}, r - 1]$. The edges violated by f_1' are a subset of the low edges, since SQUASH repairs all high violated edges and adds no new violated edges by part 1. The edges violated by f_2' are a subset of the high edges, since CLEAR used to form f_2 repairs all low violated edges by parts 3 and 4, and no new violated edges are added by parts 1 and 2.

(*3*) First, we show that f_3 is monotone. Since the function f_3 is constructed from f using a sequence of three CLEAR operators, parts 3 and 4 imply that there is no edge violated by f_3 whose interval crosses any of the intervals $[\tfrac{r}{2} - 1, \tfrac{r}{2}]$, $[0, \tfrac{r}{2} - 1]$, or $[\tfrac{r}{2}, r - 1]$. Therefore, f_3 violates no edges at all and is monotone.

Now the distance from f to the set of monotone functions is at most the distance from f to the particular monotone function f_3, and we get:

$$\epsilon_M(f) \leq \text{dist}(f, f_3) \leq \text{dist}(f, f_1) + \text{dist}(f_1, f_2) + \text{dist}(f_2, f_3)$$
$$= \epsilon_M(f') + \epsilon_M(f_1') + \epsilon_M(f_2'). \quad \text{[By part 5]} \quad \square$$

4.3 Proof of Theorem 3

In this subsection we use the results of the preceding lemma to prove Theorem 3. The proof is by induction on s with the inductive hypothesis that for every function $f : S \mapsto \Xi$ where $|\Xi| = 2^s$,

$$\epsilon_M(f) \leq C \cdot \text{DETECT}(f, D) \cdot s.$$

In the base case where $s = 1$, the hypothesis holds by the assumption stated in the theorem. Now assume that the hypothesis holds for $s - 1$ to prove that it holds for s. We can reason as follows:

$$\epsilon_M(f) \leq \epsilon_M(f') + \epsilon_M(f_1') + \epsilon_M(f_2')$$
$$\leq C \cdot \text{DETECT}(f, D) + C \cdot \text{DETECT}(f_1', D) \cdot (s-1) + C \cdot \text{DETECT}(f_2', D) \cdot (s-1)$$
$$\leq C \cdot (\text{DETECT}(f, D) + \text{DETECT}(f, D)(s - 1))$$
$$= C \cdot \text{DETECT}(f, D) \cdot s$$

The first inequality was proved in part 3 of Lemma 9. The second inequality uses the induction hypothesis; recall that the range of f' has size 2^1, and the ranges of f_1' and f_2' have size $r/2 = 2^{s-1}$. The third step uses parts 1 and 2 of Lemma 9, and the final step is simplification. This completes the proof.

Acknowledgments

We would like to thank Michael Krivelevich, Michael Sipser and Madhu Sudan for helpful discussions.

References

1. S. Arora, C. Lund, R. Motwani, M. Sudan, and M. Szegedy. Proof verification and intractability of approximation problems. *JACM*, 45(3):501–555, 1998.
2. S. Arora and S. Sudan. Improved low degree testing and its applications. In *Proceedings of STOC97*, pages 485–495, 1997.
3. M. Bellare, D. Coppersmith, J. Håstad, M. Kiwi, and M. Sudan. Linearity testing in characteristic two. In *Proceedings of FOCS95*, pages 432–441, 1995.
4. M. Bellare, S. Goldwasser, C. Lund, and A. Russell. Efficient probabilistically checkable proofs and applications to approximation. In *Proceedings of STOC93*, pages 294–304, 1993.
5. M. Blum, M. Luby, and R. Rubinfeld. Self-testing/correcting with applications to numerical problems. *JACM*, 47:549–595, 1993.
6. Y. Dodis, O. Goldreich, E. Lehman, S. Raskhodnikova, D. Ron, and A. Samorodnitsky. Improved testing algorithms for monotonocity. Available from *ECCC*, http://www.eccc.uni-trier.de/eccc/, 1999.
7. F. Ergun, S. Kannan, S. R. Kumar, R. Rubinfeld, and M. Viswanathan. Spot-checkers. In *Proceedings of STOC98*, pages 259–268, 1998.
8. P. Gemmell, R. Lipton, R. Rubinfeld, M. Sudan, and A. Wigderson. Self-testing/correcting for polynomials and for approximate functions. In *Proceedings of STOC91*, pages 32–42, 1991.
9. O. Goldreich, S. Goldwasser, E. Lehman, and D. Ron. Testing monotonicity. In *Proceedings of FOCS98*, 1998.
10. O. Goldreich, S. Goldwasser, and D. Ron. Property testing and its connection to learning and approximation. *JACM*, 45(4):653–750, 1998. An extended abstract appeared in the proceedings of FOCS96.
11. O. Goldreich and D. Ron. Property testing in bounded degree graphs. In *Proceedings of STOC97*, pages 406–415, 1997.
12. O. Goldreich and D. Ron. A sublinear bipartite tester for bounded degree graphs. In *Proceedings of STOC98*, pages 289–298, 1998. To appear in *Combinatorica*, 1999.
13. R. Raz and S. Safra. A sub-constant error-probability low-degree test, and a sub-constant error-probability PCP characterization of NP. In *Proceedings of STOC97*, pages 475–484, 1997.
14. R. Rubinfeld and M. Sudan. Robust characterization of polynomials with applications to program testing. *SIAM Journal on Computing*, 25(2):252–271, 1996.

Linear Consistency Testing

Yonatan Aumann*, Johan Håstad**, Michael O. Rabin***, and Madhu Sudan[†]

Abstract. We extend the notion of linearity testing to the task of checking linear-consistency of multiple functions. Informally, functions are "linear" if their graphs form straight lines on the plane. Two such functions are "consistent" if the lines have the same slope. We propose a variant of a test of Blum, Luby and Rubinfeld [8] to check the linear-consistency of three functions f_1, f_2, f_3 mapping a finite Abelian group G to an Abelian group H: Pick $x, y \in G$ uniformly and independently at random and check if $f_1(x) + f_2(y) = f_3(x + y)$. We analyze this test for two cases: (1) G and H are arbitrary Abelian groups and (2) $G = \mathbb{F}_2^n$ and $H = \mathbb{F}_2$.

Questions bearing close relationship to linear-consistency testing seem to have been implicitly considered in recent work on the construction of PCPs (and in particular in the work of Håstad [9]). It is abstracted explicitly for the first time here. We give an application of this problem (and of our results): A (yet another) new and tight characterization of NP, namely $\forall \epsilon > 0$, $\mathrm{NP} = \mathrm{MIP}_{1-\epsilon, \frac{1}{2}}[O(\log n), 3, 1]$. I.e., every language in NP has 3-prover 1-round proof systems in which the verifier tosses $O(\log n)$ coins and asks each of the three provers one question each. The provers respond with one bit each such that the verifier accepts instance of the language with probability $1 - \epsilon$ and rejects non-instances with probability at least $\frac{1}{2}$. Such a result is of some interest in the study of probabilistically checkable proofs.

1 Introduction

The study of linearity testing was initiated by Blum, Luby and Rubinfeld in [8]. A function f mapping a finite Abelian group G to an Abelian group H is "linear" (or more conventionally, a homomorphism) if for every $x, y \in G$,

* Department of Mathematics and Computer Science, Bar-Ilan University, Ramat-Gan, 52900, Israel. Email: aumann@cs.biu.ac.il.
** Department of Numerical Analysis and Computing Science, Royal Institute of Technology, Stockholm, Sweden. Email: johanh@nada.kth.se.
*** DEAS, Harvard University, Cambridge, MA 02138, USA and Institute of Computer Science, Hebrew University, Jerusalem, Israel. Email: rabin@deas.harvard.edu. Research supported, in part, by NSF Grant NSF-CCR-97-00365.
† Department of Electrical Engineering and Computer Science, MIT, 545 Technology Square, Cambridge, MA 02139, USA. Email: madhu@mit.edu. Research supported in part by a Sloan Foundation Fellowship an MIT-NEC Research Initiation Grant and an NSF Career Award.

$f(x) + f(y) = f(x + y)$. Blum, Luby and Rubinfeld showed that if a function f satisfies the identity above for a large fraction of pairs $x, y \in G$, then f is close to being linear. This seminal result played a catalytic role in the study of program checking/self-testing [7,8]. It is also a crucial element in the development of efficient PCP characterizations of NP and in particular occupies a central role in the results of [1,6,5].

In this paper we extend this study to testing the consistency of multiple functions. Given a triple of functions $f_1, f_2, f_3 : G \rightarrow H$, we say that they are "linear-consistent" if they satisfy: $\forall x, y, \in G$, $f_1(x) + f_2(y) = f_3(x + y)$. [1] At first glance this definition does not seem to enforce any structural property in f_1, f_2 or f_3. We show, however, that if f_1, f_2, f_3 are linear-consistent, then they are: (1) Affine: i.e., there exists $a_1, a_2, a_3 \in H$ such that for every $i \in \{1, 2, 3\}$ and $\forall x, y \in G$, $f_i(x) + f_i(y) = f_i(x+y) + a_i$; and (2) Consistent: i.e., $a_1 + a_2 = a_3$ and for every $i, j \in \{1, 2, 3\}$ and $\forall x \in G$, $f_i(x) - a_i = f_j(x) - a_j$.

We go on to study triples of functions f_1, f_2, f_3 that do not satisfy the identity $f_1(x) + f_2(y) = f_3(x + y)$ everywhere, but do satisfy this identity with high probability over a random choice of x and y. We provide two analyses for this case. The first is a variant of the analysis of [8] for linearity testing over arbitrary Abelian groups. We obtain the following result:

If $f_1, f_2, f_3 : G \rightarrow H$ satisfy $\delta \triangleq \Pr_{x,y \in G}[f_1(x) + f_2(y) \neq f_3(x + y)] < \frac{2}{9}$, then there exists a triple of linear-consistent functions $\tilde{f}_1, \tilde{f}_2, \tilde{f}_3 : G \rightarrow H$ such that for every $i \in \{1, 2, 3\}$, $\Pr_{x \in G}[f_i(x) \neq \tilde{f}_i(x)] \leq \delta$.

The second variant we study is when $G = \mathbb{F}_2^n$ and $H = \mathbb{F}_2$, where $\mathbb{F}_2$ is the finite field of two elements. This special case is of interest due to its applicability in the construction of efficient "probabilistically checkable proofs" and has been extensively studied due to this reason — see the work of Bellare et al. [4] and the references therein. Bellare et al. [4] give a nearly tight analysis of the linearity test in this case and show, among other things, that if a function f fails the linearity test with probability at most δ then it is within a distance of δ from some linear function. We extend their analysis to the case of linear-consistency testing and show an analogous result for this test:

If $f_1, f_2, f_3 : \mathbb{F}_2^n \rightarrow \mathbb{F}_2$ and $\gamma > 0$, satisfy $\Pr_{x,y \in \mathbb{F}_2^n}[f_1(x) + f_2(y) \neq f_3(x + y)] = \frac{1}{2} - \gamma < \frac{1}{2}$, then there exists a triple of linear-consistent functions $\tilde{f}_1, \tilde{f}_2, \tilde{f}_3 : \mathbb{F}_2^n \rightarrow \mathbb{F}_2$ such that for every $i \in \{1, 2, 3\}$, $\Pr_{x \in \mathbb{F}_2}[f_i(x) \neq \tilde{f}_i(x)] \leq \frac{1}{2} - \frac{2\gamma}{3}$.

---

[1] A slightly more symmetric equivalent definition would be to use: $\forall x, y, z \in G$ such that $x + y + z = 0$, $f_1(x) + f_2(y) + f_3'(z) = 0$. To see this is equivalent we set $f_3'(z) = -f_3(-z)$.

Motivation: We believe that the linear-consistency test is a natural variant of the linearity test and will potentially find similar applications in general. In fact, our original motivation came from the analysis of a variant of a protocol for deniable encryption proposed by Aumann and Rabin [3]. However, at this point we do not have any concrete applications to this case. One scenario where the linear-consistency test does appear naturally, and where we do have a concrete application, is the study of "multiple-prover one-round proof systems for NP".

An (r, p, a)-restricted MIP verifier V (for a p-prover one-round proof system) is one that acts as follows: On input $x \in \{0, 1\}^n$, V tosses $r(n)$ random coins and generates one question each for each of the p provers. The provers respond with a bits each. The response of the ith prover is allowed to be an arbitrary function of x and the query to the i prover, but is independent of the queries to the other provers. The verifier then outputs a verdict "accept/reject" based on the input x, its random coins and the answers of the p-provers. V is said to verify membership of a language L with completeness c and soundness s, if for every $x \in L$, there exist p-provers that are accepted by V with probability at least c; and for every $x \notin L$, for every p-provers, the verifier accepts with probability at most s. The class of all languages with p-prover one-round proof systems, in which the provers respond with a bits and the verifier is $r(\cdot)$ restricted and has completeness c and soundness s is denoted $\mathrm{MIP}_{c,s}[r, p, a]$.

Multiple prover interactive proof systems (MIPs) are a special case of the more familiar case of probabilistically checkable proof systems (PCPs). The difference is that in a PCP, all questions are sent to one "oracle-prover". The two main parameters of interest are the "randomness-parameter" (same as in MIP) and the "query-parameter", which counts the total number of bits of response from the oracle-prover. Thus the following containment is obtained easily $\mathrm{MIP}_{c,s}[r, p, a] \subseteq \mathrm{PCP}_{c,s}[r, p \cdot a]$ (where the second parameter is the number of queries). However, a converse of the form $\mathrm{PCP}_{c,s}[r, q] \subseteq \mathrm{MIP}_{c,s}[r, q, 1]$ is not known to be true and is a subject of some interest. Most strong PCP constructions today are obtained from some strong MIP construction. It is generally believed that MIP is a more restrictive model, but no results are known separating p-prover 1-bit MIPs from p-query PCPs. In view of the recent tight analysis of 3-query proof systems by Håstad [9] showing $\mathrm{NP} = \mathrm{PCP}_{1-\epsilon, \frac{1}{2}}[\log, 3]$, it was conceivable that one could separate 3-query PCPs from 3-prover 1-bit proof systems. However, our analysis of the linear-consistency tests leads us to an equally tight characterization of NP with MIPs. We show:

$$\forall \epsilon > 0, \mathrm{NP} = \mathrm{MIP}_{1-\epsilon, \frac{1}{2}}[O(\log n), 3, 1].$$

In fact in view of our analysis we believe that there may be no separation between p-prover 1-bit MIPs and p-query PCPs for any constant p.

Outline of this paper. In Section 2 we present some basic definitions of linear-consistency. In Section 3 we provide the analysis of linear-consistency tests over arbitrary Abelian groups. In Section 4 we consider the special case where the groups are vector spaces over $\mathbb{F}_2$. In Section 5 we sketch the MIP construction.

2 Definitions

For groups G, H, let $\text{HOM}_{G \to H}$ denote the set of homomorphisms from G to H. I.e.,

$$\text{HOM}_{G \to H} \overset{\triangle}{=} \{\phi : G \to H | \forall x, y \in G, \phi(x) + \phi(y) = \phi(x + y)\}.$$

For groups G, H, let $\text{AFF}_{G \to H}$ denote the set of *affine* functions from G to H. I.e.,

$$\text{AFF}_{G \to H} \overset{\triangle}{=} \{\psi : G \to H | \exists a \in H, \phi \in \text{HOM}_{G \to H} \text{ s.t. } \forall x \in G, \psi(x) = \phi(x) + a\}.$$

A triple of functions (f_1, f_2, f_3) is defined to be *linear-consistent* if there exists a homomorphism $\phi \in \text{HOM}_{G \to H}$ and $a_1, a_2, a_3 \in H$ such that $a_1 + a_2 = a_3$ and for every $i \in \{1, 2, 3\}$ and $x \in G$, $f_i(x) = \phi(x) + a_i$.

The following proposition gives an equivalent characterization of linear-consistent functions.

Proposition 1 *Functions $f_1, f_2, f_3 : G \to H$ are linear-consistent if and only if for every $x, y \in G$, $f_1(x) + f_2(y) = f_3(x + y)$.*

Proof: Let f_1, f_2, f_3 be linear-consistent, and let $\phi \in \text{HOM}_{G \to H}$ and $a_1, a_2, a_3 \in H$ be as guaranteed to exist by the definition of linear-consistency. Then, for every $x, y \in G$, $f_1(x) + f_2(y) - f_3(x+y) = \phi(x) + \phi(y) - \phi(x+y) + a_1 + a_2 - a_3 = 0$ as required. This gives one direction of the proposition.

Now suppose f_1, f_2, f_3 satisfy $\forall x, y, f_1(x) + f_2(y) = f_3(x + y)$. Using $x = y = 0$, we get

$$f_1(0) + f_2(0) = f_3(0) \tag{1}$$

Next we notice that $f_1(x) + f_2(0) = f_3(x)$ (using $y = 0$). Subtracting $f_1(0) + f_2(0) = f_3(0)$ from both sides we get $f_1(x) - f_1(0) = f_3(x) - f_3(0)$. Similarly we get $f_2(x) - f_2(0) = f_3(x) - f_3(0)$. Thus we may define $\phi(x) = f_1(x) - f_1(0) = f_2(x) - f_2(0) = f_3(x) - f_3(0)$. We now verify that $\phi \in \text{HOM}_{G \to H}$. For arbitrary $x, y \in G$, $\phi(x) + \phi(y) - \phi(x+y) = f_1(x) - f_1(0) + f_2(y) - f_2(0) - (f_3(x+y) - f_3(0)) = (f_1(x) + f_2(y) - f_3(x + y)) - (f_1(0) + f_2(0) - f_3(0)) = 0$. Thus for $a_i = f_i(0)$ and ϕ as above, we see that f_1, f_2, f_3 satisfy the definition of linear-consistency. ∎

For $x, y \in G$, the *linear-consistency test* through x and y is the procedure which accepts iff $f_1(x) + f_2(y) = f_3(x + y)$. Our goal in the remaining sections is to derive relationships between the probability with which a triple f_1, f_2, f_3 is rejected by the linear-consistency tests when x and y are chosen at random, and the proximity of f_1, f_2 and f_3 to linear-consistent functions.

3 Linear-consistency over arbitrary Abelian groups

In this section we consider the case of G and H being arbitrary finite Abelian groups. We extend the analysis of Blum, Luby and Rubinfeld [8] to this case. We show that if the test rejects with probability $\delta < \frac{2}{9}$, then by changing the value of each of the f_i's on at most δ fraction on the inputs, we get a triple of linear-consistent functions. In what follows, we use $\Delta(f, g)$ to denote the distance of f from g, i.e., $\Pr_{x \in G}[f(x) \neq g(x)]$.

Theorem 2 *Let G, H be finite Abelian groups and let $f_1, f_2, f_3 : G \rightarrow H$. If*

$$\delta \triangleq \Pr_{x,y \in G}[f_1(x) + f_2(y) \neq f_3(x + y)] < \frac{2}{9},$$

then there exists a triple of linear-consistent functions g_1, g_2, g_3 such that for every $i \in \{1, 2, 3\}$, $\epsilon_i \triangleq \Delta(f_i, g_i) \leq \delta$. Furthermore, $\epsilon \triangleq \frac{\epsilon_1 + \epsilon_2 + \epsilon_3}{3}$ satisfies $3\epsilon(1 - 2\epsilon) \leq \delta$.

Remark 3 *1. If $f_1 = f_2 = f_3$, then we recover the linearity testing theorem of [8] (see also [4]).*

2. The proof actually shows that $\epsilon_1 + \epsilon_2 + \epsilon_3 - 2(\epsilon_1\epsilon_2 + \epsilon_2\epsilon_3 + \epsilon_3\epsilon_1) \leq \delta$. Tightness of this and other aspect of the theorem are discussed in Section 3.1.

Proof: For $f : G \rightarrow H$, define $\text{CORR}^f(x; y)$ to be $f(x + y) - f(y)$. Define $\tilde{f}(x) = \text{PLURALITY}_{i \in \{1,2,3\}, y \in G}\{\text{CORR}^{f_i}(x; y)\}$ (where $\text{PLURALITY}(S)$ for a multiset S is the most commonly occurring element in S, with ties being broken arbitrarily). For $i \in \{1, 2, 3\}$ and $x \in G$, let $\gamma_i(x) \triangleq \Pr_{y \in G}\left[\tilde{f}(x) \neq \text{CORR}^{f_i}(x; y)\right]$. Let $\gamma_i = \mathbf{E}_x[\gamma_i(x)]$. Let $\gamma(x) = \frac{1}{3}[\gamma_1(x) + \gamma_2(x) + \gamma_3(x)]$ and let $\gamma = \mathbf{E}_x[\gamma(x)]$.

Our plan is to show that the $\gamma_i(x)$'s are all small and then to use this in two ways: First we use it to show that $\tilde{f}$ is a homomorphism. Then we show that the functions f_i's within a distance of γ_i from affine functions that are in the orbit of $\tilde{f}$.

Claim 4 *For every $x \in G$, and $i \neq j \in \{1, 2, 3\}$,*

$$\Pr_{y_1, y_2}\left[\text{CORR}^{f_i}(x; y_1) \neq \text{CORR}^{f_j}(x; y_2)\right] \leq 2\delta.$$

Proof: We prove the claim only for the case $i = 1, j = 2$. Other cases are proved similarly.

Over the choice of y_1 and y_2, consider two possible "bad" events: (A) $f_1(x + y_1) + f_2(y_2) \neq f_3(x + y_1 + y_2)$ and (B) $f_1(y_1) + f_2(x + y_2) \neq f_3(x + y_1 + y_2)$.

Observe first that if neither of the bad events listed above occur, then we have

$$\begin{aligned}
&\text{CORR}^{f_1}(x; y_1)\\
&= f_1(x + y_1) - f_1(y_1)\\
&= (f_3(x + y_1 + y_2) - f_2(y_2)) - f_1(y_1) \text{ ((A) does not occur)}\\
&= (f_3(x + y_1 + y_2) - f_2(y_2)) - (f_3(x + y_1 + y_2) - f_2(x + y_2))\\
&\qquad\qquad\qquad\qquad\qquad\qquad\qquad \text{((B) does not occur)}\\
&= f_2(x + y_2) - f_2(y_2)\\
&= \text{CORR}^{f_2}(x; y_2).
\end{aligned}$$

Now notice that the event listed in (A) has probability exactly δ (in particular, this event is independent of x). Similarly probability of the event in (B) is also δ. Thus the probability that (A) or (B) occurs may be bounded from above by 2δ (by the union bound). The claim follows. $\blacksquare$

The claim above allows us to prove upper bounds on the quantities $\gamma_i(x)$ for every x. This implies, in particular, that the function $\tilde{f}$ is defined at every point x by an overwhelming majority; a fact that is critical in proving that $\tilde{f}$ is a homomorphism.

Claim 5 *For every $x \in G$, and $i \in \{1, 2, 3\}$ and $j \neq i \in \{1, 2, 3\}$, the following hold:*

1. $\gamma_i(x) \leq 2\delta$.
2. $\gamma_i(x) + \gamma_j(x) - 2\gamma_i(x)\gamma_j(x) \leq 2\delta$.
3. $\gamma(x) < \frac{1}{3}$.

Proof: Let $p_\alpha = \Pr_{y \in G}[\text{CORR}^{f_i}(x; y) = \alpha]$ and $q_\alpha = \Pr_{y \in G}[\text{CORR}^{f_j}(x; y) = \alpha]$. We start by showing that $\max_{\alpha \in H}\{p_\alpha\}$ is very large. Observe that

$$\Pr_{y_1, y_2}[\text{CORR}^{f_i}(x; y_1) = \text{CORR}^{f_j}(x; y_2)] = \sum_{\alpha \in H} p_\alpha q_\alpha \leq \max_{\alpha \in H}\{p_\alpha\}.$$

Using Claim 4 the left-hand side of the inequality above is at least $1 - 2\delta$. Thus we establish that $\max_\alpha\{p_\alpha\} \geq 1 - 2\delta > 5/9$. Similarly we can show that $\max_\alpha\{q_\alpha\} > 5/9$.

Next we show that these maxima occur for the same value of $\alpha \in H$. Assume otherwise. Let $\tilde{p} = \max_\alpha\{p_\alpha\}$ and $\tilde{q} = \max_\alpha\{q_\alpha\}$. By the above $\tilde{p}, \tilde{q} > 5/9 > 1/2$. Since the maxima occur for distinct values of α, we may upper bound the quantity $\Pr_{y_1, y_2}[\text{CORR}^{f_i}(x; y_1) = \text{CORR}^{f_j}(x; y_2)]$ by $\tilde{p}(1-\tilde{q}) + (1-\tilde{p})\tilde{q}$. With some manipulation, the latter quantity is seen to be equal to $\frac{1}{2} - 2(\tilde{p} - \frac{1}{2})(\tilde{q} - \frac{1}{2}) < \frac{1}{2}$, which contradicts Claim 4.

Thus we find that $\text{PLURALITY}_y\{\text{CORR}^{f_i}(x; y)\}$ points to the same value for every $i \in \{1, 2, 3\}$; and this value is the value of $\tilde{f}(x)$. Thus we conclude $\gamma_i(x) =$

$1 - \max_\alpha \{p_\alpha\} \leq 2\delta$, yielding Part (1) of the claim. Part (2) follows by observing that

$$\Pr_{y_1,y_2} [\text{CORR}^{f_i}(x; y_1) = \text{CORR}^{f_j}(x; y_2)] \leq (1 - \gamma_i(x))(1 - \gamma_j(x)) + \gamma_i(x)\gamma_j(x)$$

and then using Claim 4 to lower bound the left-hand side by $1 - 2\delta$. Part (3) follows by some algebraic manipulation. Details omitted. ∎

The following claim now follows by a convexity argument.

Claim 6 *For every distinct* $i, j \in \{1, 2, 3\}$, $\gamma_i + \gamma_j - 2\gamma_i\gamma_j \leq 2\delta$.

Proof omitted.

Claim 7 $\tilde{f}$ *is a homomorphism. I.e.,* $\forall \ x, y \in G$, $\tilde{f}(x) + \tilde{f}(y) = \tilde{f}(x + y)$.

Proof [Sketch]: The claim is proven by showing that there exist $i \in \{1, 2, 3\}$ and $u \in G$ such that none of the following are true: (A) $\tilde{f}(x) \neq f_i(x+u) - f_i(u)$; (B) $\tilde{f}(y) \neq f_i(u) - f_i(u-y)$; and (C) $\tilde{f}(x+y) \neq f_i(x+u) - f_i(u-y)$. The existence of such i, u is shown by picking them at random and showing probability of (A) or (B) or (C) happening is bounded away from 1. It is easy to show that if none of the events occur, then $\tilde{f}(x) + \tilde{f}(y) = \tilde{f}(x + y)$. Details omitted. ∎

Claim 8 *For every* $i \in \{1, 2, 3\}$, *there exists* $\alpha_i \in H$ *such that*

$$\Pr_{x \in G} [f_i(x) \neq \tilde{f}(x) + \alpha_i] \leq \gamma_i.$$

Furthermore $\alpha_1 + \alpha_2 = \alpha_3$.

Proof: Fix $i \in \{1, 2, 3\}$. By definition of $\gamma_i(x)$, we have for every x, $\Pr_{a \in G}[\tilde{f}(x) \neq f_i(x+a) - f_i(a)] \leq \gamma_i(x)$. Thus, we get $\Pr_{x,a \in G}[\tilde{f}(x) \neq f_i(x+a) - f_i(a)] \leq \gamma_i$. In particular, there exists $a_0 \in G$ such that $\Pr_{x \in G}[\tilde{f}(x) \neq f_i(x + a_0) - f_i(a_0)] \leq \gamma_i$ or equivalently $\Pr_{x \in G}[\tilde{f}(x - a_0) \neq f_i(x) - f_i(a_0)] \leq \gamma_i$. But $\tilde{f}$ is a homomorphism, and thus we have $\tilde{f}(x - a_0) = \tilde{f}(x) - \tilde{f}(a_0)$. Thus we find that for this choice of a_0, $\Pr_{x \in G}[f_i(x) \neq \tilde{f}(x) + f_i(a_0) - \tilde{f}(a_0)] \leq \gamma_i$. The first part of the claim follows by setting $\alpha_i = f_i(a_0) - \tilde{f}(a_0)$.

The second part is shown by assuming, for contradicion, that $\alpha_1 + \alpha_2 \neq \alpha_3$ and then showing that that a random choice of x, y leads to the event "$f_i(x) = \tilde{f}(x) + \alpha_i$ for every $i \in \{1, 2, 3\}$" with probability greater than δ. But when this event happens, the test rejects, and this contradicts the fact that the rejection probability is at most δ. Details omitted. ∎

We are almost done with the proof of Theorem 2. The final claim, sharpens the bounds on the proximity of the functions f_i to the functions $\tilde{f}(x) + a_i$. Its proof is omitted from this version.

Claim 9 *The following inequalities hold:*

1. $\gamma_1 + \gamma_2 + \gamma_3 - 2(\gamma_1\gamma_2 + \gamma_2\gamma_3 + \gamma_3\gamma_1) \leq \delta$.
2. $3\gamma - 6\gamma^2 \leq \delta$.
3. $\gamma_1, \gamma_2, \gamma_3 \leq \delta$.

The theorem now follows from the above claims as follows. Set $g_i(x) = \tilde{f}(x) + \alpha_i$, where α_i's are as given by Claim 8. It follows from Claims 7 and 8 that g_1, g_2, g_3 are linear-consistent. It follows from Claim 8 that f_i is within a distance of γ_i from g_i; and the bounds on γ_i from Claim 9 bound these distances. ∎

3.1 Tightness of Theorem 2

Theorem 2 is tight in that one cannot improve the bound $\delta < \frac{2}{9}$ without significantly weakening the bound on the proximity of the nearest linear-consistent functions to f_1, f_2 and f_3. This tightness is inherited from the tightness of the linearity testing theorem of Blum, Luby and Rubinfeld, whose analysis also imposes the same upper bound on δ. For the sake of completeness, we recall the example, due to Coppersmith, here.

Let $G = H = \mathbb{Z}_{3n}$ for some large n, and let $f = f_1 = f_2 = f_3$ be the function

$$f(x) = \begin{cases} 3n - 1 & \text{if } x = -1 \bmod 3 \\ 0 & \text{if } x = 0 \bmod 3 \\ 1 & \text{if } x = 1 \bmod 3 \end{cases}$$

Then the probability that the linearity test rejects is $\frac{2}{9}$, while (for large enough n), the nearest affine functions to f are the constant functions, which disagree from f in at least $\frac{2}{3}$ of the inputs.

As we increase $\delta > 2/9$, the bounds on the proximity of the nearest linear(-consistent) functions become worse, approaching 0 as $\delta \to 1/4$ as demonstrated by the following example. For positive integers m, n let $f : \mathbb{Z}_{(2m+1)n} \to \mathbb{Z}_{(2m+1)n}$ be the function $f(x) = x \bmod (2m + 1)$ if $x \bmod (2m + 1) \in \{0, \ldots, m\}$ and $f(x) = (x \bmod (2m + 1)) + n - 2m - 1$ otherwise. It may be verified that the closest affine functions to f are the constant functions which are at a distance of at least $1 - \frac{1}{2m+1}$ from f. On the other hand the linearity test (and the hence the linear-consistency test on $f_1 = f_2 = f_3 = f$) accepts with probability at least $\frac{3}{4}$.

Thus for $\delta \geq \frac{1}{4}$ the linearity tests can not guarantee any non-trivial proximity with a linear function. In the range $\delta = [2/9, 1/4]$ we do not seem to have tight bounds. For $\delta < \frac{2}{9}$, the bounds given on ϵ_i can not be improved either, as shown in the following proposition.

Proposition 10 *For every $\epsilon_1, \epsilon_2, \epsilon_3 < \frac{1}{4}$, there exist a family of triples of functions $f_1^{(n)}, f_2^{(n)}, f_3^{(n)} : \mathbb{F}_2^n \to \mathbb{F}_2$ such that the distance of $f_i^{(n)}$ to the space of affine functions converges to ϵ_i and the probability that the linear-consistency test rejects is at most $\epsilon_1 + \epsilon_2 + \epsilon_3 - 2(\epsilon_1\epsilon_2 + \epsilon_2\epsilon_3 + \epsilon_3\epsilon_1)$.*

Proof: Let S_i be any subset of $\lfloor \epsilon_i 2^n \rfloor$ vectors from $\mathbb{F}_2^n$ with first coordinate being 1. Let $f_i^{(n)}(x) = 1 \Leftrightarrow x \in S_i$. Then, since $\epsilon_i < \frac{1}{4}$, the nearest affine function is the zero function, thus establishing the claim on distance. By the nature of the S_i's it is not possible that $x \in S_1$, $y \in S_2$ and $x + y \in S_3$. Therefore, the linear-consistency test rejects if and only if exactly one of $x, y, x + y$ fall in S_1, S_2, S_3 respectively. If we let ρ_i denote $2^{-n}|S_i|$, then the probability of this event is easily shown to be (exactly) $\rho_1 + \rho_2 + \rho_3 - 2(\rho_1\rho_2 + \rho_2\rho_3 + \rho_3\rho_1)$ which in turn is at most $\epsilon_1 + \epsilon_2 + \epsilon_3 - 2(\epsilon_1\epsilon_2 + \epsilon_2\epsilon_3 + \epsilon_3\epsilon_1)$. ∎

4 Linear-consistency tests over $\mathbb{F}_2$

In this section we consider the collection of affine functions and homomorphisms from $\mathbb{F}_2^n$ to $\mathbb{F}_2$. The results obtained are stronger in that it shows that any triple of functions that are accepted by the linear-consistency tests with non-trivial probability[2] are non-trivially close to a triple of linear-consistent functions.

For the purposes of this section it is better to think of the elements of $\mathbb{F}_2$ as $\{+1, -1\}$. Thus multiplication (over the reals) replaces addition modulo two in this representation. The set of homomorphisms HOM_n mapping $\{+1, -1\}^n \rightarrow \{+1, -1\}$ is given by $\text{HOM}_n = \{\ell_\alpha | \alpha \subseteq [n]\}$, where $\ell_\alpha(x) = \prod_{i \in \alpha} x_i$. The set of affine functions is given by $\text{AFF}_n = \{\ell_\alpha | \alpha \subseteq [n]\} \cup \{-\ell_\alpha | \alpha \subseteq [n]\}$. The homomorphisms now preserve $\ell_\alpha(x)\ell_\alpha(y) = \ell_\alpha(x \cdot y)$, where $x \cdot y$ represents the coordinate-wise product of the two vectors.

Let $\langle f, g \rangle$, the inner product between $f, g : \{+1, -1\}^n \rightarrow \{+1, -1\}$, be given by

$$\langle f, g \rangle = \frac{1}{2^n} \sum_{x \in \{+1, -1\}^n} f(x)g(x).$$

Then $\langle \ell_\alpha, \ell_\alpha \rangle = 1$ and $\langle \ell_\alpha, \ell_\beta \rangle = 0$ if $\alpha \neq \beta$. Thus the homomorphisms form a orthonormal basis over the reals for the set of functions from $\{+1, -1\}^n \rightarrow \mathbb{R}$. I.e. every function $f : \{+1, -1\}^n \rightarrow \mathbb{R}$ is given by $f(x) = \sum_{\alpha \subseteq [n]} \hat{f}_\alpha \ell_\alpha(x)$, where $\hat{f}_\alpha = \langle f, \ell_\alpha \rangle$ is the α-th Fourier coefficient of f. It is easily verified that the following (Parseval's identity) holds: $\langle f, f \rangle = \sum_{\alpha \subseteq [n]} \hat{f}_\alpha^2$. For functions $f : \{+1, -1\}^n \rightarrow \{+1, -1\}$, $\langle f, f \rangle = 1$. The Fourier coefficients are of interest due to the following easily verified fact.

Proposition 11 *For every function* $f : \{+1, -1\}^n \rightarrow \{+1, -1\}$:

$$- \epsilon_{\text{HOM}}(f) \stackrel{\triangle}{=} \min_{\alpha \subseteq [n]}\{\Delta(f, \ell_\alpha)\} = \min_{\alpha \subseteq [n]}\{\tfrac{1 - \hat{f}_\alpha}{2}\}.$$

---

[2] Since a triple of random functions would pass the linear-consistency tests with probability $\frac{1}{2}$, we consider the passing probability to be non-trivial if it is strictly larger than $\frac{1}{2}$.

$$- \epsilon_{\text{AFF}}(f) \overset{\Delta}{=} \min_{g \in \text{AFF}_n} \{\Delta(f,g)\} = \min_{\alpha \subseteq [n]} \{\tfrac{1-|\hat{f}_\alpha|}{2}\}.$$

Our result is the following:

Theorem 12 *Given functions $f_i : \{+1,-1\}^n \to \{+1,-1\}$, for $i \in \{1,2,3\}$, such that*

$$\Pr_{\boldsymbol{x},\boldsymbol{y}} [f_1(\boldsymbol{x})f_2(\boldsymbol{y}) \neq f_3(\boldsymbol{x}\cdot\boldsymbol{y})] = \delta,$$

for every $i \in \{1,2,3\}$, $\epsilon_{\text{AFF}}(f_i) \leq \delta$. Furthermore, there exists a triple of linear-consistent functions g_1,g_2,g_3 such that for every $i \in \{1,2,3\}$, $\Delta(f_i,g_i) \leq \frac{1}{2} - \frac{2\gamma}{3}$, where $\gamma = \frac{1}{2} - \delta$.

Remark 13 *Notice that even when $G = \mathbb{F}_2^n$ and $H = \mathbb{F}_2$, Theorem 12 does not subsume Theorem 2. In particular the error bounds given by Theorem 2 are stronger, when $\delta < 2/9$. However for $\delta > 2/9$, and in particular for $\delta \to \frac{1}{2}$, Theorem 12 is much stronger.*

Proof [Sketch]: The proof is obtained by modifying a proof of [4]. We omit the details, but the main steps are as follows. By arithmetizing the acceptance condition of the test we show that the rejection probability equals

$$\mathbb{E}_{\boldsymbol{x},\boldsymbol{y} \in_R \{+1,-1\}^n} \left[\frac{1}{2}(1 - f_1(\boldsymbol{x}) \cdot f_2(\boldsymbol{y}) \cdot f_3(\boldsymbol{x}\cdot\boldsymbol{y})) \right]$$

We then use the orthogonality of the Fourier basis to show that $1 - 2\delta = \sum_{\alpha \subseteq [n]} \hat{f}_{1,\alpha}\hat{f}_{2,\alpha}\hat{f}_{3,\alpha}$. Some algebraic manipulation, using in particular Parseval's identity, yields $\max_\alpha |\hat{f}_{i,\alpha}| \geq 1 - 2\delta$ and $\max_\alpha \{\min\{|\hat{f}_{1,\alpha}|, |\hat{f}_{2,\alpha}|, |\hat{f}_{3,\alpha}|\}\} \geq \frac{2}{3}(1 - 2\delta)$. Applying Proposition 11 to these two bounds yields the two conclusions of the theorem. ∎

5 3-prover 1-bit proof systems

For integers p,a and function $r : \mathbb{Z}^+ \to \mathbb{Z}^+$, an MIP verifier V is (r,p,a) restricted if on input $x \in \{0,1\}^n$, V tosses $r(n)$ coins and issues p queries $q_1,\dots,q_p$ to p-provers $P_1,\dots,P_p$ and receives a bit responses $a_1,\dots,a_p$ from the p provers. (The prover P_i is a function mapping q_i to some a bit string a_i.) The verifier then outputs a Boolean verdict accept/reject based on x, its random coins and the responses $a_1,\dots,a_p$. An (r,p,a)-restricted MIP verifier V achieves completeness c and soundness s for a language L if for every $x \in L$ there exists a collection of p-provers that force the V to accept with probability at least c, while for $x \notin L$ V does not accept any tuple of p-provers with probability greater than s. $\text{MIP}_{c,s}[r,p,a]$ is the collection of all languages L that have (r,p,a) restricted MIP verifiers achieving completeness c and soundness s.

We prove the following containment for NP, that is tight in that none the parameters c,s can be improved.

Theorem 14 *For every $\epsilon > 0$, $NP = MIP_{1-\epsilon,\frac{1}{2}}[O(\log n), 3, 1]$.*

We only sketch the proof here. Our verifier and analysis are simple variants of the verifier and analysis of Håstad [9]. As is usual in many of the recent PCP constructions, we start with the powerful 2-prover 1-round proof system of Raz [10] for NP, and then apply the technique of recursive proof checking [2]. To apply this technique, we define an appropriate "inner verifier system". The main point of difference in our construction from the construction of [9] is in the inner verifier that we construct and in the "decoding procedure" used in the construction. The formalism for the inner verifier system is derived from that of Trevisan [11]. Theorem 14 follows from the existence of a good inner-verifier system.

Definition 15 *An inner-verifier system consists of an $(r, 3, 1)$-restricted MIP verifier V_{inner} (for some function r); 3 encoding functions E_1, E_2 and E_3; and two (probabilistic) decoding functions D_1 and D_2. An inner-verifier system is good, if for every $\epsilon, \delta > 0$ there exists a $\gamma > 0$ such for every pair of positive integers m, n, the following hold:*

Completeness *If $a \in [n]$, $b \in [m]$ and $\pi : [m] \to [n]$ satisfy $\pi(b) = a$ then V_{inner}, on input (m, n, π, ϵ) accepts the provers $P_1 = E_1(a)$, $P_2 = E_2(b)$, and $P_3 = E_3(b)$ with probability at least $1 - \epsilon$.*

Soundness *If V_{inner} on input (m, n, π, ϵ) accepts provers P_1, P_2, P_3 with probability $\frac{1}{2} + \delta$, then $\pi(D_2(P_2, P_3)) = D_1(P_1)$ with probability at least γ (over the coin tosses of the decoding procedures D_1 and D_2).*

The inner verifier V_{inner} is derived directly from [9]. Given (n, m, π, ϵ), V_{inner} picks three functions $f : [n] \to \{+1, -1\}$, $g : [m] \to \{+1, -1\}$ and $\eta : [m] \to \{+1, -1\}$ such that $f(1) = g(1) = \eta(1) = 1$ and otherwise f and g are random and unbiased while η is random with bias $1 - \epsilon$ i.e., for every input $j \in [m]$ $\eta(j)$ is 1 with probability $1 - \epsilon$ and -1 with probability ϵ, independently. Let $b = f(\pi(1))$ and g' be the function given by $g'(j) = bf(\pi(j))g(j)\eta(j)$. The verifier sends the questions f to P_1, g to P_2 and g' to P_3. If the responses are $a_1, a_2, a_3 \in \{+1, -1\}$, then V_{inner} accepts if $a_1 a_2 a_3 = b$. (The main difference between this verifier and that of [9] is that this verifier sends the queries g and g' to two different provers, while the verifier of [9] sent it to a (single) oracle.)

The encoding functions are just the "long codes" (see [5, 9, 11]). I.e., $E_1(a)$ is the function P_1 that on input $f : [n] \to \{+1, -1\}$ responds with $f(a)$, while $E_2(b)$ (as also $E_3(b)$) is the function P_2 that on input $g : [m] \to \{+1, -1\}$ responds with $g(b)$. The completeness follows immediately.

The decoding function D_1 is from [11], which is in turn based on [9]. To describe this decoding, we notice that $f : [n] \to \{+1, -1\}$ may also be viewed as a vector $f \in \{+1, -1\}^n$. Thus P_1 may be viewed as a function from $\{+1, -1\}^n$ to $\{+1, -1\}$. (Actually, P_1 (resp. P_2, P_3) is never queried with any function f with

$f(1) = -1$. Thus we may set $P_1(-f) = -P_1(f)$ for every f, without altering the acceptance probability of V_{inner}. We assume here onwards that P_1 (resp. P_2, P_3 are such functions.) The decoding function is then based on the Fourier coefficients of P_1. $D_1(P_1)$ works as follows: Pick $\alpha \subseteq [n]$ with probability $\hat{P}_{1,\alpha}^2$, and output a random element of α (α is never empty, since $\hat{P}_{1,\phi} = 0$ for any function P_1 satisfying $P_1(f) = -P_1(-f)$).

The new element of our proof is the decoding function D_2. $D_2(P_2, P_3)$ works as follows: Pick $\alpha \subseteq [m]$ with probability $|\hat{P}_{2,\alpha} \cdot \hat{P}_{3,\alpha}|$ and output a random element of α. Notice that the probabilities of picking the sets α add up to at most 1. (If the sum is smaller, we do nothing in the remaining case.)

A proof similar to that of [9] with modifications (and, in particular, the use of the Cauchy-Schwartz inequality) as in the proof of Theorem 12 provide the analysis of the soundness condition, thus yielding Theorem 14.

References

1. S. ARORA, C. LUND, R. MOTWANI, M. SUDAN AND M. SZEGEDY. Proof verification and the hardness of approximation problems. *Journal of the ACM*, 45(3):501-555, 1998.
2. S. ARORA AND S. SAFRA. Probabilistic checking of proofs: A new characterization of NP. *Journal of the ACM*, 45(1):70-122, 1998.
3. Y. AUMANN AND M. O. RABIN. Manuscript. 1999.
4. M. BELLARE, D. COPPERSMITH, J. HÅSTAD, M. KIWI AND M. SUDAN. Linearity testing in characteristic two. *IEEE Transactions on Information Theory*, 42(6): 1781-1795, 1996.
5. M. BELLARE, O. GOLDREICH AND M. SUDAN. Free bits, PCPs, and non-approximability – towards tight results. *SIAM Journal on Computing*, 27(3):804-915, 1998.
6. M. BELLARE, S. GOLDWASSER, C. LUND AND A. RUSSELL. Efficient probabilistically checkable proofs and applications to approximation. *Proceedings of the Twenty-Fifth Annual ACM Symposium on the Theory of Computing*, pages 294-304, San Diego, California, 16-18 May 1993.
7. M. BLUM AND S. KANNAN. Designing programs that check their work. *Journal of the ACM*, 42(1):269-291, 1995.
8. M. BLUM, M. LUBY AND R. RUBINFELD. Self-testing/correcting with applications to numerical problems. *Journal of Computer and System Sciences*, 47(3):549-595, 1993.
9. J. HÅSTAD. Some optimal inapproximability results. *Proceedings of the Twenty-Ninth Annual ACM Symposium on Theory of Computing*, pages 1-10, El Paso, Texas, 4-6 May 1997.
10. R. RAZ. A parallel repetition theorem. *SIAM Journal on Computing*, 27(3):763-803, 1998.
11. L. TREVISAN. Recycling queries in PCPs and in linearity tests. *STOC*, 1998.

Improved Bounds for Sampling Contingency Tables

Ben Morris[1]

Statistics Department, University of California, Berkeley CA 94720, U.S.A.
Supported by an NSF Graduate Fellowship.
morris@stat.berkeley.edu

Abstract. This paper addresses the problem of sampling contingency tables (non-negative integer matrices with specified row and column sums) uniformly at random. We give an approximation algorithm which runs in polynomial time provided that the row and column sums satisfy $r_i = \Omega(n^{3/2} m \log(m))$, and $c_j = \Omega(m^{3/2} n \log(n))$. Our algorithm is based on a reduction to continuous sampling from a convex set. This is an approach which was taken by Dyer, Kannan and Mount in previous work. However, the algorithm we present is simpler, and has a greater range of applicability since the requirements on the row and column sums are weaker.

1 Introduction

1.1 The Problem

Given positive integer vectors $r = (r_i)_{i=1}^m$ and $c = (c_j)_{j=1}^n$, let $I(r, c)$ denote the set of non-negative integer $m \times n$ matrices with row sums $r_1, \ldots, r_m$ and column sums $c_1, \ldots, c_n$. We consider the problem of generating an element of $I(r, c)$ uniformly at random.

1.2 Motivation

We will now give a brief sketch of how this problem arises in Statistics. The interested reader should consult [1] for a comprehensive analysis.

Suppose that we perform an experiment in which N independent samples are taken and classified according to two characteristics A and B, which take the values $1, \ldots, m$, and $1, \ldots, n$, respectively. For example, A might classify each subject's blood type and B might measure cholesterol level. We then assemble the results in an $m \times n$ matrix X such that X_{ij} is equal to the number of samples having $A = i$ and $B = j$. Such a matrix is called a *contingency table*. We will be interested in measuring the amount of dependence between the two variables in a contingency table. Now, a traditional way to quantify this dependence is via the chi-squared statistic

$$\chi^2(X) = \sum_{i=1}^m \sum_{j=1}^n \frac{(X_{ij} - \frac{r_i c_j}{N})^2}{\frac{r_i c_j}{N}} . \tag{1}$$

This is the sum of (Observed − Expected)^2/Expected, where "Expected" refers to the quantity $\frac{r_i c_j}{N}$, which is the number of observations we would expect to see in a particular cell if A and B were independent with $\Pr(A = i) = \frac{r_i}{N}$ and $\Pr(B = j) = \frac{c_j}{N}$ for all i and j. The *p-value* is defined as the probability that N samples from a distribution with independent row and column variables (i.e., independent A and B) would give a chi-squared statistic which is at least as large as the observed value. Thus, when the p-value is small, there is evidence that the data in the contingency table is dependent.

Consider the following simple examples with $m = n = 2$. Let

$$X_1 = \begin{bmatrix} 10 & 20 \\ 40 & 30 \end{bmatrix} \qquad X_2 = \begin{bmatrix} 20 & 40 \\ 80 & 60 \end{bmatrix} . \qquad (2)$$

For X_1 we have $\chi^2 = 4.76$, which gives a p-value of about .03. For X_2, we have $\chi^2 = 9.52$, and a p-value of about .002. Thus, comparing p-values we would conclude that there is more dependence in the second data set than in the first. However, both data sets appear to have come from a similar underlying distribution, e.g., something like

$$\begin{bmatrix} .1 & .2 \\ .4 & .3 \end{bmatrix} . \qquad (3)$$

The second p-value is smaller only because the sample size is larger. This illustrates a problem with using the p-value alone to measure dependence. While the p-value is useful for determining the *existence* of dependence, it should not be used to measure the *amount* of dependence. Thus it is unwise to compare a p-value or chi-squared statistic from one experiment with that of another. This and other considerations led Diaconis and Efron [1] to propose the following statistic:

$$T(X) = \frac{|\{X' \in I(r, c) : \chi^2(X') < \chi^2(X)\}|}{|I(r, c)|} . \qquad (4)$$

$T(X)$ is the fraction of all contingency tables in $I(r, c)$ which have a smaller chi-squared statistic than the observed value. $T(X)$ is not highly sensitive to sample size, and is thus a better measure of dependence. In the examples above, we have $T(X_1) = \frac{9}{31} \approx \frac{19}{61} = T(X_2)$. Of course, when the contingency tables are large we cannot always calculate $T(X)$ exactly. This explains why we want an algorithm to sample uniformly from $I(r, c)$; given such an algorithm, we can estimate $T(X)$ in the following way:

1. Take a large number of independent samples from $I(r, c)$.
2. Compute the fraction of samples X' for which $\chi^2(X') < \chi^2(X)$.

1.3 Results

Our method of sampling will rely on the fact that $I(r, c)$, when viewed as a subset of $\mathbf{R}^{mn}$, is equal to the intersection of a convex set and the integer lattice. In recent years, a number of (random walk-based) polynomial-time algorithms have been developed for sampling nearly uniformly from a convex set K (see [2],[5], [6], and [8]). Thus, we can solve our discrete sampling problem by reducing it to a continuous one. Given a convex set K which contains $I(r, c)$, we can generate a random sample from $I(r, c)$ using the following algorithm:

1. Generate a random point Y in K, and "round" it to an integer point Z.
2. If $Z \notin I(r, c)$, repeat.

The two main ingredients in this sampling technique are the convex set K and the rounding method. Now, the choice of K is a delicate matter. We require that the distribution of the final sample be nearly uniform. Thus, as X varies over $I(r, c)$, we require that $\mathrm{vol}(X)$ is nearly constant, where $\mathrm{vol}(X)$ denotes the volume of points in K which round to X. Thus we must make K sufficiently large. (We could not, for example, naively set K equal to the set of non-negative, real matrices with the given row and column sums, since the resulting distribution would place too little mass on matrices with a large number of zero entries).

On the other hand, as K becomes larger, it becomes more likely that each sample Z will fall outside of $I(r, c)$. Thus, if we make K too large, then the expected number of trials taken before $Z \in I(r, c)$ will be too high. Hence there is a tradeoff between the running time of the algorithm and the uniformity of the distribution.

The approach we have just described has formed the basis for previous work on sampling contingency tables. Dyer, Kannan and Mount [4], describe an algorithm similar to the one above and show that it samples nearly uniformly from $I(r, c)$ in polynomial time, provided that the row and column sums satisfy $r_i = \Omega(n^2 m)$ for all $1 \le i \le m$, and $c_j = \Omega(m^2 n)$ for all $1 \le j \le n$. These are essentially the best known bounds for general m and n (although when $m = 2$ there is always a polynomial-time algorithm; see [3]). In this paper, we will show that the requirements can be loosened to $r_i = \Omega(n^{3/2} m \log(m))$ and $c_i = \Omega(m^{3/2} n \log(n))$. We accomplish this using the following method for rounding. For $Y \in K$, we will round Y to the integer matrix Z which has $|Z_{ij} - Y_{ij}| \le 1/2$ for all $i < m, j < n$, and has the appropriate row and column sums. This rounding method is quite simple, and it allows us to prove easy bounds on $\mathrm{vol}(X)$ for $X \in I(r, c)$. In turn, this allows us to determine the best choice for the convex set K, and leads to the improved requirements on the row and column sums.

The Contingency Tables problem is a special case of the problem of sampling from the set of integer points contained in the polytope $\{x \in \mathbf{R}^d : Ax < b\}$, where A is a non-negative matrix and b is a non-negative vector. This is a class of problems that was studied by Kannan and Vempala in [7], where they give conditions on the polytope which guarantee a polynomial-time algorithm. (When they apply their results to the Contingency Tables problem, they improve on the row and column sum requirements given in [4], but only by logarithmic factors.) We believe that the techniques in this paper may extend to other problems of this general type.

2 The Algorithm

We will now describe the algorithm for sampling from $I(r,c)$ detail. Two things are needed to specify the algorithm. First, we need to define the convex set K from which we perform continuous sampling. Second, we need to describe the rounding method which takes elements of K to integer lattice points. Since the best choice of K will depend on the rounding method, we will discuss the rounding method first.

To avoid a triviality, we will assume that $\sum_{i=1}^{m} r_i = \sum_{j=1}^{n} c_j$. We will also assume, without loss of generality, that $m \le n$. Now, note that any $m \times n$ matrix X whose row and column sums are fixed can be completely specified by $(X_{ij})_{i<m,j<n}$. It will be helpful to think of $I(r,c)$ as a subset of $\mathbf{R}^{(m-1)(n-1)}$, indexed by $\{(i,j) : i < m, j < n\}$. Thus, we will define $I(r,c)$ as the set of $(m-1) \times (n-1)$ non-negative integer matrices satisfying the constraints:

$$\sum_{j=1}^{n-1} X_{ij} \le r_i \text{ for all } i < m \qquad \sum_{i=1}^{m-1} X_{ij} \le c_j \text{ for all } j < n \qquad (5)$$

$$\sum_{i=1}^{m-1} \sum_{j=1}^{n-1} X_{ij} \ge \sum_{i=1}^{m-1} r_i - c_n . \qquad (6)$$

For $X \in I(r,c)$, we may still at times refer to the quantities X_{in} and X_{mj}, but it will be with the understanding that they are defined in terms of the other entries, via

$$X_{in} = r_i - \sum_{j=1}^{n-1} X_{ij} \text{ for all } i < m \qquad X_{mj} = r_j - \sum_{i=1}^{m-1} X_{ij} \text{ for all } j \le n . \qquad (7)$$

Using this convention, the rounding method (which we described in the Introduction) simply consists of rounding each coordinate to the nearest integer. Furthermore, the convex set K from which we perform continuous sampling will be a full-dimensional subset of $\mathbf{R}^{(m-1)(n-1)}$.

Recall that for integer matrices X, $\mathrm{vol}(X)$ is defined as the volume of points in K which round to X. Thus, $\mathrm{vol}(X)$ is equal to the volume of the intersection of K and the $(m-1)(n-1)$ dimensional unit hypercube centered at X. This leads to the following appealing characterization of $\mathrm{vol}(X)$. Let $\mathcal{E}$ be a random $(m-1) \times (n-1)$ matrix, whose entries are mutually independent and have the uniform distribution over $[-1/2, 1/2]$. Then for all integer matrices X, we have $\mathrm{vol}(X) = \Pr(X + \mathcal{E} \in K)$. Thus, $\mathrm{vol}(X)$ is equal to the probability that, if we perturb X by adding a small random variable to each entry, then the result is in K.

Now, in order for the output of our algorithm to have a nearly uniform distribution, K must be large enough so that $\mathrm{vol}(X)$ is nearly 1 for all $X \in I(r,c)$. In light of the above, this means that if we take any X in $I(r,c)$ and add $\mathcal{E}$, then the result must be in K with high probability. We are now ready to define K. Let $0 < \epsilon < 1/2$ be an error parameter, and let

$$C_1 = \frac{\log(4/\epsilon)}{2\log(m)} + 1/2 \quad C_2 = \frac{\log(4/\epsilon)}{2\log(n)} + 1/2 \quad C_3 = \frac{1}{2}\log(2/\epsilon) . \qquad (8)$$

Let K be the set of real, $(m-1) \times (n-1)$ matrices Y satisfying

$$Y_{ij} > -1/2 \qquad (9)$$

$$Y_{in} \geq -\sqrt{C_1 n \log(m)} \qquad Y_{mj} \geq -\sqrt{C_2 m \log(n)} \qquad (10)$$

$$Y_{mn} \geq -\sqrt{C_3 mn}, \qquad (11)$$

for all $i < m$ and $j < n$. The reasons behind our choices for the above parameters should become clear after we prove the following lemma.

Lemma 1. *For any $X \in I(r,c)$, we have $\mathrm{vol}(X) \geq 1 - \epsilon$.*

Proof. Let $X \in I(r,c)$ and let $X' = X + \mathcal{E}$. We want to show that $\Pr(X' \notin K)$ is less than ϵ. Now, $X'_{ij} \geq -1/2$ for all $i < m$ and $j < n$, since $|\mathcal{E}_{ij}| < 1/2$. We also have $X'_{in} = X_{in} - \sum_{j=i}^{n-1} \mathcal{E}_{ij} \geq -\sum_{j=i}^{n-1} \mathcal{E}_{ij}$ for all $i < m$. Thus, Chernoff's bounds give

$$\Pr(X'_{in} < -\sqrt{C_1 n \log(m)}) \leq e^{-2C_1 \log(m)} = m^{-2C_1}, \quad \text{for all } i < m. \qquad (12)$$

Hence the probability that some X'_{in} is too small is at most $m^{1-2C_1} \leq \epsilon/4$. Applying Chernoff's bounds again, this time to the column sums, gives

$$\Pr(X'_{mj} < -\sqrt{C_2 m \log(n)}) \leq e^{-2C_2 \log(n)} = n^{-2C_2}, \quad \text{for all } j < n. \qquad (13)$$

Hence the probability that some X'_{mj} is too small is also at most $\epsilon/4$. Finally, note that $X'_{mn} = X_{mn} + \sum_{i=1}^{m-1} \sum_{j=1}^{n-1} \mathcal{E}_{ij}$. Hence, Chernoff's bounds imply that $\Pr(X'_{mn} < -\sqrt{C_3 mn}) \leq e^{-2C_3} < \epsilon/2$. Putting this all together, we conclude that $\Pr(X' \notin K) \leq \epsilon/4 + \epsilon/4 + \epsilon/2 = \epsilon$, so $\mathrm{vol}(X) > 1 - \epsilon$. $\qquad \square$

Let Z be the random integer matrix produced by a single trial of our algorithm. The preceding lemma implies that the conditional distribution of Z, given that it is in $I(r,c)$, is nearly uniform. Thus, our algorithm will indeed generate a nearly uniform sample from $I(r,c)$.

3   Main Theorem

In order to bound the running time of the algorithm, we must bound the expected number of trials taken before the random sample Z is in $I(r,c)$. We now state the main result of this paper.

Theorem 1. *Suppose that the row and column sums satisfy $r_i = \Omega(n^{3/2} m \log(m))$ and $c_j = \Omega(m^{3/2} n \log(n))$ for all i and j. Then the expected number of trials is $O(1/\epsilon^2)$.*

Proof. We will assume for simplicity that $r_i \geq 2n^{3/2}m\log(m)$ and $c_j \geq 2m^{3/2}n\log(n)$ for all i and j. The proof for the general case is not any harder. The algorithm repeatedly samples integer matrices Z until one of them is in $I(r, c)$. Thus, the number of trials is a geometric random variable with parameter $\Pr(Z \in I(r, c))$. Let $\bar{I}(r, c)$ denote the set of integer matrices X having $\mathrm{vol}(X) > 0$, i.e. the points which have some $Y \in K$ rounding to them. Then the expected number of trials $\frac{1}{\Pr(Z \in I(r,c))}$ is equal to

$$\frac{\mathrm{vol}(K)}{\mathrm{vol}(I(r, c))} = \frac{\mathrm{vol}(\bar{I}(r, c))}{\mathrm{vol}(I(r, c))}. \tag{14}$$

Suppose that $X \in \bar{I}(r, c)$. Then by (9), X must satisfy $X_{ij} \geq 0$ for all $i < m$ and $j < n$. Thus, X is in $I(r, c)$ if and only if it satisfies $X_{in} \geq 0$ for all $1 \leq i \leq m$ and $X_{mj} \geq 0$ for all $1 \leq j \leq n$. Thus X, when thought of as a $m \times n$ matrix, is in $I(r, c)$ when it has no negative entry anywhere in its last row or column. We must show that such points form a non-neglible fraction of $\bar{I}(r, c)$.

A sketch of our argument is as follows. First, consider the random variable Z_{1n}. Since the row and column sums are large, the probabilities $\Pr(Z_{1n} = k)$ will remain roughly constant over a long interval. Thus, since the number of possible negative values for Z_{1n} is limited to about n, the probability that Z_{1n} will take a non-negative value is quite large. Of course, a similar argument will hold for the other Z_{in} and Z_{mj}, so Z will stand a good chance to be in $I(r, c)$.

Instead of working directly with the vol function, we will find it easier to work with an upper bound on vol which is based on Chernoff's bounds. Let $X \in \bar{I}(r, c)$. Then Chernoff's bounds give the following three upper bounds on $\mathrm{vol}(X)$.

$$\mathrm{vol}(X) \leq \exp\left(\frac{-2\sum_{i<m}((X_{in} + c_1)^-)^2}{n}\right) \tag{15}$$

$$\mathrm{vol}(X) \leq \exp\left(\frac{-2\sum_{j<n}((X_{mj} + c_2)^-)^2}{m}\right) \tag{16}$$

$$\mathrm{vol}(X) \leq \exp\left(\frac{-2((X_{mn} + c_3)^-)^2}{mn}\right), \tag{17}$$

where $c_1 = \sqrt{C_1 n \log(m)}$, $c_2 = \sqrt{C_2 m \log(n)}$, $c_3 = \sqrt{C_3 mn}$, and $x^- = \max(0, -x)$. Putting these together (using monotonicity of the exponential function), we get $\mathrm{vol}(X) \leq w(X)$, where $w(X)$ is defined as the exponential of the quantity

$$-\frac{2}{3}\left[\frac{\sum_{i<m}((X_{in} + c_1)^-)^2}{n} + \frac{\sum_{j<n}((X_{mj} + c_2)^-)^2}{m} + \frac{((X_{mn} + c_3)^-)^2}{mn}\right]. \tag{18}$$

We will call w the weight function. Note that for all integer points X we have $0 \leq \mathrm{vol}(X) \leq w(X) \leq 1$, and since $\epsilon < \frac{1}{2}$, we have $w(I(r, c)) \leq 2 \cdot \mathrm{vol}(I(r, c))$.

This implies that $\frac{\mathrm{vol}(\bar{I}(r,c))}{\mathrm{vol}(I(r,c))} \leq 2 \cdot \frac{w(\bar{I}(r,c))}{w(I(r,c))}$. Thus, our task reduces to giving an upper bound on the quantity $\frac{w(\bar{I}(r,c))}{w(I(r,c))}$.

Recall that $\bar{I}(r,c) - I(r,c)$ consists of the points in $\bar{I}(r,c)$ which have a negative entry somewhere their last row or column. Let $W_0 = \{X \in \bar{I}(r,c) : X_{mn} \geq 0\}$. Next, for all $1 \leq i \leq m-1$, define $W_i = \{X \in W_{i-1} : X_{in} \geq 0\}$. Finally, for all $m \leq i \leq m+n-2$, define $W_i = \{X \in W_{i-1} : X_{m,i-m+1} \geq 0\}$. We have $\bar{I}(r,c) \supset W_0 \supset W_1, \ldots, \supset W_{m+n-2} = I(r,c)$. Our strategy will be to write $\frac{w(\bar{I}(r,c))}{w(I(r,c))}$ as the product

$$\left(\frac{w(\bar{I}(r,c))}{w(W_0)} \right) \left(\frac{w(W_0)}{w(W_1)} \right) \cdots \left(\frac{w(W_{m+n-3})}{w(W_{m+n-2})} \right), \tag{19}$$

and then show that each factor is not too large.

First, we will bound $w(\bar{I}(r,c))/w(W_0)$, which amounts to showing that there isn't too much weight on the points X which have $X_{mn} < 0$. For integers s, let $V_s = \{X \in \bar{I}(r,c) : X_{mn} = s\}$. Then we want to bound

$$\frac{\sum_s w(V_s)}{\sum_{s \geq 0} w(V_s)} . \tag{20}$$

We will do this by giving an upper bound on $\frac{w(V_{s-1})}{w(V_s)}$ for each s. What we want to show is that $w(V_{s-1}) < \alpha w(V_s)$, for some α only slightly larger than 1. We will do this in a way that is reminiscent of the standard technique in Combinatorics in which one shows that two sets S_1 and S_2 satisfy $|S_1| < k|S_2|$ by giving a k-to-1 function from S_1 to S_2. What we will do here is give a *random* function (i.e., a function which is itself a random variable) from V_{s-1} to V_s with the property that for every $X \in V_s$, the expected value of the total weight of points mapping to X is less than $\alpha \cdot w(X)$.

For positive integers $a < m$ and $b < n$, let $T_{ab}(\cdot)$ be the transformation, acting on a matrix X, that increases X_{ab} and X_{mn} by 1 and reduces X_{an} and X_{mb} by 1. Let $M = m^{3/2} n \log(n)$, and for $0 \leq s \leq M$, let f_s be a random function from V_{s-1} to V_s such that

$$f_s(X) = T_{ab}(X) \text{ with probability } \frac{X_{an}^+ X_{mb}^+}{\left(\sum_{i<m} X_{in}^+ \right) \left(\sum_{j<n} X_{mj}^+ \right)}, \tag{21}$$

for all $a < m$ and $b < n$, where $x^+ = \max(0, x)$. Note that if $X \in V_s$ for $s \leq -c_3$, and $X' = T_{ab}^{-1}(X)$ for some a and b, then

$$\frac{w(X')}{w(X)} = \exp\left\{ \frac{2}{3} \cdot \frac{2(s+c_3)+1}{mn} \right\}. \tag{22}$$

Let

$$p(s) = \begin{cases} \exp\left\{ \frac{2}{3} \cdot \frac{2(s+c_3)+1}{mn} \right\} & \text{if } s \leq -c_3; \\ 1 & \text{otherwise.} \end{cases} \tag{23}$$

Then for all $s \le M$ and $X \in V_s$ we have

$$E(w(f_s^{-1}(X))) \le \sum_{a=1}^{m-1} \sum_{b=1}^{n-1} \frac{(X_{an}^+ + 1)(X_{mb}^+ + 1)}{(\sum_a X_{an}^+ + 1)(\sum_b X_{mb}^+ + 1)} \cdot p(s) \cdot w(X)$$

$$= \frac{[\sum_a X_{an}^+ + m] \cdot [\sum_b X_{mb}^+ + n]}{[\sum_a X_{an}^+ + 1] \cdot [\sum_b X_{mb}^+ + 1]} \cdot p(s) \cdot w(X)$$

$$\le (1 + \frac{1}{\sqrt{mn}\log(n)})(1 + \frac{1}{\sqrt{nm}\log(m)}) \cdot p(s) \cdot w(X)$$

$$\le (1 + \frac{3}{\sqrt{nm}\log(m)}) \cdot p(s) \cdot w(X),$$

since $[\sum_a X_{an}^+] \ge m^{3/2} n \log(n)$ and $[\sum_b X_{mb}^+] \ge n^{3/2} m \log(m)$ (recall that $X_{mn} \le m^{3/2} n \log(n)$).

It follows that for all $s \le M$ we have $\frac{w(V_{s-1})}{w(V_s)} \le (1 + \frac{3}{\sqrt{nm}\log(m)}) \cdot p(s)$. A calculation shows that this bound is less than $(1 - \frac{1}{\sqrt{mn}})$ when $s \le -c_3 - 3\sqrt{mn}$. Thus,

$$\frac{w(V_{s-1})}{w(V_s)} \le \begin{cases} (1 - 1/\sqrt{mn}) & \text{if } s \le -c_3 - 3\sqrt{mn}; \\ 1 + \frac{3}{\sqrt{nm}\log(m)} & \text{otherwise.} \end{cases} \tag{24}$$

This implies that the quantity $\frac{\sum_s w(V_s)}{\sum_{s \ge 0} w(V_s)}$ is less than

$$\frac{\sum_{j \ge 0}(1 - 1/\sqrt{mn})^j + \sum_{j=1}^{M+c_3+3\sqrt{mn}}(1 + \frac{3}{\sqrt{nm}\log(m)})^{-j}}{\sum_{j=c_3+3\sqrt{mn}}^{M+c_3+3\sqrt{mn}}(1 + \frac{3}{\sqrt{nm}\log(m)})^{-j}}, \tag{25}$$

which a straightforward calculation shows is $O(1/\epsilon)$.

Next, using calculations similar to above, we obtain the bounds

$$\frac{w(W_i)}{w(W_{i+1})} = O\left(\exp\{\frac{\log(\frac{1}{\epsilon})}{2m}\}\right), \quad \text{for } 0 \le i \le m - 2, \tag{26}$$

and[1]

$$\frac{w(W_i)}{w(W_{i+1})} = O\left(\exp\{\frac{\log(\frac{1}{\epsilon})}{2n}\}\right), \quad \text{for } m - 1 \le i \le m + n - 3. \tag{27}$$

Plugging these bounds into (19), we get $\frac{w(I(r,c))}{w(\bar{I}(r,c))} = O(1/\epsilon^2)$. □

We conclude that the running time of the entire algorithm is of the form

$$q(m, n) \cdot O(1/\epsilon^2), \tag{28}$$

where $q(m, n)$ is the (polynomial) running time of a single trial.

---

[1] The reader who wishes to verify (27) should note that it is necessary here to assume w.l.o.g that $r_m \ge r_1, \ldots, r_{m-1}$.

References

1. P. Diaconis and B. Efron. Testing for independence in a two-way table: new interpretations of the chi-squared statististic. *Annals of Statistics* **13** (1985), pp. 845–913.
2. M. E. Dyer, A. M. Frieze, and R. Kannan. A random polynomial time algorithm for approximating the volume of convex bodies. *Journal of the ACM* **38** (1991), pp. 1–17.
3. M. E. Dyer and C. Greenhill. A genuinely polynomial-time algorithm for sampling two-rowed contingency tables. *Proceedings of the 25th International Colloquium on Automata, Languages and Programming*, 1998, pp. 339–350.
4. M. E. Dyer, R. Kannan and J. Mount. Sampling contingency tables. *Random Structures & Algorithms* **10** (1997), pp. 487–506.
5. R. Kannan and L. Lovász. A logarithmic Cheeger inequality and mixing in random walks. *Proceedings of ACM Symposium on Theory of Computing*, 1999, to appear.
6. R. Kannan, L. Lovász and M. Simonovits. Random walks and an $O^*(n^5)$ volume algorithm for convex bodies. *Random Structures & Algorithms* **11** (1997), pp. 1–50.
7. R. Kannan and S. Vempala. Sampling Lattice Points. *Proceedings of the 29th Annual Symposium on the Theory of Computing*, 1997, pp. 696–700.
8. L. Lovász. *Hit and run mixes fast*. Preprint, Yale University, 1998.

Probabilistic and Deterministic Approximations of the Permanent

Avi Wigderson

The Hebrew University, Jerusalem and
The Institute for Advanced Study, Princeton

Abstract. The exact computation of the permanent of a matrix is #*P*-complete. Many efforts have been made to efficiently approximate the permanent. In this talk we will survey some of these methods, both probabilistic and deterministic.

The papers below and the references within them serve as a good source of information on this topic.

References

1. A. I. Barvinok, Computing Mixed Discriminants, Mixed Volumes, and Permanents, *Discrete & Computational Geometry* , 18, 205-237, 1997.
2. P. Dagum, M. Luby, Approximating the Permanent of Graphs with Large Factors *Theretical Computer Science*, Part A, Vol. 102, pp. 283-305, 1992.
3. U. Feige and C. Lund. On the hardness of computing the permanent of random matrices, *STOC* 24, 643-654, 1992.
4. M. Jerrum and A. Sinclair, Approximating the permanent, *SIAM J. Comput.*, 18, 1149-1178, 1989.
5. M. Jerrum and U. Vazirani, A mildly exponential approximation algorithm for the permanent, *Algorithmica*, 16(4/5), 392-401, 1996.
6. P. W. Kasteleyn, The statistics of dimers on a lattice 1. The number of dimer arrangements on a quadratic lattice. *Physica*, 27, 1209–1225, 1961.
7. N. Karmarkar, R. Karp, R. Lipton, L. Lovasz and M. Luby, A Monte-Carlo algorithm for estimating the permanent, *SIAM Journal on Computing*, 22(2), 284-293, 1993.
8. L. G. Valiant, The complexity of computing the permanent, *Theoretical Computer Science*, 8(2), 189-201, 1979.

Improved Derandomization of BPP Using a Hitting Set Generator

Oded Goldreich[1] and Avi Wigderson[2]

[1] Department of Computer Science, Weizmann Institute of Science, Rehovot, ISRAEL.
oded@wisdom.weizmann.ac.il
[2] Institute of Computer Science, The Hebrew University of Jerusalem, Givat-Ram,
Jerusalem, ISRAEL. avi@cs.huji.ac.il

Abstract. A hitting-set generator is a deterministic algorithm which generates a set of strings that intersects every dense set recognizable by a small circuit. A polynomial time hitting-set generator readily implies $\mathcal{RP} = \mathcal{P}$. Andreev et. al. (ICALP'96, and JACM 1998) showed that if polynomial-time hitting-set generator in fact implies the much stronger conclusion $\mathcal{BPP} = \mathcal{P}$. We simplify and improve their (and later) constructions.

Keywords: Derandomization, $\mathcal{RP}$, $\mathcal{BPP}$, one-sided error versus two-sided error,

1 Introduction

The relation between randomized computations with one-sided error and randomized computations with two-sided error is one of the most interesting questions in the area. Specifically, we refer to the relation betwen $\mathcal{RP}$ and $\mathcal{BPP}$. In particular, does $\mathcal{RP} = \mathcal{P}$ imply $\mathcal{BPP} = \mathcal{P}$?

The breakthrough paper of Andreev et. al. [1] (and its sequel [2]) gave a natural setting in which the answer is YES. The setting is a specific natural way to prove $\mathcal{RP} = \mathcal{P}$, namely via "hitting-set generators" (see exact definition below). Intuitively, such a generator outputs a set of strings, that hits every large efficiently-recognizable set (e.g., the witness set of a positive input of an $\mathcal{RP}$ language). Having such a generator which runs in polynomial time enables the trivial deterministic simulation of an $\mathcal{RP}$ algorithm using each of its outputs as the random pad of the given algorithm.

The main result of [1] was that such a generator for 1-sided error algorithms already suffices to derandomize 2-sided error algorithms: the existence of polynomial-time hitting set generators imply $\mathcal{BPP} = \mathcal{P}$.

Definition 1 (hitting set generator): *An algorithm, G, is called a* **hitting set generator** *for circuits if for every $n, s \in \mathsf{N}$ (given in unary) generates as output a set of n-bit strings $G(n, s)$ with the following property: every circuit of size s on n input bits, which accepts at least half its inputs, accepts at least one element from the set $G(n, s)$.*[3]

---

[3] Usually generators are defined to output only one string; in terms of the above

Since $s = s(n)$ is the essential complexity parameter ($n \leq s$), we let $t_G(s)$ denote the running time of the generator G on input (n, s), and $N_G(s)$ denote the size of its output set. Clearly $N_G(s) \leq t_G(s)$. The result of Andreev *et. al.* [1] is

Theorem 2 [1]: *If there exists a hitting-set generator G running in time t_G then $BPP \subseteq DTime(\text{poly}(t_G(\text{poly}(n))))$.*

With the most important special case (i.e., $t_G(s) = \text{poly}(s)$)

Corollary 3 [1]: *If G runs in polynomial time then $BPP = P$.*

Our main result is a simple proof of Theorem 2. To explain what simple means is not so simple, and we have to explain how the given generator assumed in the theorem is used to enable the derandomization of BPP, in the proof of [1] and in later proofs. Indeed later proofs (of [2] and then [3]) were much simpler, but while proving Corollary 3, they fell short of proving Theorem 2.

The reader is warned that the following discussion is on an intuitive level and some things cannot easily be made precise. If you don't like such discussions, you are welcome to skip to the formal proof in the next two sections.

The proof in [1] uses the generator in two ways. Once, literally as a producer of a hitting set for all large efficient sets. Second, and more subtly, as a hard function. Observe that the existence of such a generator G immediately implies the existence of a function in E on $O(\log t_G(s))$ bits which cannot be computed by circuits of size s. These two ways are combined in a rather involved way for the derandomization of BPP.

It is interesting to note that for the case $t_G(s) = \text{poly}(s)$, the resulting hard function mentioned above can be plugged into the pseudo–random generator of [6], to yield $BPP = P$ as in Corolarry 3. However, [6] was unavailable to the authors of [1] at the time (the two papers are independent). Moreover, [6] is far from "simple", it does use the computational consequence which we are trying to avoid, and anyway it is not strong enough to yield Theorem 2.

A considerably simpler proof was given in [2]. There the generator is used only in its "original capacity", as a hitting set generator, without explicitly using any computational consequence of its existence. In some sense, this proof is more clearly a "black-box" use of the output set of the generator. However, something was lost. The running time of the derandomization is replaced by $\text{poly}(t_G(t_G(\text{poly}(n))))$.

On the one hand, this is not too bad. For the interesting case of $t_G(s) = \text{poly}(s)$ (which implies $RP = P$), they still get the consequence of 3 $BPP = P$ (as iterating a polynomial function twice results in a polynomial). On the other hand, if the function t_G grows moderately so that $t_G(t_G(n)) = 2^n$, then we have as assumption a highly nontrivial derandomization of RP, but the consequence is a completely trivial derandomization of BPP.

---

definition it means that on input an index $i \in \{1, ..., |G(n, s)|\}$, the generator outputs the ith string in $G(n, s)$. However, we find the current convention simpler to work with in the current context.

The best (to our taste) way to understand the origin of the iterated application of the function t_G in the result above, is explained in the recent paper [3], which further simplifies the proof of [2]. They remind the reader that Sipser's proof [8] putting $\mathcal{BPP}$ in $\Sigma^2 \cap \Pi^2$ actually gives much more. In fact, viewed appropriately, it almost begs (with hindsight) the use of hitting sets!

The key is, that in both the $\forall\exists$ and $\exists\forall$ expressions for the $\mathcal{BPP}$ language, the "witnesses" for the existential quantifier are abundant. Put differently, $\mathcal{BPP} \subseteq RP^{pr\mathcal{RP}}$, (where $pr\mathcal{RP}$ is the promise-problem version of $\mathcal{RP}$). But if you have a hitting set, you can use it first to derandomize the "oracle" part or the right hand side. This leaves us with an $RTime(t_G(\text{poly}(n)))$ machine, which can again be derandomized (using hitting sets for $t_G(\text{poly}(n))$ size circuits).

In short, the "two quantifier" representation of $\mathcal{BPP}$, leads to a two-level recursive application of the generator. It seems hopeless to reduce the number of quantifiers to one in Sipser's result. So another route has to be taken to prove Theorem 2 in a similar "direct" (or "black-box") as above, without incurring the penalty arising from this two level recursion.

We eliminate the recursion to have only one-level use of the hitting set, by "increasing the dimension to two": We view the possible random strings of the $\mathcal{BPP}$ algorithm as elements in a matrix. This is inspired by another, recent proof (strengthening Sipser's result) that $\mathcal{BPP} \subseteq \mathcal{MA}$, due to Goldreich and Zuckerman [5]. There and here strong extractors (cf., [10] or [9]) are used to ensure that in this matrix, the "non-witnesses" are not only few, but actually miss most rows and columns. The hitting set is used to select a small subset of the rows and a small subset of the columns, and the entries of this submatrix determine the result. Specifically we will look for "enough" (yet few) rows which are monochromatic, and decide accordingly. The correctness and efficiency of the test is spelled out is Lemma 6. It is essentially captured by the following simple Ramsey-type result, which is seemingly new and may be of independent interest.

Proposition 4 *Let $n \leq 2^k$. Then for every n-vertex graph, either the graph or its complement has a dominating set of size k. Furthermore, one can find such a set in polynomial time.*

We end by observing that (like the previous results) our result holds in the context of promise problems. Hence, the existence of hitting set generators provide an efficient way for approximately counting the fraction of inputs accepted by a given circuit within additive polynomial fraction. Formalizing this is standard and we leave it to the reader.

2 The Derandomization Procedure

Given $L \in \mathcal{BPP}$ we first use strong results regarding extractors (cf., [10] or [9]) to obtain a probabilistic polynomial-time algorithm, A, which on inputs of length n

uses $2\ell = \text{poly}(n)$ many random bits and errs with probability at most $2^{-(\ell+1)}$.[4] Let $A(x,r)$ denote the output of algorithm A on input $x \in \{0,1\}^n$ and random-tape contents $r \in \{0,1\}^{2\ell}$, and p be some fixed polynomial so that the computation of A on inputs of length n can be implemented by circuits of size $p(\ell)/\ell$. Our derandomization procedure, described below, utilizes a hitting-set generator H as defined above (cf., Def. 1).

Derandomization procedure: On input $x \in \{0,1\}^n$, letting A and ℓ be as above.

1. Invoking the hitting-set generator G obtain $H \leftarrow G(\ell, p(\ell))$. That is, H is a hitting set for circuits of size $p(\ell)$ and input length ℓ. Denote the elements of H by $e_1, ..., e_N$, where $N \overset{\text{def}}{=} N_G(p(\ell))$ and each e_i is in $\{0,1\}^\ell$.
2. Construct an N-by-N matrix, $M = (v_{i,j})_{i,j}$, so that $v_{i,j} = A(x, e_i e_j)$. That is, we run A with all possible random-pads composed of pairs of strings in H.
3. Using a procedure to be specified below, determine whether for every ℓ columns there exists a row on which all these columns have 1-value. If the procedure accepts then **accept** else **rejects**. That is, we accept if and only if

$$\forall c_1, ..., c_\ell \in [N] \; \exists r \in [N] \text{ s.t. } \wedge_{i=1}^\ell (v_{c_i,r} = 1) \tag{1}$$

We first show that if $x \in L$ then Eq. (1) holds, and analogously if $x \notin L$ then

$$\forall r_1, ..., r_\ell \in [N] \; \exists c \in [N] \text{ s.t. } \wedge_{i=1}^\ell (v_{r_i,c} = 0) \tag{2}$$

Note that this by itself does not establish the correctness of the procedure. Neither did we specify how to efficiently implement the procedure. To that end we use a general technical lemma which implies that it cannot be the case that both Eq. (1) and Eq. (2) hold, and in fact efficiently decides at least one which does not hold. These are defered to the next section. But first we prove the above implications.

Proposition 5 *If $x \in L$ (resp., $x \notin L$) then Eq. (1) (resp., Eq. (2)) holds,*

Proof. We shall prove a more general statement. That is, let χ_L be the characteristic function of L (i.e., $\chi_L(x) = 1$ if $x \in L$ and $\chi_L(x) = 0$ otherwise). Then we prove that for every $x \in \{0,1\}^n$, for every ℓ rows (resp., columns) there exists a column (resp., row) on which the value of the matrix is $\chi_L(x)$.

Fixing the input $x \in \{0,1\}^n$ to algorithm A, we consider the circuit C_x which takes an 2ℓ-bit input r and outputs $A(x,r)$ (i.e., evaluates A on input x and coins r). By the above hypothesis (regarding the error probability of A), we have

$$\Pr_{r \in \{0,1\}^{2\ell}}[C_x(r) \neq \chi_L(x)] \leq 2^{-(\ell+1)}$$

Thus, at least half the values of $z \in \{0,1\}^\ell$ satisfy $\forall y \, C_x(y,z) = \chi_L(x)$. We will use a much weaker consequence, namely, that the above holds for every set of ℓ values of y (and this weakness is the key to our more efficient reduction).

---

[4] We note that using [10], ℓ is linear in the randomness of the original BPP-algorithm, and the polynomial p below is quite large. Using the extractors in [9, 7], one may be able to obtain more favorite bounds.

1. Fix any sequence $\bar{y} = (y_1, ..., y_\ell)$ so that $y_1, ..., y_\ell \in \{0,1\}^\ell$. Then,

$$\Pr_{z \in \{0,1\}^\ell}[(\forall i)\, C_x(y_i z) = \chi_L(x)] \geq 1/2 \tag{3}$$

Consider the circuit $C_{x,\bar{y}}(z) \stackrel{\text{def}}{=} \wedge_{i=1}^\ell (C_x(y_i z) = \chi_L(x))$. Then, by the above $\Pr_z[C_{x,\bar{y}}(z) = \chi_L(x)] \geq 1/2$. On the other hand, the size of $C_{x,\bar{y}}$ is merely ℓ times the size of C_x, which was at most $p(\ell)/\ell$. Thus, by definition of the hitting-set generator G, the set $H = G(\ell, p(\ell))$ must contain a string z so that $C_{x,\bar{y}}(z) = \chi_L(x)$. By definition of $C_{x,\bar{y}}$ it follows that $C_x(y_i z) = \chi_L(x)$ holds for every $i \in [\ell]$.

The above holds for any $\bar{y} = (y_1, ..., y_\ell)$. Thus, for every $y_1, ..., y_\ell \in \{0,1\}^\ell$ there exists $z \in H$ so that $A(x, y_i z) = C_x(y_i z) = \chi_L(x)$ for every $i \in [\ell]$. Thus we have proved that for every ℓ rows in M there exists a column on which the value of the matrix is $\chi_L(x)$.

2. A similar argument applies to sets of ℓ columns in M. Specifically, for every $z_1, ..., z_\ell \in \{0,1\}^\ell$

$$\Pr_{y \in \{0,1\}^\ell}[(\forall i)\, C_x(y z_i) = \chi_L(x)] \geq \frac{1}{2}$$

Again, we conclude that for every $z_1, ..., z_\ell \in \{0,1\}^\ell$, there exists $y \in H$ so that $C_x(y z_i) = \chi_L(x)$ for every $i \in [\ell]$. Thus, for every ℓ columns in M there exists a row on which the value of the matrix is $\chi_L(x)$.

The proposition follows.

3   Correctness and Efficiency of the Derandomization

Proposition 5 shows that for every x either Eq. (1) or Eq. (2) holds. But, as stated above, it is not even clear that Eq. (1) and Eq. (2) cannot hold simultaneously. This is asserted next.

Lemma 6 *Every n-by-n Boolean matrix, with $n \leq 2^k$, either has k rows whose OR is the all 1's row, or k columns whose AND is the all 0's column. Moreover, there is a (deterministic) polynomial-time algorithm that given such a matrix find such a set.*

We prove the lemma momentarily. But first let use show that Eq. (1) and Eq. (2) cannot hold simultaneously. We first note that in our case $n = N = N_G(\ell, p(\ell))$ (which is smaller than 2^ℓ by the hypothesis of Theorem 2) and $k = \ell$. Then we just apply the following corollary.

Corollary 7 *For every n-by-n Boolean matrix, with $n \leq 2^k$, it is impossible that both*

1. *For every k rows there exists a column so that all the k rows have a 0-entry in this column.*

2. *For every k columns there exists a row so that all the k columns have a 1-entry in this row.*

Furthermore, assuming one of the above holds, we can decide which holds in (deterministic) polynomial-time.

Proof (of Corollary 7): Suppose Item (1) holds. Then, the OR of every k rows contains a 0-entry, and so cannot be the all 1's row. Likewise, if Item (2) holds then the AND of every k columns contains a 1-enrty, and so cannot be the all 0's column. Thus, the case where both items holds stands in contradiction to Lemma 6. Furthermore, finding a set as in the lemma yields which of the two items does not hold. ∎

Proof of Lemma 6: Let $S_0 = [n]$, $R = \emptyset$, and repeart for $i = 1, 2, ...$: Take a row j not in R which has at least $|S_i|/2$ 1's in S_i. Add j to R, and let S_{i+1} be the part of S_i that had 0's in row j. We get stuck if for any i, no row in current $[n] - R$ has at least $|S_i|/2$ 1's in S_i. Otherwise, we terminate when $S_i = \emptyset$

If we never get stuck, then we generated at most $\log_2 n \leq k$ rows whose OR is the all 1's row (as the ith row has 1-entries in every column in $S_{i-1} - S_i$, and the last S_i is empty). On the other hand, if we got stuck at iteration i, let $S = S_i$. Note that every row has at least $S/2$ 0's in the columns S. (This includes the rows in the current R which have only 0's in the columns in $S \subset S_{i-1} \subset \cdots \subset S_0$.) But now picking greedily columns from S in sequence so as to contain the largest number of 0's in the remaining rows will clearly pick a 0 from every row after a set T of at most k columns from S were chosen.

Turning to the algorithmics, note that the above procedure for constructing R, S and T is implementable in polynomial-time. Thus, in case the "row" procedure was completed successfully, we may output the set of rows R, and otherwise the set T of columns. ∎

Proof of Theorem 2: Proposition 5 shows that for every x either Eq. (1) or Eq. (2) holds, and furthermore that the former (resp., latter) holds whenever $x \in L$ (resp., $x \notin L$). By applying Corollary 7 as indicated above it follows that only one of these equation may hold. Using the decision procedure gauarnteed by this corollary, we implement Step 3 in our derandomized procedure, and Theorem 2 follows. ∎

References

1. A.E. Andreev, A.E.F. Clementi, and J.D.P. Rolim. A new general derandomization method. *Journal of the Association for Computing Machinery (J. of ACM)*, 45(1), pages 179–213, 1998.
   Hitting Sets Derandomize BPP. In *XXIII International Colloquium on Algorithms, Logic and Programming (ICALP'96)*, 1996.

2. A.E. Andreev, A.E.F. Clementi, J.D.P. Rolim and L. Trevisan, Weak Random Sources, Hitting Sets, and BPP Simulations. To appear in *SIAM J. on Comput.*. Preliminary version in *38th FOCS*, pages 264–272, 1997.

3. H. Buhrman and L. Fortnow. One-sided versus two-sided randomness. In Proceedings of the *16th Symposium on Theoretical Aspects of Computer Science*. Lecture Notes in Computer Science, Springer, Berlin, 1999.

4. S. Even, A.L. Selman, and Y. Yacobi. The Complexity of Promise Problems with Applications to Public-Key Cryptography. *Inform. and Control*, Vol. 61, pages 159–173, 1984.

5. O. Goldreich and D. Zuckerman. Another proof that BPP subseteq PH (and more). *ECCC*, TR97-045, 1997.

6. R. Impagliazzo, A. Wigderson, P=BPP unless E has Subexponential Circuits: Derandomizing the XOR Lemma. *29th STOC*, pages 220–229, 1997.

7. R. Raz, O. Reingold and S. Vadhan. Extracting all the Randomness and Reducing the Error in Trevisan's Extractors In *31st STOC*, pages 149–158, 1999.

8. M. Sipser. A complexity-theoretic approach to randomness. In *15th STOC*, pages 330–335, 1983.

9. L. Trevisan. Constructions of Near-Optimal Extractors Using Pseudo-Random Generators. In *31st STOC*, pages 141–148, 1999.

10. D. Zuckerman. Simulating BPP Using a General Weak Random Source. *Algorithmica*, Vol. 16, pages 367–391, 1996.

Probabilistic Construction of Small Strongly Sum-Free Sets via Large Sidon Sets

Andreas Baltz, Tomasz Schoen, and Anand Srivastav

Mathematisches Seminar, Christian-Albrechts-Universität zu Kiel,
Ludewig-Meyn-Str. 4, D-24098 Kiel, Germany aba@numerik.uni-kiel.de
tos@numerik.uni-kiel.de asr@numerik.uni-kiel.de

Abstract. We give simple randomized algorithms leading to new upper bounds for combinatorial problems of Choi and Erdős: For an arbitrary additive group G let $\mathcal{P}_n(G)$ denote the set of all subsets S of G with n elements having the property that 0 is not in $S + S$. Call a subset A of G admissible with respect to a set S from $\mathcal{P}_n(G)$ if the sum of each pair of distinct elements of A lies outside S. For $S \in \mathcal{P}_n(G)$ let $h(S)$ denote the maximal cardinality of a subset of S admissible with respect to S. In particular we show $h(n) := \min\{h(S) \mid G \text{ group}, S \in \mathcal{P}_n(G)\} = \mathcal{O}((\ln n)^2)$. The methodical innovation of the whole approach is the use of large Sidon sets.

Keywords. Strongly Sum-Free Sets, Sidon Sets, Independent Sets in Hypergraphs

1 Introduction

1.1 Statement of the Problems and Previous Results

In [4] Erdős asked the following interesting question:

Let $a_1, a_2, \ldots, a_n$ be n distinct real numbers. A subset $a_{i_1}, \ldots, a_{i_k}$ is called *strongly sum-free* if $a_{i_j} + a_{i_l} \neq a_r$ for all $1 \leq j < l \leq k$, $1 \leq r \leq n$. Let $\Phi(n)$ be the maximum cardinality of a strongly sum-free set. How large is $\Phi(n)$?

The best known bounds so far have been given by Choi [3] who proved that

$$\Phi(n) \geq \ln n$$

and, using sieve methods, showed

$$\Phi(n) = \mathcal{O}(n^{\frac{2}{5}+\epsilon}) \ .$$

Moreover Choi observed that Erdős' problem can be equivalently formulated in terms of non-negative integers: $\Phi(n)$ equals $g(n)$, the largest number such that, from every set of n distinct non-negative integers, one can always select a strongly sum-free subset of $g(n)$ integers.

Choi also considered the following variant of the problem:

Let us call a set A of non-negative integers *admissible* with respect to a set S of non-negative integers if the sum of each pair of distinct elements of A lies outside S. Let $n \in \mathbb{N}$, and suppose that S is a subset of the interval $[2n, 4n)$. Denote by $f(S)$ the number of elements in a maximal subset of $[n, 2n)$ admissible with respect to S, and define $f(n)$ by

$$f(n) := \min\{|S| + f(S) \mid S \subseteq [2n, 4n)\} .$$

How large is $f(n)$?

It is easy to see that $f(n) \geq \sqrt{n}$: Given $|S| < \sqrt{n}$ we can construct an admissible set A by successively selecting $a_i \in [n, 2n) \backslash D_i$, where $D_1 := \emptyset$ and $D_{i+1} := -a_i + S$. In each step we remove at most $|S|$ elements, so the procedure can be carried out at least $\frac{n}{|S|} > \sqrt{n}$ times yielding an admissible set of the claimed size.

For an upper bound Choi proved that $f(n) = \mathcal{O}(n^{\frac{3}{4}})$ and conjectured $f(n) = \mathcal{O}(n^{\frac{1}{2}+\varepsilon})$.

1.2 Results

In this article we improve Choi's upper bound by showing that

$$f(n) = \mathcal{O}(n^{\frac{2}{3}} \ln^{\frac{2}{3}} n) .$$

As an easy consequence we get an improved upper bound for the problem of Erdős: $g(n) = \mathcal{O}(n^{\frac{2}{5}} \ln^{\frac{2}{5}} n)$. The proof is a probabilistic construction based on a deep theorem of Komlós, Sulyok, and Szemerédi [6] who showed that every set $A \subseteq \mathbb{N}$ contains a Sidon set of size $\Theta(\sqrt{|A|})$. Note that Sidon sets and independent sets in hypergraphs are closely related [8].

Finally, we study a more general setting (see [4] and [5]): For an arbitrary additive group G that contains at least n elements let $\mathcal{P}_n(G)$ denote the set of all subsets S of G satisfying $|S| = n$ and $0 \notin S + S$. (The latter condition prevents us from taking S as a subgroup of G.) If the maximal cardinality of a subset A of $S \in \mathcal{P}_n(G)$ admissible with respect to S is $h(S)$, how large is

$$h(n) := \min\{h(S) \mid G \text{ group}, \ S \in \mathcal{P}_n(G)\} ?$$

It is shown in [7] that $h(n) \geq 3$ for abelian groups. We give the first upper bound by proving that $h(n) = \mathcal{O}(\ln^2 n)$.

We leave open the problem of efficient, i.e. polynomial-time derandomization of our randomized algorithms.

1.3 Some Notations

As we consider only intervals of positive integers we abbreviate $[a, b] \cap \mathbb{N}$, $(a, b] \cap \mathbb{N}$, and $[a, b) \cap \mathbb{N}$ (for positive real numbers a and b) by $[a, b]$, $(a, b]$, and $[a, b)$.

If z is an integer and S, T are sets of integers we define:

$$- z + S := \{z + s \mid s \in S\}$$
$$- z - S := \{z - s \mid s \in S\}$$
$$- z \cdot S := \{z \cdot s \mid s \in S\}$$
$$- S + T := \{s + t \mid s \in S, \ t \in T\}$$
$$- S \dotplus T := \{s + t \mid s \in S, \ t \in T, \ s \neq t\}$$

In our approach Sidon sets play a key role.

A *Sidon set* is a set of integers with the property that all pairwise sums of its elements are distinct. For us the crucial property of a Sidon set S is

$$|S \dotplus S| = \binom{|S|}{2}. \tag{1}$$

The terms c, c', c_1, c_2, etc. mysteriously appearing and disappearing in our proofs should always be absolute constants.

2 Strongly Sum-Free Sets in $\mathbb{N}$

Komlós, Sulyok, and Szemerédi proved the following remarkable theorem generalizing the celebrated Erdős/Turan theorem that the size of a Sidon set in $[1, n]$ is $\Theta(\sqrt{n})$.

Lemma 1 (Komlós, Sulyok, Szemerédi). *There is an absolute constant $c > 0$, such that each finite set A of positive integers contains a Sidon set with at least $c \cdot |A|^{\frac{1}{2}}$ elements.*

Theorem 1. $f(n) = \mathcal{O}(n^{\frac{2}{3}} \ln^{\frac{2}{3}} n)$.

Proof. Choose a random subset $S \subseteq [2n, 4n)$ by picking each element independently with probability $p = \left(\frac{\ln^2 n}{n} \right)^{\frac{1}{3}}$. Let

$$r := \lceil 2(n \ln n)^{\frac{1}{3}} \rceil$$

and define

$$\mathcal{S}_r := \{ R \subseteq [n, 2n) \mid R \text{ Sidon set}, \ |R| = r \} \ .$$

For every $R \in \mathcal{S}_r$ we consider the indicator random variable

$$X_R := \begin{cases} 1, & \text{if } R \dotplus R \cap S = \emptyset \\ 0 & \text{otherwise} \end{cases} .$$

Then $X := \sum_{R \in \mathcal{S}_r} X_R$ denotes the random variable counting the number of Sidon sets $R \subseteq [n, 2n)$ with $|R| = r$ and $R \dotplus R \cap S = \emptyset$. We have

$$\mathbb{E}(X) = \sum_{R \in \mathcal{S}_r} \mathbb{E}(X_R) = \sum_{R \in \mathcal{S}_r} \mathbb{P}(R \dotplus R \cap S = \emptyset)$$
$$= \sum_{R \in \mathcal{S}_r} \mathbb{P}(a + b \notin S \text{ for all } a, b \in R \text{ where } a \neq b) \ .$$

As R is a Sidon set, all of the sums $a + b$ are distinct. So (using (1)) for each R we are facing $|R \dot{+} R| = \binom{|R|}{2} = \frac{r^2 - r}{2}$ independent events, each of which happens with probability $(1 - p)$. This yields

$$\mathbb{E}(X) = \sum_{R \in \mathcal{S}_r} (1 - p)^{\frac{r^2 - r}{2}} \leq \binom{n}{r}(1 - p)^{\frac{r^2 - r}{2}} \leq \left(\frac{en}{r}\right)^r \left[(1 - p)^{\frac{1}{p}}\right]^{\frac{r^2 - r}{2}p}$$

$$\leq \left(\frac{en}{re^{\frac{rp - p}{2}}}\right)^r \leq \left(\frac{en}{re^{\frac{rp}{2}}}\right)^r \leq \frac{en}{2(n \ln n)^{\frac{1}{3}}n} .$$

For large n the latter expression becomes arbitrarily small, so that by Markov's inequality

$$\mathbb{P}(|S| \geq 4(n \ln n)^{\frac{2}{3}}) + \mathbb{P}(X \geq 1) \leq \frac{1}{2} + \mathbb{E}(X) < 1 .$$

Hence there exists $S \subseteq [2n, 4n)$ of size $\mathcal{O}(n^{\frac{2}{3}} \ln^{\frac{2}{3}} n)$ such that every Sidon set R of size *at least* r fulfills $R \dot{+} R \cap S \neq \emptyset$.

Let A be a (maximal) subset of $[n, 2n)$ with $A \dot{+} A \cap S = \emptyset$. From Lemma 1 we know that A contains a Sidon set R with cardinality $c \cdot \sqrt{|A|}$. Obviously, $R \dot{+} R \cap S = \emptyset$ and thus $|A| = \frac{1}{c^2}|R| < \frac{1}{c^2}r^2 = \mathcal{O}(n^{\frac{2}{3}} \ln^{\frac{2}{3}} n)$. We conclude that $f(n) \leq |S| + |A| = \mathcal{O}(n^{\frac{2}{3}} \ln^{\frac{2}{3}} n)$.

Corollary 1. $g(n) = \mathcal{O}(n^{\frac{2}{5}} \ln^{\frac{2}{5}} n)$.

Proof. Let $m := \lfloor n^{\frac{3}{5}} \rfloor$. From Theorem 1 we know that there exists $S' \subseteq [2m, 4m)$ of size at most $c_1(m \ln m)^{\frac{2}{3}}$ such that any subset $A' \subseteq [m, 2m)$ admissible with respect to S' has no more than $c_2(m \ln m)^{\frac{2}{3}}$ elements. Obviously, for any $k \in \mathbb{N}$ the set $2^{k-1} \cdot S'$ has the property that no subset of $2^{k-1} \cdot [m, 2m)$ consisting of more than $c_2(m \ln m)^{\frac{2}{3}}$ elements is admissible with respect to S'.

Now choose

$$k := \frac{n - |S'|}{m}$$

and define

$$S := \left(\bigcup_{i=1}^{k} 2^{i-1} \cdot [m, 2m)\right) \cup 2^{k-1} \cdot S' .$$

We have

$$|S| = k \cdot m + |S'| = n .$$

Let $A \subseteq S$ be a set of maximal cardinality admissible with respect to S. Clearly, $2^{k-1} \cdot S' \subseteq A$. A contains at most 2 elements from each set $2^{i-1} \cdot [m, 2m)$, $i \in \{1, \ldots, k-1\}$ and at most $c_2(m \ln m)^{\frac{2}{3}}$ elements from $2^{k-1} \cdot [m, 2m)$. Thus $|A| \leq 2(k-1) + (c_1 + c_2)(m \ln m)^{\frac{2}{3}} = \mathcal{O}(n^{\frac{2}{5}} \ln^{\frac{2}{5}} n)$.

3 Strongly Sum-Free Sets in Arbitrary Groups

Theorem 2. $h(n) = \mathcal{O}(ln^2 n)$.

Proof. We show:

There exists $S \in \mathcal{P}_n(\mathbb{Z}_{2n+1})$ such that each $A \subseteq \mathbb{Z}_{2n+1}$ admissible with respect to S has no more than $\mathcal{O}(\ln^2 n)$ elements.

This is slightly stronger than the claim, since we do not restrict A to be a subset of S.

Choose a random subset $T \subseteq [1, n]$ by selecting each element with probability $p = \frac{1}{2}$. Set

$$S := T \cup \{[n+1, 2n]\backslash(2n+1-T)\} \ .$$

Clearly, $0 \notin S + S$ and $|S| = |T| + (n - |T|) = n$.

Let $X^1{}_r$, $X^1{}_r$, $X^3{}_r$, and $X^4{}_r$ be random variables counting the number of Sidon sets R of size r in $[1, \frac{n}{2}]$, $(\frac{n}{2}, n]$, $(n, \frac{3n}{2}]$ and $(\frac{3n}{2}, 2n]$ respectively, where R satisfies $R \dot{+} R \cap S = \emptyset$. (Note that any such R is a Sidon set in $\mathbb{Z}_{2n+1}$ iff it is a Sidon set in $\mathbb{N}$.)

As in the proof of Theorem 1 we estimate

$$\mathbb{E}(X^i{}_r) \leq \binom{\frac{n}{2}}{r}(1-p)^{\binom{r}{2}} \leq \left(\frac{en}{2re^{\frac{r-1}{4}}}\right)^r, \ i \in \{1, 3\}$$

and

$$\mathbb{E}(X^i{}_r) \leq \binom{\frac{n}{2}}{r}p^{\binom{r}{2}} \leq \left(\frac{en}{2re^{\frac{r-1}{4}}}\right)^r, \ i \in \{2, 4\} \ .$$

Choosing

$$r := 4\ln(en)$$

we get

$$\mathbb{E}(X^i{}_r) \leq \frac{e^{\frac{1}{4}}}{8\ln(en)} < \frac{1}{4}$$

and hence by Markov's inequality

$$\mathbb{P}(X^1{}_r \geq 1) + \mathbb{P}(X^2{}_r \geq 1) + \mathbb{P}(X^3{}_r \geq 1) + \mathbb{P}(X^4{}_r \geq 1) < 1 \ .$$

Thus there exists $S \in \mathcal{P}_n(\mathbb{Z}_{2n+1})$ such that every Sidon set R in $[1, \frac{n}{2}]$, $(\frac{n}{2}, n]$, $(n, \frac{3n}{2}]$ or $(\frac{3n}{2}, 2n]$ of size *at least* $4\ln(en)$ has the property $R \dot{+} R \cap S \neq \emptyset$.

Let A be a subset of $[1, 2n]$ admissible with respect to S and let

$$A_1 := A \cap [1, \frac{n}{2}], \ A_2 := A \cap (\frac{n}{2}, n], \ A_3 := A \cap (n, \frac{3n}{2}], \ A_4 := A \cap (\frac{3n}{2}, 2n] \ .$$

Pigeon-hole principle gives

$$|A_j| \geq \frac{|A|}{4}$$

for some $j \in \{1, 2, 3, 4\}$. From Lemma 1, $c\sqrt{|A_j|}$ elements in A_j form a Sidon set, and we conclude that $|A| \leq 4 \cdot |A_j| \leq \frac{4}{c^2} \cdot r^2 = \mathcal{O}(\ln^2 n)$.

Note that Theorem 2 is similar to Theorem 2.13 in Agarwal, Alon, Aronov and Suri [1], and in fact it can be proved by the method given there which does not use the result of Komlós, Sulyok and Szemerédi. Also, Theorem 6 in Alon and Orlitsky [2] is related to Theorem 2.

4 An open Problem

Can we efficiently find a set S with the properties guaranteed by our theorems? Exhaustive search takes time $\mathcal{O}(e^{n^{\frac{2}{3}}\ln^{\frac{5}{3}}n-\frac{2}{3}\ln\ln n})$ and $\mathcal{O}(e^{\ln^3 n-2\ln\ln n})$ respectively. The method of conditional probabilities seems hardly applicable since we do not know how to compute the number of Sidon sets with fixed cardinality that contain a certain subset without testing all Sidon sets.

References

1. P.K. Agarwal, N. Alon, B. Aronov, S. Suri, Can visibility graphs be represented compactly?, *Discrete and Computational Geometry* 12 (1994), 347-365.
2. N. Alon, A. Orlitsky, Repeated communication and Ramsey graphs, *IEEE Transactions on Information Theory* 41 (1995), 1276-1289.
3. S.L.G. Choi, On a combinatorial problem in number theory, *Proc. London Math. Soc.* (3)23 (1971), 629-642.
4. P. Erdős, Extremal problems in number theory, *Proc. Sympos. Pure Math., vol.8, Amer. Math. Soc., Providence, R.I.* (1965), 181-189.
5. R.F. Guy, *Unsolved problems in number theory*, Springer, New York (1994), 128-129.
6. J. Komlós, M. Sulyok, E. Szemerédi, Linear problems in combinatorial number theory, *Acta Math. Acad. Sci. Hungar.* 26 (1975), 113-121.
7. T. Łuczak, T. Schoen, On strongly sum-free subsets of abelian groups, *Coll. Math.* 71 (1996), 149-151.
8. T. Thiele, Geometric Selection Problems and Hypergraphs, *PhD thesis*, Freie Universität Berlin (October 1995).

Stochastic Machine Scheduling: Performance Guarantees for LP-based Priority Policies[*]

(Extended Abstract)

Rolf H. Möhring[1], Andreas S. Schulz[2], and Marc Uetz[1]

[1] Technische Universität Berlin, Fachbereich Mathematik
Sekr. MA 6–1, Straße des 17. Juni 136, 10623 Berlin, Germany
{moehring,uetz}@math.tu-berlin.de
[2] MIT, Sloan School of Management and Operations Research Center
E53-361, 30 Wadsworth St, Cambridge, MA 02139
schulz@mit.edu

Abstract. We consider the problem to minimize the total weighted completion time of a set of jobs with individual release dates which have to be scheduled on identical parallel machines. The durations of jobs are realized on-line according to given probability distributions, and the aim is to find a scheduling policy that minimizes the objective in expectation. We present a polyhedral relaxation of the corresponding performance space, and then derive the first constant-factor performance guarantees for priority policies which are guided by optimum LP solutions, thus generalizing previous results from deterministic scheduling. In the absence of release dates, our LP-based analysis also yields an additive performance guarantee for the WSEPT rule which implies both a worst-case performance ratio and a result on its asymptotic optimality.

1 Introduction

The model. Let $J = \{1, \ldots, n\}$ be a set of jobs which have to be processed non-preemptively on m identical parallel machines so as to minimize the total weighted completion time. That is, each job has a nonnegative weight w_j and one wants to minimize $\sum_{j \in J} w_j C_j$, where C_j denotes the completion time of job j. Any machine can process at most one job at a time, and every job has to be processed on one of the m machines. We consider scenarios where jobs may, or may not have individual release dates.

The crucial assumption is that processing times of jobs are not known in advance, but are instead given by a random variable $\boldsymbol{p} = (\boldsymbol{p}_1, \ldots, \boldsymbol{p}_n)$. Here, $\boldsymbol{p}_j$ denotes the random variable for the processing time of job j. (All random variables are typeset in bold face.) Throughout the paper job durations are supposed to be stochastically independent, and first as well as second moments are finite. It is usually assumed that these distributions are known from the outset, but

[*] Research partially supported by the German–Israeli Foundation for Scientific Research and Development (G.I.F.), grant I 246–304.02/97.

for our approach it suffices that their expected processing times (and an upper bound on their coefficients of variation) are given. Using the classification scheme for scheduling problems introduced by Graham, Lawler, Lenstra, and Rinnooy Kan [7], the problem may be written as P $\mid p_j \sim$ stoch, $r_j \mid E[\sum w_j C_j]$.

Due to the lack of beforehand information on processing times, the jobs have to be allocated to the machines "on-line". This dynamic allocation of jobs to machines is the task of a *scheduling policy*. It specifies which job(s) should be started at any given time t. The decisions of such a policy may only depend on the "past up to t", which is given by the sets of jobs already finished or being performed at t, their start times, and the conditional distribution of remaining processing times of jobs. For a detailed account of the theoretical foundations of the stochastic model considered in this paper, we refer to Möhring, Radermacher, and Weiss [13, 14]. Of special importance for our work is the class of *priority policies* which implement a given priority order on the set of jobs. They will be more formally defined in Sect. 4. Simple examples show that in general one cannot expect to find a non-anticipative scheduling policy that minimizes the objective point-wise for any realization of processing times. Therefore, one aims to minimize the objective in expectation. If $E[C_j^\Pi]$ denotes the expected completion time of job j under policy Π, the problem reads

$$\text{minimize } \{\sum_{j\in J} w_j E[C_j^\Pi] \mid \Pi \text{ policy}\}.$$

Furthermore, let $Z^{OPT} := \inf\{ \sum_{j\in J} w_j E[C_j^\Pi] \mid \Pi \text{ policy} \}$ denote the corresponding optimum value. It follows from [13, Sect. 4] that in the present setting there exists an *optimum* policy with expected performance Z^{OPT}. Note that an optimum policy is not necessarily work conserving. It may involve deliberate idling of machines, even in the absence of job release dates.

The model considered in this paper is somewhat related to certain on-line scenarios, which recently have received some attention. These scenarios are also based on the assumption that the scheduler learns the input piece by piece over time and has to make decisions based on partial knowledge only. In contrast to the present model, on-line algorithms are usually analyzed with respect to optimum off-line solutions. We refer to Sgall [19] for an overview of recent achievements in this direction. Note that stochastic scheduling is more moderate in the sense that one supposes that the number of jobs as well as (at least) their expected processing times are known in advance. Our approach also differs from the probabilistic analysis of parallel machine scheduling problems as considered, e.g., in [21] where it is implicitly assumed that the instance, including processing times of jobs, is known from the outset.

Related work. Stochastic machine scheduling problems have been considered, among others, by Möhring, Radermacher, and Weiss [13, 14], Weber, Varaiya, and Walrand [22], Kämpke [10], and Weiss [23, 24]. For a survey and more bibliographic references we refer to Sect. 16 of the survey by Lawler, Lenstra, Rinnooy Kan, and Shmoys [12]. Except for the mentioned work of Möhring, Radermacher, and Weiss [13, 14] and Weiss [23, 24], research mainly concentrated on identifying conditions that guarantee optimality of simple priority policies such as SEPT, LEPT (shortest/longest expected processing times first), or WSEPT (schedule jobs with highest ratio of weight to expected processing time first). Already for the deterministic case without release dates, the problem under consideration is NP-hard, even for fixed $m \geq 2$ [2], and the WSPT rule (weighted shortest processing time first) is known to achieve a worst-case performance ratio of $\frac{1}{2}(\sqrt{2} + 1)$ [11]. For the special case of a single machine, WSPT is known to be optimal [20], and this result easily generalizes to stochastic processing times [17]. However, results for parallel machines are more complex. For unit weights, the SEPT rule is optimal whenever the processing time distributions of the jobs are stochastically comparable in pairs [22], but it fails to be optimal in general. For arbitrary weights, the WSEPT rule is optimal whenever processing times are exponentially distributed and additionally the job weights are compliant with the ratios of weight to expected processing time [10]. In the general case, Weiss [23, 24] has analyzed the optimality gap of WSEPT, and he proved that WSEPT is asymptotically optimal under mild assumptions on the input parameters of the problem. To the best of our knowledge, no results were previously known for problems where jobs are released over time. Our work also relates to recent developments in the optimal control of stochastic systems [1, 6, 3], and we will discuss this issue more detailed in Sect. 4.3.

Results. Our approach to stochastic machine scheduling is LP-based, and motivated by the success of polyhedral approaches to deterministic scheduling problems. The driving idea is to exploit a polynomially solvable LP-relaxation of the performance space of the problem in order to get both a lower bound on the performance of an optimum policy as well as a clue to design a corresponding LP-based priority policy with provable good performance. Most relevant for our work in this respect is the paper by Hall, Schulz, Shmoys, and Wein [8], where several approximation algorithms are derived on the basis of LP-relaxations in completion time variables. For related and previous work in deterministic scheduling, we refer to the bibliographic references in [8]. We extend this methodology to the stochastic setting. For the model with release dates, we derive an LP-based priority policy with a performance guarantee of $3 - \frac{1}{m} + \max\{1, \frac{m-1}{m}\Delta\}$, where Δ is an upper bound on the squared coefficients of variation of the occurring probability distributions. The underlying polyhedral relaxation of the performance space generalizes previous relaxations that have been used in the deterministic setting. It is further shown in [15] that this LP-relaxation can be solved in polynomial time by a purely combinatorial algorithm.

We also analyze the performance of the WSEPT rule for the model without job release dates, and we derive a worst-case performance guarantee of $1 + \frac{(\Delta+1)(m-1)}{2m}$. Examples show that the performance ratio of $\frac{1}{2}(\sqrt{2}+1)$ of the WSPT rule in deterministic scheduling does not generalize to the stochastic setting. The LP-based analysis also yields an additive bound of the performance of WSEPT which implies a result on its asymptotic optimality, thus complementing previous results by Weiss [23]. Furthermore, our LP lower bound generalizes a previous lower bound on the cost of any deterministic schedule by Eastman, Even, and Isaacs [4], and one obtains a lower bound on the expected cost of any scheduling policy in terms of the optimum cost for a corresponding single-machine problem.

Organization of the paper. Section 2 introduces the basic concept of LP-based priority policies in stochastic scheduling, while in Sect. 3 a new class of valid inequalities for the performance space in stochastic parallel machine scheduling is presented. In Sect. 4.1, these are used to prove constant-factor worst-case performance guarantees for LP-based priority policies within the model where jobs have nontrivial release dates. The analysis of the performance of WSEPT for the model without release dates is presented in Sect. 4.2. We conclude with some remarks and open problems in Sect. 5. Due to space limitations, some proofs and details have been omitted; a complete version can be obtained from the authors.

2   LP-based approximation in stochastic scheduling

A policy is called an α-*approximation* if its expected performance is always within a factor of α of the optimum value, and if it can be determined and executed in polynomial time with respect to the input size of the problem. To cope with the input size of a stochastic scheduling problem which includes non-discrete data in general, we assume that the input is specified by the number of jobs, the number of machines, and the encoding lengths of weights w_j, release dates r_j, expected processing times $E[\boldsymbol{p}_j]$, and, as only additional stochastic information, an upper bound on the *coefficients of variation* of all processing time distributions $\boldsymbol{p}_j$, $j = 1, \ldots, n$. (The coefficient of variation of a given random variable X is defined as the ratio $\sqrt{\mathrm{Var}[X]}/E[X]$.) Thus, it is particularly sufficient if all second moments $E[\boldsymbol{p}_j^2]$ are given. Note, however, that the performance guarantees we derive actually hold with respect to optimal policies that make use of the *complete* knowledge of the distributions of processing times.

In most cases optimal policies and the corresponding optimum value Z^{OPT} are unknown. Hence, in order to prove performance guarantees for simple priority policies we use lower bounds on the optimum value Z^{OPT}. The problem we consider reads

$$\text{minimize } \{\sum_{j \in J} w_j \, C_j \mid C \in \mathcal{C} \},$$

where $\mathcal{C} := \{ \, (E[C_1^\Pi], \ldots, E[C_n^\Pi]) \mid \Pi \text{ policy} \, \} \subseteq \mathbb{R}_+^n$ denotes the *performance space*. Since one cannot hope to completely characterize the performance space in general, we approximate $\mathcal{C}$ by a polyhedron $\mathcal{P}$ and solve the LP-relaxation

$$\text{minimize } \{ \sum_{j \in J} w_j \, C_j \mid C \in \mathcal{P} \},$$

and denote by $C^{LP} = (C_1^{LP}, \ldots, C_n^{LP})$ some optimal solution. If the LP captures sufficient structure of the original problem, the ordering of jobs according to nondecreasing values of C_j^{LP} is a promising candidate for a priority policy (see Sect. 4 for a formal definition).

Clearly, $\sum_{j\in J} w_j C_j^{LP} \leq Z^{OPT} \leq \sum_{j\in J} w_j E[C_j^\Pi]$ if Π denotes this policy, and the goal is to prove $\sum_{j\in J} w_j E[C_j^\Pi] \leq \alpha \sum_{j\in J} w_j C_j^{LP}$ for some $\alpha \geq 1$. This leads to a performance guarantee of α for the priority policy Π and also to a (dual) guarantee for the quality of the LP lower bound:

$$\sum_{j\in J} w_j E[C_j^\Pi] \leq \alpha\, Z^{OPT} \quad \text{and} \quad \sum_{j\in J} w_j C_j^{LP} \geq \frac{1}{\alpha}\, Z^{OPT}.$$

3 Valid inequalities for stochastic machine scheduling

In deterministic scheduling, Schulz [18, Lemma 7] proved that for any feasible schedule on m machines the following inequalities are valid:

$$\sum_{j\in A} p_j\, C_j \;\geq\; \frac{1}{2m}\left(\sum_{j\in A} p_j\right)^2 + \frac{1}{2}\sum_{j\in A} p_j^2 \quad \text{for all } A \subseteq J. \tag{1}$$

Here p_j and C_j denote the deterministic processing and completion times of jobs, resp. The following class of valid inequalities extends (1) to stochastic parallel machine scheduling. They are crucial for all our subsequent results.

$$\begin{aligned} \sum_{j\in A} E[p_j]\, E[C_j^\Pi] \;\geq\; & \frac{1}{2m}\left(\sum_{j\in A} E[p_j]\right)^2 + \frac{1}{2}\sum_{j\in A}(E[p_j])^2 \\ & - \frac{m-1}{2m}\sum_{j\in A} \mathrm{Var}[p_j] \quad \text{for all } A \subseteq J. \end{aligned} \tag{2}$$

Theorem 1. *Let Π be any policy for stochastic parallel machine scheduling. Then inequalities (2) are valid for the corresponding vector of expected completion times $E[C^\Pi]$.*

The proof is based on the idea that the corresponding deterministic inequalities hold pointwise for any realization of processing times. The result then follows by taking expectations, and by exploiting the fact that policies are non-anticipative. Weiss [25] has communicated to us that an alternate proof of the above inequalities can be obtained on the basis of [23], where an exact formula for the left-hand side of (2) is derived for nonidling (work conserving) policies.

With an additional assumption on the second moments of all processing time distributions, one can rewrite (2) more conveniently. Therefore, assume that the squared coefficients of variation of all processing times p_j are bounded by some constant Δ, that is, $\mathrm{Var}[p_j]/(E[p_j])^2 \leq \Delta$ for all jobs $j \in J$. Then, the following inequalities are valid for the performance space $\mathcal{C}$:

$$\sum_{j \in A} E[\boldsymbol{p}_j] E[C_j^\Pi] \geq \frac{1}{2m} \left(\left(\sum_{j \in A} E[\boldsymbol{p}_j] \right)^2 + \sum_{j \in A} E[\boldsymbol{p}_j]^2 \right)$$
$$- \frac{(m-1)(\Delta-1)}{2m} \left(\sum_{j \in A} E[\boldsymbol{p}_j]^2 \right) \quad \text{for all } A \subseteq J. \tag{3}$$

Note that an upper bound on the coefficients of variation of the $\boldsymbol{p}_j$ is a quite natural assumption for scheduling problems. For instance, if job processing times follow NBUE distributions (i.e., the expected remaining processing time of a job in process never exceeds its total expected processing time), it follows from the work of Hall and Wellner [9] that $\mathrm{Var}[\boldsymbol{p}_j]/(E[\boldsymbol{p}_j])^2 \leq 1$.

4 Performance guarantees for (LP-based) priority policies

In this section, we present the first constant-factor worst-case performance guarantees for LP-based priority policies in stochastic machine scheduling.

Let us first give a formal definition of priority policies. A job j is called *available* at time t if $r_j \geq t$. A policy is called a *priority policy* or *priority rule* or *list scheduling policy* if at any time t a maximal number of available jobs is scheduled according to a given priority order on the set of jobs. More precisely, we are given a linear order on J, and when a machine is or becomes idle at time t, the available job with highest priority is started at t. Widely used priority policies are, e.g., LEPT and SEPT as well as WSEPT, also known as Smith's ratio rule. In the presence of release dates, a priority policy may schedule jobs with low priority prior to jobs with higher priority. If this is not desired, we additionally enforce that jobs with low priority are scheduled only if all jobs with higher priority have already been started. In this case, we call a job available at t if $r_j \geq t$ and all its predecessors with respect to the given priority order have already been started by t. The corresponding priority policy is then called *job-based*. Note that this may yield idling of machines although there are jobs waiting that in principle could have been started.

4.1 Parallel machine scheduling with release dates

We now consider the problem $P \mid \boldsymbol{p}_j \sim \text{stoch}, r_j \mid E[\sum w_j C_j]$. The first ingredient is an upper bound on the expected completion times whenever the jobs are scheduled according to a (job-based) priority rule. The following lemma is a generalization of a corresponding bound for the deterministic case [16, 8]. (For the deterministic case without release dates, a similar bound also appears in [4].)

Lemma 1. *Let Π be a job-based priority policy which schedules the jobs in the order $1 < \cdots < n$. Then,*

$$E[C_j^\Pi] \leq \max_{k=1,\dots,j} r_k + \frac{1}{m} \left(\sum_{k=1}^{j-1} E[\boldsymbol{p}_k] \right) + E[\boldsymbol{p}_j] \quad \text{for all } j \in J. \tag{4}$$

To prove the lemma, we crucially need job-based priority rules instead of ordinary priority policies if release dates are present. If no release dates are present, clearly $\max_{k=1,...,j} r_k = 0$, and the claim also holds for ordinary priority policies.

The second ingredient establishes the critical linkage between the LP solution and the value obtained from an LP-based priority policy; it is again a generalization of a corresponding result in deterministic scheduling [8, 18].

Lemma 2. *Let $m \geq 1$ and $C \in \mathbb{R}^n$ be any point which satisfies $C_j \geq E[p_j]$ for all $j \in J$ as well as inequalities (3) for some $\Delta \geq 0$. Assume without loss of generality that $C_1 \leq \cdots \leq C_n$, then*

$$\frac{1}{m}\sum_{k=1}^{j} E[p_k] \leq \left(1 + \max\{1, \frac{m-1}{m}\Delta\}\right) C_j \quad \text{for all } j = 1,\ldots,n.$$

Now we are ready to analyze the following LP-based approximation algorithm for stochastic parallel machine scheduling with release dates. Suppose that the squared coefficient of variation of processing times is bounded from above by some $\Delta \geq 0$. Then inequalities (3) are valid for any scheduling policy Π. Moreover, every vector of expected completion times corresponding to Π additionally fulfills

$$E[C_j^\Pi] \geq r_j + E[p_j] \quad \text{for all } j \in J. \tag{5}$$

We thus consider the linear programming relaxation

$$\min\{ \sum_{j \in J} w_j C_j \mid \text{(3) and (5)} \}, \tag{6}$$

and let C^{LP} denote an optimum solution to (6). Define Π to be a job-based priority policy according to the order given by nondecreasing values of C_j^{LP}.

Theorem 2. *Let $Var[p_j]/(E[p_j])^2 \leq \Delta$ for all jobs j and some $\Delta \geq 0$, and let Π be the job-based priority policy corresponding to an optimal solution to the LP-relaxation (6). Then Π is a $(3 - \frac{1}{m} + \max\{1, \frac{m-1}{m}\Delta\})$-approximation.*

It can be shown that linear program (6) can be solved in $\mathcal{O}(n^2)$ time by purely combinatorial methods [15]. Thus, the corresponding priority list can be computed efficiently.

Proof. First assume without loss of generality that $C_1^{LP} \leq C_2^{LP} \leq \cdots \leq C_n^{LP}$. We apply Lemma 1 to Π, and observe that $\max_{k=1,...,j} r_k \leq C_j^{LP}$ for all $j = 1,\ldots,n$. This holds, since from inequalities (5) we get $C_k^{LP} \geq r_k$, and because $C_j^{LP} \geq C_{j-1}^{LP} \geq \cdots \geq C_1^{LP}$. Moreover, $E[p_j] \leq C_j^{LP}$, thus Lemma 1 yields

$$E[C_j^\Pi] \leq \left(2 - \frac{1}{m}\right) C_j^{LP} + \frac{1}{m}\left(\sum_{k=1}^{j} E[p_k]\right)$$

for all jobs $j \in J$. Since C^{LP} fulfills the conditions of Lemma 2, we now obtain

$$E[C_j^\Pi] \leq \left(3 - \frac{1}{m} + \max\{1, \frac{m-1}{m}\Delta\}\right) C_j^{LP}$$

for all jobs $j \in J$. The fact that linear program (6) is a relaxation of the scheduling problem concludes the proof. □

Theorem 2 particularly yields a worst-case performance guarantee of $(4 - \frac{1}{m})$ whenever $Var[p_j]/(E[p_j])^2 \leq m/(m-1)$ for the given processing time distributions. This bound is already known for deterministic scheduling [8].

4.2 Parallel machine scheduling without release dates

We now consider the problem P | $p_j \sim$ stoch | $E[\sum w_j C_j]$. Using the framework of the preceding section, one easily obtains an LP-based priority policy which has a performance guarantee of $2 - \frac{1}{m} + \max\{1, \frac{m-1}{m}\Delta\}$. However, for this case we can do better by considering the WSEPT rule and a different LP-relaxation. Recall that WSEPT works as follows: Whenever a machine becomes available, schedule the job(s) with highest ratio $w_j/E[p_j]$.

Theorem 3. *Let* $Var[p_j]/(E[p_j])^2 \leq \Delta$ *for all jobs j and some $\Delta \geq 0$, then the WSEPT priority policy is a* $(1 + \frac{(\Delta+1)(m-1)}{2m})$-*approximation.*

Proof. First assume without loss of generality that $w_1/E[p_1] \geq w_2/E[p_2] \geq \cdots \geq w_n/E[p_n]$. Now consider the linear programming relaxation

$$\min\{ \sum_{j \in J} w_j C_j \mid (3) \},\tag{7}$$

and let C^{LP} denote an optimum solution with optimum value Z^{LP}. Since inequalities (3) define a supermodular polyhedron, the solution to LP-relaxation (7) is given by Edmonds' greedy algorithm for supermodular polyhedra, which yields $C_j^{LP} = \frac{1}{m} \sum_{k=1}^{j} E[p_k] - \frac{(\Delta-1)(m-1)}{2m} E[p_j]$ for $j = 1, \ldots, n$. We now apply Lemma 1 to the WSEPT priority policy and obtain

$$Z^{WSEPT} \leq \frac{1}{m} \sum_{j=1}^{n} w_j \sum_{k=1}^{j} E[p_k] + (1 - \frac{1}{m}) \sum_{j=1}^{n} w_j E[p_j]$$

$$= Z^{LP} + \frac{(\Delta+1)(m-1)}{2m} \sum_{j=1}^{n} w_j E[p_j].$$

Since LP (7) is a relaxation for the scheduling problem, and since $\sum_{j \in J} w_j E[p_j]$ is a lower bound on the optimum value Z^{OPT}, the claim follows. □

It is clear from the proof of Theorem 3 that, apart from the above worst-case ratio, an additive performance guarantee for WSEPT can be derived as well.

Corollary 1. *Let* $Var[p_j]/(E[p_j])^2 \leq \Delta$ *for all jobs* j *and some* $\Delta \geq 0$, *then*

$$Z^{WSEPT} - Z^{OPT} \leq \frac{(\Delta+1)(m-1)}{2m} \sum_{j \in J} w_j \, E[p_j] \; .$$

With some additional conditions on weights and expected processing times, we obtain asymptotic optimality for the performance of WSEPT.

Corollary 2. *If* $Var[p_j]/(E[p_j])^2 \leq \Delta$ *for all jobs* j *and some* $0 \leq \Delta < \infty$, *and if there exists some* $\varepsilon > 0$ *such that* $\varepsilon \leq w_j \leq 1/\varepsilon$ *and* $\varepsilon \leq E[p_j] \leq 1/\varepsilon$ *for all jobs* j, *and* $m/n \xrightarrow{n \to \infty} 0$, *then*

$$(Z^{WSEPT} - Z^{OPT})/Z^{OPT} \xrightarrow{n \to \infty} 0.$$

(Subject to the same conditions, also LP-relaxation (7) is asymptotically tight.)

Corollary 1 complements a previous result by Weiss [23, 24], who showed that $Z^{WSEPT} - Z^{OPT} \leq \frac{(m-1)}{2} \cdot \Omega \cdot \max_{j=1,\ldots,n} \{w_j/E[p_j]\}$. Here Ω is an upper bound on the second moment of the remaining processing time of any uncompleted job. Subject to some assumptions on the input parameters of the problem which assure that the right-hand side remains bounded, Weiss [23] thus has proved asymptotic optimality of WSEPT for a wide class of processing time distributions. In fact, since one can construct examples which show that either of the two above additive bounds can be favorable, Corollary 1 complements Weiss' analysis of the quality of the WSEPT heuristic in stochastic machine scheduling.

Theorem 3 particularly includes as a special case a performance guarantee of $3/2 - \frac{1}{2m}$ for the WSPT rule in deterministic scheduling, a result which already follows from the lower and upper bounds given by Eastman, Even, and Isaacs [4]. Moreover, the performance ratio of the WSPT rule in the deterministic setting is $\frac{1}{2}(\sqrt{2}+1)$ [11]. Again, these bounds do not hold in the stochastic setting, as will become clear in the following example.

Example 1. Consider a set of 4 jobs $J = \{1, \ldots, 4\}$ and $m = 2$ machines. All jobs have weight 1, i.e., the objective is the total expected completion time $\sum_{j=1}^{4} E[C_j]$. Let $0 < \varepsilon < 1$. Jobs 1 and 2 have processing time ε with probability $1 - \varepsilon$ and $1/\varepsilon$ with probability ε, independent of each other. Then, $E[p_1] = E[p_2] = 1 + \varepsilon - \varepsilon^2$, which we choose to be the (deterministic) processing time of jobs 3 and 4. Hence, the expected total completion time on a single machine is $Z_1^{OPT} = 10$ for $\varepsilon \to 0$, for any priority policy. For the parallel (two) machine case, elementary calculations show that the optimum policy is to schedule according to the priority list $1 < 2 < 3 < 4$ if ε is small enough, and we obtain an expected total completion time of $Z_m^{OPT} = 4$ for $\varepsilon \to 0$. Thus, in contrast to the above mentioned bound by Eastman, Even, and Isaacs [4], we obtain $\frac{1}{m} Z_1^{OPT} > Z_m^{OPT}$. Moreover, any priority policy is SEPT (or WSEPT, respectively) in this example. Scheduling according to the priority list $3 < 4 < 1 < 2$ yields an expected total completion time of 6 for $\varepsilon \to 0$. Thus, in the stochastic case SEPT (or WSEPT) may differ from the optimum value by a factor of $3/2$. $\qquad \square$

However, the proof of Theorem 3 yields the following generalization of the lower bound by Eastman, Even, and Isaacs [4] to stochastic machine scheduling.

Corollary 3. *If $Var[p_j]/(E[p_j])^2 \leq \Delta$ for all processing times p_j, then*

$$\frac{1}{m} Z_1^{OPT} - \frac{(\Delta - 1)(m - 1)}{2m} \sum_{j=1}^{n} w_j E[p_j] \leq Z_m^{OPT}, \qquad (8)$$

where Z_m^{OPT} is the optimum value for a parallel machine problem on m machines, and Z_1^{OPT} is the optimum value of the same problem on a single machine.

This particularly shows that for $\Delta \leq 1$ the optimum value for a single machine problem with an m-fold faster machine is a relaxation for the corresponding problem on m parallel machines. Moreover, Example 1 not only reveals that the condition $\Delta \leq 1$ is necessary to use the fast single-machine relaxation, but it also shows that — in contrast to the deterministic case — a negative term in the right-hand side of inequalities (2) is necessary as well.

4.3 LP-based priority policies and the achievable region approach

The LP-based approach presented in this paper is closely related to recent and independent developments in the optimal control of stochastic systems via characterizing or approximating "achievable regions". For instance, Bertsimas and Niño-Mora [1] show that previous results on the optimality of Gittins indexing rules can alternatively be derived by a polyhedral characterization of corresponding performance spaces as (extended) polymatroids. Subsequently, Glazebrook and Niño-Mora [6] have proved approximate optimality of Klimov's index rule in multiclass queueing networks with parallel servers. Their work is based on *approximate conservation laws* for the performance of Klimov's index rule (which corresponds to the WSEPT rule for the model we consider here). Since from the bounds (7) and (4) one can obtain an approximate conservation law for the performance of WSEPT, Theorem 3 (respectively Corollary 1) of the present paper can also be derived within their framework [5].

There is, however, an interesting difference between the techniques employed in their work and those of the present paper. For the case where release dates are present (Sect. 4.1), we explicitly make use of an optimum *primal* solution to LP-relaxation (6) in order to obtain a priority policy with provable good performance. (Note that in this case the performance of WSEPT can be arbitrarily bad.) While the achievable region approach as proposed in [6] and [3, Sect. 3] is also based on the concept of LP-relaxations, the *dual* of the corresponding LP-relaxation is solved in order to derive Klimov's index rule and to analyze its performance for the case of parallel servers. Primal and dual solutions, however, can in fact lead to substantially different priority policies.

5 Concluding remarks

With this work we extend the concept of LP-based approximation algorithms from deterministic scheduling to a more general stochastic setting. We have generalized several previous results, including LP-relaxations for parallel machine scheduling and corresponding LP-based performance guarantees. For the model without release dates, our work complements previous work on the performance of the WSEPT heuristic, and extends a previous lower bound on the value of optimum schedules to the stochastic setting.

In particular, LP relaxations of scheduling problems are shown to be a quite powerful tool for producing not only good lower bounds, but also to obtain high-quality priority policies. It is one of the outcomes of our studies that successful combinatorial methods from deterministic machine scheduling also bear on the algorithm design and analysis for stochastic machine scheduling problems. Moreover, another advantage of using LP relaxations is that one not only obtains "a priori" worst-case bounds, but also "a posteriori" guarantees depending on the particular instance. This aspect adds to the practical appeal of this approach.

Altogether, the presented results underline the potential of the polyhedral approach to scheduling problems – in both the deterministic and the stochastic setting, and we hope that this methodology may lead to progress also in other stochastic systems rather than scheduling.

Acknowledgements. We are grateful to Kevin Glazebrook and Gideon Weiss for stimulating discussions on a previous version of this paper [15]. In particular, Kevin Glazebrook pointed out that our analysis also yields an additive bound for the WSEPT rule.

References

1. D. BERTSIMAS AND J. NIÑO-MORA, *Conservation laws, extended polymatroids and multi-armed bandit problems: A polyhedral approach to indexable systems*, Mathematics of Operations Research, 21 (1996), pp. 257–306.
2. J. L. BRUNO, E. G. COFFMAN JR., AND R. SETHI, *Scheduling independent tasks to reduce mean finishing time*, Communications of the Association for Computing Machinery, 17 (1974), pp. 382 – 387.
3. M. DACRE, K. D. GLAZEBROOK, AND J. NIÑO-MORA, *The achievable region approach to the optimal control of stochastic systems*, to appear in: Journal of the Royal Statistical Society.
4. W. L. EASTMAN, S. EVEN, AND I. M. ISAACS, *Bounds for the optimal scheduling of n jobs on m processors*, Management Science, 11 (1964), pp. 268–279.
5. K. D. GLAZEBROOK, *Personal communication*, January 1999.
6. K. D. GLAZEBROOK AND J. NIÑO-MORA, *Scheduling multiclass queueing networks on parallel servers: Approximate and heavy-traffic optimality of Klimov's rule*, in Algorithms – ESA'97, R. Burkard and G. Woeginger, eds., vol. 1284 of Lecture Notes in Computer Science, Springer, 1997, pp. 232–245. Proceedings of the 5th Annual European Symposium on Algorithms.
7. R. L. GRAHAM, E. L. LAWLER, J. K. LENSTRA, AND A. H. G. RINNOY KAN, *Optimization and approximation in deterministic sequencing and scheduling: A survey*, Annals of Discrete Mathematics, 5 (1979), pp. 287–326.
8. L. A. HALL, A. S. SCHULZ, D. B. SHMOYS, AND J. WEIN, *Scheduling to minimize average completion time: Off-line and on-line approximation algorithms*, Mathematics of Operations Research, 22 (1997), pp. 513–544.

9. W. J. HALL AND J. A. WELLNER, *Mean residual life*, in Statistics and Related Topics, Proceedings of the International Symposium on Statistics and Related Topics, M. Csörgö, D. A. Dawson, J. N. K. Rao, and A. K. Md. E. Saleh, eds., North-Holland, 1981, pp. 169–184.

10. T. KÄMPKE, *On the optimality of static priority policies in stochastic scheduling on parallel machines*, Journal of Applied Probability, 24 (1987), pp. 430–448.

11. T. KAWAGUCHI AND S. KYAN, *Worst case bound on an LRF schedule for the mean weighted flow-time problem*, SIAM Journal on Computing, 15 (1986), pp. 1119–1129.

12. E. L. LAWLER, J. K. LENSTRA, A. H. G. RINNOY KAN, AND D. B. SHMOYS, *Sequencing and scheduling: Algorithms and complexity*, in Logistics of Production and Inventory, vol. 4 of Handbooks in Operations Research and Management Science, North-Holland, Amsterdam, 1993, pp. 445–522.

13. R. H. MÖHRING, F. J. RADERMACHER, AND G. WEISS, *Stochastic scheduling problems I: General strategies*, ZOR - Zeitschrift für Operations Research, 28 (1984), pp. 193–260.

14. ———, *Stochastic scheduling problems II: Set strategies*, ZOR - Zeitschrift für Operations Research, 29 (1985), pp. 65–104.

15. R. H. MÖHRING, A. S. SCHULZ, AND M. UETZ, *Approximation in stochastic scheduling: The power of LP-based priority policies*, Tech. Rep. 595/1998, Department of Mathematics, Technical University of Berlin, 1998.

16. C. A. PHILLIPS, C. STEIN, AND J. WEIN, *Minimizing average completion time in the presence of release dates*, Mathematical Programming, 82 (1998), pp. 199–223. A preliminary version of this paper (*Scheduling jobs that arrive over time*) appeared in vol. 955 of Lecture Notes in Computer Science, Springer, 1995, pp. 86–97.

17. M. H. ROTHKOPF, *Scheduling with random service times*, Management Science, 12 (1966), pp. 703–713.

18. A. S. SCHULZ, *Scheduling to minimize total weighted completion time: Performance guarantees of LP-based heuristics and lower bounds*, in Integer Programming and Combinatorial Optimization, W. H. Cunningham, S. T. McCormick, and M. Queyranne, eds., vol. 1084 of Lecture Notes in Computer Science, Springer, 1996, pp. 301–315. Proceedings of the 5th International IPCO Conference.

19. J. SGALL, *On-line scheduling*, in Online Algorithms: The State of the Art, A. Fiat and G. J. Woeginger, eds., vol. 1442 of Lecture Notes in Computer Science, Springer, 1998, pp. 196–231. Proceedings of the Dagstuhl Workshop on On-Line Algorithms.

20. W. E. SMITH, *Various optimizers for single-stage production*, Naval Research and Logistics Quarterly, 3 (1956), pp. 59–66.

21. A. M. SPACCAMELA, W. S. RHEE, L. STOUGIE, AND S. VAN DE GEER, *Probabilistic analysis of the minimum weighted flowtime scheduling problem*, Operations Research Letters, 11 (1992), pp. 67–71.

22. R. R. WEBER, P. VARAIYA, AND J. WALRAND, *Scheduling jobs with stochastically ordered processing times on parallel machines to minimize expected flowtime*, Journal of Applied Probability, 23 (1986), pp. 841–847.

23. G. WEISS, *Approximation results in parallel machines stochastic scheduling*, Annals of Operations Research, 26 (1990), pp. 195–242.

24. ———, *Turnpike optimality of Smith's rule in parallel machines stochastic scheduling*, Mathematics of Operations Research, 17 (1992), pp. 255–270.

25. ———, *Personal communication*, January 1999.

Efficient Redundant Assignments under Fault-Tolerance Constraints[*]

DIMITRIS A. FOTAKIS[1,2]   PAUL G. SPIRAKIS[1,2]

[1] Computer Engineering and Informatics Department
University of Patras, 265 00 Rion, Patras, Greece
[2] Computer Technology Institute — CTI
Kolokotroni 3, 262 21 Patras, Greece
Email: fotakis@cti.gr, spirakis@cti.gr

Abstract. We consider the problem of computing minimum congestion, fault-tolerant, redundant assignments of messages to faulty parallel delivery channels. In particular, we are given a set M of faulty channels, each having an integer capacity c_i and failing independently with probability f_i. We are also given a set of messages to be delivered over M, and a fault-tolerance constraint $(1 - \epsilon)$, and we seek a redundant assignment ϕ that minimizes congestion $\mathrm{Cong}(\phi)$, i.e. the maximum channel load, subject to the constraint that, with probability no less than $(1 - \epsilon)$, all the messages have a copy on at least one active channel. We present a 4-approximation algorithm for identical capacity channels and arbitrary message sizes, and a $2\left\lceil \frac{\ln(|M|/\epsilon)}{\ln(1/f_{\max})} \right\rceil$-approximation algorithm for related capacity channels and unit size messages.

Both algorithms are based on computing a collection of disjoint channel subsets such that, with probability no less than $(1 - \epsilon)$, at least one channel is active in each subset. The objective is to maximize the sum of the minimum subset capacities. Since the exact version of this problem is $\mathcal{NP}$-complete, we present a 2-approximation algorithm for identical capacities, and a $(8 + o(1))$-approximation algorithm for arbitrary capacities.

1   Introduction

In many practical applications involving design with faulty components (e.g. fault-tolerant network design, fault-tolerant scheduling), a combinatorial structure, such as a graph, should be optimized to best tolerate random and independent faults with respect to a given property, such as connectivity or non-existence of isolated points (e.g. [10]).

For instance, let us consider some messages to be delivered over a set of faulty parallel delivery channels. Each channel has an integer capacity c, fails independently with probability f_i, and, in case of failure, it delivers no message to the

[*] This work was partially supported by ESPRIT LTR Project no. 20244—ALCOM–IT.

destination. Then, redundancy can be employed to increase the probability that all the messages have a copy on at least one active channel, i.e. all the messages are delivered to the destination. On the other hand, redundancy also increases the channel congestion, that is the maximum number of messages assigned to each channel.

A natural question arising from this setting is whether it is possible to deliver, with probability no less than $(1 - \epsilon)$, at least one copy of all the messages to the destination, without assigning more than c messages to any channel. The answer requires the construction of the most reliable, redundant assignment $\phi^\star$ not violating the capacity constraint, and the computation of $\phi^\star$'s reliability, that is the probability all the messages to get a copy on at least one active channel. Then, we investigate how fault-tolerant redundant assignments can be efficiently computed, and how the structure of the most reliable assignment $\phi^\star$ looks like.

In this paper, we consider a set M of faulty parallel channels, each having an integer capacity c_i and failing independently with probability f_i, for some rational $1 > f_i > 0$. We are also given a set J of messages, each of an integer size s_j, and a rational fault-tolerance constraint $(1 - \epsilon)$. Each message $j \in J$ of size s_j causes a load s_j/c_i when assigned to a channel $i \in M$ of capacity c_i. We seek a redundant assignment $\phi : J \mapsto 2^M$ that tolerates the faults (i.e. all the messages get a copy on at least one active channel) with probability no less than $(1 - \epsilon)$, and minimizes channel congestion $\mathsf{Cong}(\phi)$ (i.e. the maximum channel load caused by ϕ). Then, we distinguish the case of identical capacity channels, where all the capacities c_i are equal, and the case of related capacity channels, where the capacities c_i can be arbitrary integer numbers.

A natural strategy for constructing a redundant, $(1 - \epsilon)$-fault-tolerant assignment is to compute a collection $\mathcal{M} = \{M_1, \ldots, M_\kappa\}$ of disjoint channel subsets such that, with probability no less than $(1 - \epsilon)$, at least one channel is active in each subset M_j. Then, each subset M_j can be thought as a single, reliable channel of effective capacity $C(M_j) = \min_{i \in M_j}\{c_i\}$, and any algorithm for non-redundant scheduling on $\mathcal{M}$ can be used for computing ϕ. A reasonable objective for a reliable collection $\mathcal{M}$ is to maximize the total effective capacity $C(\mathcal{M}) = \sum_{j=1}^{\kappa} C(M_j)$.

We observe that the problem of computing an optimal, $(1 - \epsilon)$-fault-tolerant collection of disjoint subsets is $\mathcal{NP}$-complete even for identical capacity channels, and we present a polynomial-time 2-approximation algorithm for partitioning a set of identical capacity channels. In case of related capacities, we obtain a simple $2 \left\lceil \frac{\ln(|M|/\epsilon)}{\ln(1/f_{max})} \right\rceil$-approximation algorithm, and, for any constant $\delta > 0$, a polynomial-time $(8 + \delta)$-approximation algorithm.

As for the approximability of Fault-Tolerant Congestion, in case of identical capacities, we prove that near optimal assignments can be obtained by partitioning M into an optimal number of reliable effective channels. This proof is based on a technical lemma of independent interest that provides a tight upper bound on the reliability of any redundant assignment. This lemma also bounds from above the probability that isolated nodes do not appear to a not necessarily connected hypergraph, whose edges fail randomly and independently. Then, we show that Minimum Fault-Tolerant Congestion can be approximated within a factor of 4 in polynomial-time. In case of related capacities, we restrict our attention to unit size messages, and we present a simple $2\left\lceil\frac{\ln(|M|/\epsilon)}{\ln(1/f_{\max})}\right\rceil$-approximation algorithm.

Up to the best of our knowledge, similar optimization problems concerning the computation of minimum congestion, redundant assignments, subject to the constraint the assignments to tolerate random and independent faults with a given probability, have not been studied so far. Minimum Fault-Tolerant Congestion does not assume any upper bound on the number of faulty channels, and since any kind of reaction to channel failures is not allowed, a tight lower bound on the optimal congestion should identify the most reliable redundant assignments. On the other hand, unlike other on-line fault-tolerant scheduling problems (e.g. [7]), Fault-Tolerant Congestion is an off-line problem, and any algorithm with sufficient computational power will eventually come with the optimal solution.

The fault-tolerant versions of some routing problems, such as minimizing congestion of Virtual Path layouts in a complete ATM network [4], have been studied in an off-line setting. In [4], redundancy is also employed to overcome faulty links, but mainly the worst-case fault model, where each layout must tolerate any configuration of at most f faulty links, is considered. In case of random and independent faults, they only provide a trivial logarithmic upper bound.

Due to lack of space, we only provide proof sketches for most of the results. The reader is referred to [2] for a full version of this paper.

1.1 Notation and Definitions

Let M be a set of faulty parallel channels. Each channel $i \in M$ has an integer capacity $c_i \geq 1$ and fails independently with probability f_i, for some rational $1 > f_i > 0$. For any subset $M' \subseteq M$, let $\Pr[M']$ denote the reliability of M', that is the probability at least one channel of M' to be active, $\Pr[M'] = 1 - \prod_{i \in M'} f_i$. Also, $f_{\max} = \max_{i \in M}\{f_i\}$ denotes the failure probability of the least reliable channel.

Let $J = \{1, \ldots, n\}$ be a set of messages to be delivered over M. Each message $j \in J$ has an integer size s_j, $S_{tot} = \sum_{j \in J} s_j$ denotes the total size of J, and $s_{\max} = \max_{j \in J}\{s_j\}$.

A redundant assignment $\phi : J \mapsto 2^M$ is a function that assigns each message $j \in J$ to a non-empty set of channels $\phi(j) \subseteq M$. An assignment ϕ is feasible for a set of channels $M' \subseteq M$, if, for all $j \in J$, $\phi(j) \cap M' \neq \emptyset$. Given an assignment ϕ, $\Pr[\phi]$ denotes the reliability of ϕ, that is the probability ϕ to be feasible over the channel availability distribution defined by the probabilities f_i over M.

$$\Pr[\phi] = \sum_{\substack{M' \subseteq M \\ \phi \text{ is feasible for } M'}} \left(\prod_{i \in M'} (1 - f_i) \prod_{i \in M - M'} f_i \right)$$

Given a redundant assignment $\phi : J \mapsto 2^M$, a *minimal feasible set of channels for ϕ* is any subset $M' \subseteq M$ such that ϕ is feasible for M', but ϕ is not feasible for any $M'' \subset M'$. A *minimum feasible set of channels for ϕ* is a minimal feasible set for ϕ of minimum cardinality. Moreover, $\mathsf{MF}(\phi)$ denotes the cardinality of any minimum feasible set for ϕ.

Given an assignment ϕ that is feasible for M, $\mathsf{Cong}(\phi)$ denotes the channel congestion, that is the maximum load assigned by ϕ to the channels of M,
$$\mathsf{Cong}(\phi) = \max_{i \in M} \left\{ \sum_{j : i \in \phi(j)} \frac{s_j}{c_i} \right\}$$

Definition 1. (Minimum Fault-Tolerant Congestion)
INSTANCE: A set of faulty parallel channels $M = \{(f_1, c_1), \ldots, (f_m, c_m)\}$. Each channel $i \in M$ has an integer capacity $c_i \geq 1$ and fails independently with probability f_i, for some rational $1 > f_i > 0$.
A set of messages $J = \{s_1, \ldots, s_n\}$ to be delivered over M. Each message $j \in J$ has an integer size $s_j \geq 1$.
A fault-tolerance constraint $(1 - \epsilon)$, for some rational $1 > \epsilon \geq \prod_{i=1}^{m} f_i$.
SOLUTION: A $(1 - \epsilon)$-fault-tolerant redundant assignment $\phi : J \mapsto 2^M$, i.e. an assignment of each message $j \in J$ to a non-empty set of channels $\phi(j) \subseteq M$, such that $\Pr[\phi] \geq 1 - \epsilon$.
OBJECTIVE: Minimize $\mathsf{Cong}(\phi) = \max_{i \in M} \left\{ \sum_{j : i \in \phi(j)} \frac{s_j}{c_i} \right\}$.

As usual, we distinguish the identical capacities case, where all the channels have unit capacity (i.e. $c_i = 1$), and the related capacities case, where each channel can have an arbitrary integer capacity. In the related capacities case, we further assume that the channels are numbered in non-increasing order of their capacities, i.e. $c_1 \geq \cdots \geq c_m \geq 1$.

The definition of Maximum Fault-Tolerant Partition arises from the following natural strategy for computing a $(1 - \epsilon)$-fault-tolerant, redundant assignment ϕ.

(1) Compute a collection of reliable effective channels by partitioning some $M' \subseteq M$ into disjoint groups $\mathcal{M} = \{M_1, \ldots, M_\kappa\}$, such that the probability of at least one channel being active in each group is at least $(1 - \epsilon)$.

(2) Use an appropriate algorithm for scheduling the message set J on the set $\mathcal{M}$ of reliable effective channels. For all messages $j \in J$ scheduled on the effective channel M_l, set $\phi(j) = M_l$.

The first step of this approach actually determines an upper bound on the amount of redundancy being necessary for satisfying the fault-tolerance constraint. Moreover, if we set the effective capacity $C(M_l)$ equal to the minimum capacity of the corresponding group, $C(M_l) = \min_{i \in M_l} \{c_i\}$, $1 \leq l \leq \kappa$, then the makespan of the non-redundant schedule obtained in the second step equals the congestion of the redundant assignment ϕ.

Since there exist many efficient algorithms for the implementation of the second step (e.g. see Chapter 1 of [5]), we focus on the design and analysis of approximation algorithms for the first step, that is the computation of a $(1 - \epsilon)$-fault-tolerant collection of disjoint channel subsets.

Definition 2. (Maximum Fault-Tolerant Partition)
INSTANCE: A set of channels/items $M = \{(f_1, c_1), \ldots, (f_m, c_m)\}$. Each item $i \in M$ has an integer capacity/profit $c_i \geq 1$ and fails independently with probability f_i, for some rational $1 > f_i > 0$.
A fault-tolerance constraint $(1 - \epsilon)$, for some rational $1 > \epsilon \geq \prod_{i=1}^{m} f_i$.
SOLUTION: A partition of a subset $M' \subseteq M$ into disjoint groups $\mathcal{M} = \{M_1, \ldots, M_\kappa\}$ such that:

$$\Pr[\mathcal{M}] = \Pr[M_1 \wedge \ldots \wedge M_\kappa] = \prod_{l=1}^{\kappa} \Pr[M_l] = \prod_{l=1}^{\kappa} \left(1 - \prod_{i \in M_l} f_i\right) \geq 1 - \epsilon$$

OBJECTIVE: Maximize the total effective capacity of the partition $\mathcal{M}$:

$$C(\mathcal{M}) = \sum_{l=1}^{\kappa} C(M_l) = \sum_{l=1}^{\kappa} \min_{i \in M_l} \{c_i\}$$

The term "partition" is somewhat abused, because the definition of Fault-Tolerant Partition allows $M' = \bigcup_{l=1}^{\kappa} M_l \subset M$. This is crucial in the related capacities case, because there exist many instances, where the optimal solution is a partition of a strict $M' \subset M$. However, in the identical capacities case, where the objective is simply to maximize the number of groups κ, we can always assume that $\bigcup_{l=1}^{\kappa} M_l = M$.

The Fault-Tolerant Partition problem can be thought as a version of Bin Covering [1], where, instead of a threshold T on the total size of each separate bin, we have to cover a constraint on the product of the total bin sizes.
Complexity Issues. Since it is $\mathcal{NP}$-complete to determine the minimum makespan for scheduling a set of jobs on reliable identical machines, it is $\mathcal{NP}$-hard to determine the optimal Fault-Tolerant Congestion, even for channels of identical capacity and failure probability.

On the other hand, Minimum Fault-Tolerant Congestion, especially in case of related capacity channels, does not seem to belong to the class $\mathcal{NP}$, because, for an arbitrary redundant assignment ϕ, it is #$\mathcal{P}$-complete to exactly compute the reliability of ϕ. In particular, given a set M of channels, each failing independently with probability $f = \frac{1}{2}$, a set J of unit size messages, and a redundant assignment $\phi : J \mapsto 2^M$, computing $\Pr[\phi]$ is equivalent to #Monotone-Sat originally shown #$\mathcal{P}$-complete by Valiant [13].

Obviously, Minimum Fault-Tolerant Congestion is in $\mathcal{PSPACE}$, but we do not know if it belongs to the Polynomial Hierarchy $\mathcal{PH}$ (e.g., see [11]). Minimum Fault-Tolerant Congestion can be included in a class containing all the languages L that can be decided by a polynomial-time non-deterministic Turing machine T reducing L to a single call of a function $g \in$ #$\mathcal{P}$. Moreover, after calling the #$\mathcal{P}$-oracle once, the only additional computation that T needs to perform is an arithmetic comparison involving the outcome of g. We denote this class by $\mathcal{NP}^{\#\mathcal{P}[1,\mathrm{comp}]}$. Additionally, a stochastic version of Knapsack defined in [9] belongs to this class. Using the results of Toda and Watanabe [12], it is easy to show that $\mathcal{NP}^{\#\mathcal{P}[1,\mathrm{comp}]}$ contains the whole Polynomial Hierarchy $\mathcal{PH}$.

Algorithm Next Fit Decreasing – NFD

Input: $M = \{f_1, \ldots, f_m\}$, failure probabilities $1 > f_i > 0$.

         Fault-tolerance constraint $(1 - \epsilon)$, $1 > \epsilon \geq \prod_{i \in M} f_i$.

Output: A $(1 - \epsilon)$-fault-tolerant partition of M into λ disjoint groups.

(1) Sort M so that $f_1 \leq f_2 \leq \cdots \leq f_m$, i.e. in non-increasing order of reliability.

(2) Compute the first index l, $0 \leq l < m$, such that $f_{l+1} > \alpha_l$, where $\alpha_l = F_l^{1/x_l}$, and x_l, F_l are defined by the following equation:

$$\left(1 - F_l^{1/x_l}\right)^{x_l} = \frac{1 - \epsilon}{P_l} \; , \;\; F_l = \prod_{i=l+1}^{m} f_i \; , \;\; P_l = \prod_{i=1}^{l} (1 - f_i) \qquad (1)$$

(3) For $j = 1, \ldots, l$, $M_j = \{f_j\}$, i.e. M_j only consists of the item f_j.

(4) The set $\{f_{l+1}, \ldots, f_m\}$ is partitioned using Next Fit [1] with threshold $(1 - \alpha_l)$.

Fig. 1. The Algorithm Next Fit Decreasing (NFD).

As for the complexity of Maximum Fault-Tolerant Partition, there exists a direct transformation from Subset Product (problem SP14, [3]) showing that, given a set $M = \{f_1, \ldots, f_m\}$, for rational $1 > f_i > 0$, and a rational $\epsilon > 0$, it is $\mathcal{NP}$-complete to decide if M can be partitioned into two sets M_1, M_2 such that $\Pr[M_1] \Pr[M_2] \geq 1 - \epsilon$.

2 Partition of Identical Capacity Channels

The Next Fit Decreasing - NFD algorithm (Figure 1) initially computes l over-reliable items placed into the single-item groups $M_1, \ldots, M_l$ (steps (1) - (3)). The remaining items are partitioned using Next Fit with threshold $(1 - \alpha_l)$.

Theorem 3. Next Fit Decreasing *runs in time* $\mathcal{O}(\log m(m + \sum \log(1/f_i)))$ *and is a 2-approximation algorithm for Maximum Fault-Tolerant Partition of identical capacity channels.*

Proof sketch. It is not hard to verify that the partition $\{M_1, \ldots, M_\lambda\}$ is always $(1 - \epsilon)$-fault-tolerant. As for the performance, if non-integral groups and placement of items were allowed, an optimal solution would consist of x_0 groups, each of reliability $(1 - \alpha_0)$, x_0, α_0 defined by (1). Therefore, if $f_1 \leq \alpha_0$, f_1 is an over-reliable item, because it contributes at least 1 to x_0. Moreover, if f_1 is placed to a single-item group, the average group reliability $(1 - \alpha_0)$ of the optimal non-integral partition decreases to $(1 - \alpha_1)$. This goes on, until no more over-reliable items exist, i.e. $f_{l+1} > \alpha_l$. Thus, NFD computes a collection of l over-reliable items placed to the single-item groups $M_1, \ldots, M_l$.

The remaining items form the set $M_r = M - \bigcup_{i=1}^{l} M_i$ and are partitioned using Next Fit [1] with threshold $(1 - \alpha_l)$. A standard argument implies that $\lambda - l > \frac{1}{2}(x_l - 1)$, and since the number of groups must be integer, $\lambda \geq l + \lfloor \frac{x_l}{2} \rfloor$.

We can show by contradiction that any $M_r' \subseteq M_r$ cannot contribute to any optimal partition more than x_l groups. This implies an upper bound of $l + \lfloor x_l \rfloor$ on

---

Algorithm Safe Partition – SP

Input: $M = \{(f_1, c_1) \ldots, (f_m, c_m)\}$, failure probabilities $1 > f_i > 0$,
             capacities c_i, $c_1 \geq \cdots \geq c_m$.
             Fault-tolerance constraint $(1 - \epsilon)$, $1 > \epsilon \geq \prod_{i \in M} f_i$.

Output: A $(1 - \epsilon)$-fault-tolerant partition of M into disjoint groups.

(1) Compute a partition $\mathcal{M} = \{M_1, \ldots, M_\lambda\}$ using Next Fit with threshold $\left(1 - \frac{\epsilon}{m}\right)$.
      $C(\mathcal{M}) = \sum_{l=1}^{\lambda} C(M_l) = \sum_{l=1}^{\lambda} \min_{i \in M_l} \{c_i\}$.

(2) Let $d \geq 0$ be the last index such that $\prod_{i=1}^{d} f_i > \epsilon$.
      If $c_{d+1} \geq C(\mathcal{M})$ then return $\{\{(f_1, c_1), \ldots, (f_{d+1}, c_{d+1})\}\}$.
      Otherwise, return $\mathcal{M} = \{M_1, \ldots, M_\lambda\}$.

Fig. 2. The Algorithm Safe Partition (SP).

the cardinality of any optimal $(1 - \epsilon)$-fault-tolerant partition. There also exists a family of instances for which the approximation ratio of NFD is exactly 2. $\square$

In the sequel, we extensively use the upper bound $(l + x_l)$ on the number of reliable groups that can be obtained from a set of identical capacity channels. In particular, given a set M of identical capacity channels and a fault-tolerance constraint $(1-\epsilon)$, $\mathsf{IUB}(M, 1-\epsilon) = l+x_l$ bounds from above the number of groups that can be produced by M with constraint $(1 - \epsilon)$. The bound $\mathsf{IUB}(M, 1 - \epsilon) = l + x_l$ consists of the integer l denoting the number of over-reliable items, and the real x_l denoting the optimal non-integral number of groups that would be obtained from the instance $\left(M_r, \frac{1-\epsilon}{P_l}\right)$, if non-integral placement of items was allowed.

3 Partition of Related Capacity Channels

3.1 A Simple Logarithmic Approximation Algorithm

The Safe Partition – SP algorithm (Figure 2) starts by applying Next Fit with threshold $\left(1 - \frac{\epsilon}{m}\right)$. Since any feasible solution cannot have than m groups, the resulting partition is always $(1 - \epsilon)$-fault-tolerant. Then, Safe Partition computes the largest effective capacity, $(1-\epsilon)$-fault-tolerant group consisting of the first $d+1$ channels, where d is the largest index such that $\prod_{i=1}^{d} f_i > \epsilon$. The Safe Partition algorithm returns the best of these solutions. The analysis of Safe Partition is simple and based on the facts that (i) all the groups of both the aforementioned partitions have cardinality at most $\left\lceil \frac{\ln(m/\epsilon)}{\ln(1/f_{\max})} \right\rceil$, and (ii) any $(1-\epsilon)$-fault-tolerant partition cannot have effective capacity more than $\sum_{i=d+1}^{m} c_i$.

Lemma 4. Safe Partition *is a polynomial-time* $2\left\lceil \frac{\ln(m/\epsilon)}{\ln(1/f_{\max})} \right\rceil$-*approximation algorithm for Maximum Fault-Tolerant Partition of related capacity channels, where* $f_{\max} = \max_{i \in M} \{f_i\}$.

3.2 A Constant Factor Approximation Algorithm

The Capacity Class Partition – CCP algorithm (Figure 3) divides the original instance into classes I_j of almost identical capacity channels, and allocates a

---

Algorithm Capacity Class Partition – CCP

Input: $M = \{(f_1, c_1) \ldots, (f_m, c_m)\}$, failure probabilities $1 > f_i > 0$, capacities c_i, $c_1 \geq \cdots \geq c_m$.

Fault-tolerance constraint $(1 - \epsilon)$, $1 > \epsilon \geq \prod_{i \in M} f_i$.

Output: A $(1 - \epsilon)$-fault-tolerant partition of M into disjoint groups.

(1) Let $d \geq 0$ be the last index such that $\prod_{i=1}^{d} f_i > \epsilon$, and $\psi = \lfloor \log c_{d+1} \rfloor$. For all $i = 1, \ldots, d$, set $c_i = c_{d+1}$. $I_j = \{f_i : (f_i, c_i) \in M \wedge 2^j \leq c_i < 2^{j+1}\}$, for all $j = 0, \ldots, \psi$.

(2) For each class I_j, compute a set Γ_j of pairs $(\kappa_j(i), \beta_j(i))$, where $1 \geq \beta_j(i) \geq 0$, and $\kappa_j(i)$'s are defined by $\kappa_j(i) = \mathsf{IUB}(I_j, (1 - \epsilon)^{\beta_j(i)})$. Each Γ_j must contain pairs $(\nu, \beta_j(i))$, for all integers $\nu = 0, 1, \ldots, \lfloor \mathsf{IUB}(I_j, 1 - \epsilon) \rfloor$, and, for all $i \geq 0$, $1 \geq \kappa_j(i) - \kappa_j(i+1) \geq 0$.

(3) For each class I_j, compute a set $\hat{\Gamma}_j$ of pairs $(\lambda_j(i), \beta_j(i))$ as follows: For all $(\kappa_j(i), \beta_j(i)) \in \Gamma_j$, $\lambda_j(i) = \mathsf{NFD}(I_j, (1 - \epsilon)^{\beta_j(i)})$.

(4) For each class I_j, select exactly one pair (λ_j, β_j) from $\hat{\Gamma}_j$ so as to

$$\text{maximize} \quad \sum_{j=0}^{\psi} 2^j \lambda_j \ , \quad \text{subject to} \quad \sum_{j=0}^{\psi} \beta_j \leq 1$$

(5) If $c_{d+1} \geq \sum_{j=0}^{\psi} 2^j \lambda_j$ then return a single group $\{(f_1, c_1), \ldots, (f_{d+1}, c_{d+1})\}$. Otherwise, return $\mathcal{M} = \bigcup_{j=0}^{\psi} \mathcal{M}^j$, where $\mathcal{M}^j = \{M_1^j, \ldots, M_{\lambda_j}^j\}$ is the partition produced by $\mathsf{NFD}(I_j, (1 - \epsilon)^{\beta_j})$.

Fig. 3. The Algorithm Capacity Class Partition (CCP).

portion $(1 - \epsilon)^{\beta_j}$ of the fault-tolerance constraint to each class according to the solution of a generalized Knapsack instance.

Theorem 5. *For any constant* $\delta > 0$, Capacity Class Partition *is a polynomial-time* $(8 + \delta)$*-approximation algorithm for Maximum Fault-Tolerant Partition of related capacity channels. The running time of* CCP *is polynomial in* $\frac{1}{\delta}$.

Proof sketch. At first, we observe that the partitions produced by CCP are always $(1 - \epsilon)$-fault-tolerant. Also, we can set $c_i = c_{d+1}$, for all $i = 1, \ldots, d$, without affecting the optimal effective capacity, and we can assume that the capacities of all the channels belonging to the class I_j are equal to 2^j, by only losing a factor of 2 in the approximation ratio. The performance analysis of the CCP algorithm is based on the following technical lemma.

Lemma 6. *There exist reals* $0 \leq \beta_j^\star \leq 1$, $j = 0, \ldots, \psi$, $\sum_{j=0}^{\psi} \beta_j^\star \leq 1$, *such that the effective capacity* $C(\mathcal{M}^\star)$ *of an optimal partition* $\mathcal{M}^\star$ *fulfills the inequality:*

$$C(\mathcal{M}^\star) \leq \sum_{j=0}^{\psi} 2^j \ \mathsf{IUB}\left(I_j, (1 - \epsilon)^{\beta_j^\star}\right) \ .$$

Lemma 6 states that there exists an allocation of portions $(1 - \epsilon)^{\beta_j^\star}$ of the fault-tolerance constraint to the classes I_j, $\sum_j \beta_j^\star \leq 1$, so that the total effective capacity of an optimal solution can be bounded from above by the sum, over all

j, of I_j's capacity times the upper bound $\mathsf{IUB}\left(I_j, (1-\epsilon)^{\beta_j^\star}\right)$ on the number of groups obtained from the class I_j with fault-tolerance constraint $(1-\epsilon)^{\beta_j^\star}$.

In order to approximate the values $\beta_j^\star$, at the steps (2) and (3), CCP computes an appropriately selected set of samples $\beta_j(i)$ on the $[0, 1]$-interval, and for each sample $\beta_j(i)$, evaluates the number of groups $\lambda_j(i) = \mathsf{NFD}\left(I_j, (1-\epsilon)^{\beta_j(i)}\right)$ produced by NFD from the capacity class I_j with fault-tolerance constraint $(1-\epsilon)^{\beta_j(i)}$. In particular, $\beta_j(i)$'s are computed so as the corresponding $\kappa_j(i) = \mathsf{IUB}\left(I_j, (1-\epsilon)^{\beta_j(i)}\right)$ to be integers, and not to differ too much from each other, i.e. $1 \geq \kappa_j(i) - \kappa_j(i+1) \geq 0$.

For all the classes I_j, the profit-size pairs $(\lambda_j(i), \beta_j(i))$ form an instance of generalized Knapsack, whose optimal solution suggests a near optimal allocation of portions of the fault-tolerance constraint to each class I_j. It is not hard to verify that the FPTAS for ordinary Knapsack (e.g., see Section 9.3 of [5], or [6]) can be generalized to an FPTAS for the generalized Knapsack instance of the step (4). Then, using Lemma 6 and the properties of the samples $\beta_j(i)$, we can show that, for any constant $\delta > 0$, we can compute exactly one pair $(\lambda_j, \beta_j) \in \hat{\Gamma}_j$, for each class I_j, $j = 0, \ldots, \psi$, such that

$$C(\mathcal{M}^\star) \leq 2c_{d+1} + 2(1+\delta) \sum_{j=0}^{\psi} 2^j \lambda_j \ .$$

Moreover, such (λ_j, β_j)-pairs can be computed in time polynomial in the input size and $\frac{1}{\delta}$. A partition $\mathcal{M}$ of objective value $C(\mathcal{M}) = \sum_{j=0}^{\psi} 2^j \lambda_j$ is formed by the union of the partitions produced by NFD on instances $\left(I_j, (1-\epsilon)^{\beta_j}\right)$. CCP returns the best of $\mathcal{M}$ and a single group consisting of the first $d+1$ channels. $\square$

4 Assignments on Identical Capacity Channels

Then, we present NFD-LS, that is a polynomial-time 4-approximation algorithm for Minimum Fault-Tolerant Congestion on identical capacity channels. Given an instance $(M, J, 1-\epsilon)$, NFD-LS calls $\mathsf{NFD}(M, 1-\epsilon)$ to obtain a $(1-\epsilon)$-fault-tolerant partition into λ disjoint groups, $\mathcal{M} = \{M_1, \ldots, M_\lambda\}$, and uses List Scheduling – LS (e.g. see Section 1.1 of [5]) to compute a non-redundant schedule ϕ' of J on λ reliable, identical capacity channels. Finally, for all messages $j \in J$, if $\phi'(j) = l$, for some integer $1 \leq l \leq \lambda$, NFD-LS sets $\phi(j) = M_l$, and returns ϕ.

It should be clear, that the reliability of the resulting assignment ϕ equals the reliability of the underlying partition $\mathcal{M}$, $\Pr[\phi] = \Pr[\mathcal{M}] \geq 1 - \epsilon$. Additionally, since all the channels are of identical capacity, the congestion of ϕ equals the makespan of the non-redundant assignment ϕ', $\mathsf{Cong}(\phi) = \mathsf{Makespan}(\phi')$.

The analysis of NFD-LS is based on the following combinatorial lemma, which applies to identical channels, that are channels of both identical capacity and failure probability. For integers κ, μ, ν, given a set of $\kappa\mu$ identical channels, a set of $\nu\kappa$ unit size messages, and an upper bound ν on the congestion, the most reliable κ-partition assignment ϕ_κ is obtained by partitioning the channels into

κ disjoint groups each of cardinality μ, and assigning exactly ν messages to each group. The following lemma states that the reliability of ϕ_κ bounds from above the reliability of any redundant assignment ϕ, $\text{Cong}(\phi) \leq \nu$, of $\nu\kappa$ unit size messages to $\kappa\mu$ identical channels.

Lemma 7. *Given an integer $\kappa \geq 1$, for any integer $\nu \geq 1$, let ϕ be any redundant assignment of $\nu\kappa$ unit size messages to a set M of identical channels, each of failure probability f. If each message is assigned to exactly μ channels, $|M| \geq \kappa\mu$, and $\text{Cong}(\phi) \leq \nu$, then $\Pr[\phi] \leq (1 - f^\mu)^\kappa = \Pr[\phi_\kappa]$.*

Proof sketch. Clearly, the most reliable κ-partition assignment ϕ_k has $\Pr[\phi_k] = (1 - f^\mu)^\kappa$. Then, we show that ϕ_κ is at least as reliable as any other assignment ϕ that fulfills the hypothesis. Wlog., we assume that $\text{MF}(\phi) = \kappa$. We adopt a time-evolving representation of random channel subsets. Similar representations of random edge subsets are used in [10, 8] for proving lower bounds on graph reliability.

We introduce non-negative time t and define a time-dependent random subset $M(t) \subseteq M$ representing the set of active channels by the time t. Initially, $M(0) = \emptyset$. Each channel is given an arrival time chosen independently from the exponential distribution with mean 1, and each time a channel arrives, it is included in $M(t)$. Each time a channel i is included in $M(t)$, all the copies of the messages assigned to i are removed from the remaining channels $M - M(t)$. Then $\phi(t)$ denotes the assignment at time t, that is obtained from ϕ by removing all the copies of the messages assigned to some channel of $M(t)$ by ϕ. Also, all the channels of $M - M(t)$ assigned no messages by $\phi(t)$ can be included in $M(t)$, because they do not further affect the stochastic process. Then, $m(t) = |M - M(t)|$ denotes the number of channels assigned non-zero number of messages by $\phi(t)$.

The random variables $\theta_r(\phi)$, $r = 1, \ldots, \kappa$, denote the length of the time interval for which $\text{MF}(\phi(t))$ is equal to r. Obviously, $\theta_r(\phi)$ is the time it takes for decreasing the cardinality of a minimum feasible set for $\phi(t)$ from r to $r-1$. The random variables $\theta_r(\phi)$, $r = 1, \ldots, \kappa$, are mutually independent. Since a channel arrival can decrease $\text{MF}(\phi(t))$ by at most one, in order to prove the lemma, it suffices to show that, for all $r = 1, \ldots, \kappa$, $\theta_r(\phi)$ stochastically dominates $\theta_r(\phi_\kappa)$.

Due to the memoryless property of the exponential distribution, the time it takes for the next channel arrival is independent of the previous arrivals and the time t. Therefore, once we have conditioned on the value of $M(t)$, the time for the next arrival has probability distribution function $1 - e^{-m(t)t}$ (see also [8]). Thus, $\theta_r(\phi_\kappa)$ is exponentially distributed with mean $1/r\mu$.

If at some time t, $\text{MF}(\phi(t)) = r$ and $m(t) \leq r\mu$, then the time for the next channel arrival, that may cause $\text{MF}(\phi(t))$ to become $r - 1$, stochastically dominates an exponentially distributed random variable with mean $1/r\mu$. However, it may be $\text{MF}(\phi(t)) = r$ and $m(t) > r\mu$. In this case, we can show by a combinatorial argument that, with probability no more than $\frac{r\mu}{m(t)}$, the next channel arrival causes $\text{MF}(\phi(t))$ to become $r-1$, while, with probability at least $1 - \frac{r\mu}{m(t)}$, $\text{MF}(\phi(t))$ remains equal to r. Since

$$\frac{m(t) - r\mu}{m(t)} \left(1 + \frac{r\mu\, e^{-m(t)t}}{m(t) - r\mu} - \frac{m(t)\, e^{-r\mu t}}{m(t) - r\mu}\right) + \frac{r\mu}{m(t)} \left(1 - e^{-m(t)t}\right) = 1 - e^{-r\mu t} ,$$

a simple induction shows that $\theta_r(\phi)$ always stochastically dominates an exponentially distributed random variable with mean $1/r\mu$. □

Remark. Lemma 7 also applies to the probability that isolated nodes do not appear to a not necessarily connected hypergraph with $\kappa\mu$ "faulty" hyperedges of cardinality ν. In particular, assume that $|M| = \mu\kappa$, ϕ assigns each message to exactly μ channels, and the load of each channel is exactly ν. Given such an assignment ϕ, we can construct a hypergraph $H_\phi(N, E)$, where N consists of $\nu\kappa$ nodes corresponding to the unit size messages, and E consists of $\mu\kappa$ hyperedges corresponding to the channels of M. Moreover, each $e \in E$ consists of the messages assigned to the corresponding channel of M by ϕ.

An assignment ϕ is feasible for a $M' \subseteq M$, iff the removal of the hyperedges corresponding to $M - M'$ does not create any isolated nodes in H_ϕ, that are nodes of degree 0. Lemma 7 implies that the hypergraph corresponding to the most reliable κ-partition assignment ϕ_κ achieves the greatest probability of not having any isolated points under random and independent hyperedge faults. □

Lemma 7 can be easily generalized to hold if the total number of message copies is equal to $\nu\kappa\mu$. We can also handle different failure probabilities, by choosing a sufficiently small real number δ, and replacing each channel i of failure probability f_i by a "bundle" of $m_i = \left\lceil \frac{-\ln f_i}{\delta} \right\rceil$ parallel channels each of failure probability $f = 1 - \delta$ (see also [8]). Using this technique, we can prove the following lower bound on the optimal Fault-Tolerant Congestion.

Lemma 8. *Given a set $M = \{f_1, \ldots, f_m\}$ of identical capacity channels, n unit size messages, and a fault-tolerance constraint $(1-\epsilon)$, the optimal Fault-Tolerant Congestion cannot be less than $n \cdot \lceil \mathsf{IUB}(M, 1-\epsilon) \rceil^{-1}$.*

Based on Lemma 8 and the analyses of Next Fit Decreasing and List Scheduling, we can show that the partition assignment ϕ produced by NFD-LS approximates Minimum Fault-Tolerant Congestion within a factor of 4.

Theorem 9. *The redundant assignment ϕ produced in polynomial-time by NFD-LS approximates within a factor of 4 Minimum Fault-Tolerant Congestion on identical capacity channels.*

5 Assignments on Related Capacity Channels

The SP-OPT algorithm is a logarithmic approximation algorithm for Minimum Fault-Tolerant Congestion, in case of unit size messages and related capacity channels. The redundant assignments of SP-OPT are obtained by an optimal non-redundant assignment of the unit size messages on the related capacity, reliable, effective channels obtained by the Safe Partition algorithm. The analysis of SP-OPT is based on the analysis of Safe Partition, and the fact that any $(1 - \epsilon)$-fault-tolerant assignment must have at least one copy of all the messages to some channel of index greater than d, where d is the largest index such that $\prod_{i=1}^{d} f_i > \epsilon$.

Theorem 10. SF-OPT *is a polynomial-time* $2 \left\lceil \frac{\ln(m/\epsilon)}{\ln(1/f_{max})} \right\rceil$*-approximation algorithm for Minimum Fault-Tolerant Congestion, in case of unit size messages and related capacity channels.*

6 Open Problems

A challenging open problem is to investigate whether Minimum Fault-Tolerant Congestion, especially in case of related capacity channels, is a complete problem for the complexity class $\mathcal{NP}^{\#\mathcal{P}[1,\text{comp}]}$, that contains the whole $\mathcal{PH}$.

Another direction for further research is to derive a non-trivial lower bound for Minimum Fault-Tolerant Congestion in case of related capacity channels. This lower bound may be combined with the CCP algorithm in order to obtain a constant factor approximation algorithm.

Additionally, the fault-tolerant generalizations of some fundamental graph optimization problems, such as shortest path or connectivity, have not been studied so far under random and independent faults. In particular, the fault-tolerant generalization of connectivity is, given a graph $G(V, E)$, where each edge $e \in E$ fails independently with probability f_e, and a fault-tolerance constraint $(1-\epsilon)$, to compute the minimum (w.r.t. the number of edges) subgraph $G'(V, E')$, $E' \subseteq E$, that remains connected with probability at least $(1 - \epsilon)$.

References

1. S.F. Assmann, D.S. Johnson, D.J. Kleitman, and J.Y.-T. Leung. On a Dual Version of the One-Dimensional Bin Packing Problem. *Journal of Algorithms* 5, pp. 502–525, 1984.
2. D.A. Fotakis and P.G. Spirakis. Minimum Congestion Redundant Assignments to Tolerate Random Faults. http://students.ceid.upatras.gr/~fotakis, 1999.
3. M.R. Garey and D.S. Johnson, *Computers and Intractability: A Guide to the Theory of NP-Completeness*, Freeman, San Francisco, 1979.
4. L. Gasieniec, E. Kranakis, D. Krizanc, A. Pelc. Minimizing Congestion of Layouts for ATM Networks with Faulty Links. *Proc. of the 21st Mathematical Foundations of Computer Science*, pp. 372–381, 1996.
5. D.S. Hochbaum (ed.). *Approximation Algorithms for NP-hard problems*. PWS Publishing, 1997.
6. O.H. Ibarra and C.E. Kim. Fast Approximation Algorithms for the Knapsack and Sum of Subset Problems. *Journal of the Association for Computing Machinery* 22, pp. 463–468, 1975.
7. B. Kalyanasundaram and K.R. Pruhs. Fault-Tolerant Scheduling. *Proc. of the 26th ACM Symposium on Theory of Computing*, pp. 115–124, 1994.
8. D.R. Karger. A Randomized Fully Polynomial Time Approximation Scheme for the All Terminal Network Reliability Problem. *Proc. of the 27th ACM Symposium on Theory of Computing*, pp. 11–17, 1995.
9. J. Kleinberg, Y. Rabani, E. Tardos. Allocating Bandwidth for Bursty Connections. *Proc. of the 29th ACM Symposium on Theory of Computing*, pp. 664–673, 1997.
10. M.V. Lomonosov. Bernoulli Scheme with Closure. *Problems of Information Transmission* 10, pp. 73–81, 1974.
11. C.H. Papadimitriou. *Computational Complexity*. Addison-Wesley, 1994.
12. S. Toda and O. Watanabe. Polynomial-time 1-Turing reductions from $\#\mathcal{PH}$ to $\#\mathcal{P}$. *Theoretical Computer Science* 100, pp. 205–221, 1992.
13. L.G. Valiant. The Complexity of Enumeration and Reliability Problems. *SIAM Journal on Computing*, 8(3), pp. 410–421, 1979.

Scheduling with Machine Cost *

Csanád Imreh[1,2] and John Noga[3]

[1] József Attila University, Department of Informatics, Szeged, Hungary.
[2] Stochastic Research Group, Hungarian Academy of Sciences, Technical University Budapest, Hungary. csanad@inf.u-szeged.hu
[3] Technical University Graz, Mathematics Department, Graz, Austria. noga@opt.math.tu-graz.ac.at

Abstract. For most scheduling problems the set of machines is fixed initially and remains unchanged for the duration of the problem. We consider two basic online scheduling problems with the modification that initially the algorithm possesses no machines, but that at any point additional machines may be purchased. Upper and lower bounds on the competitive ratio are shown for both problems.

1 Introduction

In machine scheduling, we typically have a fixed set of machines. The scheduling algorithm makes no decision regarding the initial set of machines nor is it allowed to later change the set of machines. It is usually assumed that the provided machines can be utilized without cost.

Our intent is to begin an investigation of how scheduling problems change when machine costs are considered. We have several reasons for studying this idea. Most obviously, real machines have cost. If we do not have the necessary machines then they must be obtained. Even if we already possess machines we may still incur a fixed start up or conversion cost proportional to the number of machines used. Also, we still have an "opportunity cost". By this we mean that if we use the machines for a given problem we lose the chance to use them for something else. Further, in many cases it is desirable to buy or lease additional machines. A second reason we might allow the number of machines to be determined by the algorithm is that the performance of an algorithm on a given input can be highly dependent on the number of machines. This seems particularly true when considering worst case measures (such as competitive analysis). A third reason is that by considering such a variant we may find other interesting problems and/or gain insight into the original.

Machine scheduling problems can be classified as either offline or online. For an offline problem the entire input is known to the algorithm prior to any decision. Although many scheduling problems are NP-complete, in principle an offline algorithm can find the optimal solution. We contrast this with an online

* Research supported by the START program Y43-MAT of the Austrian Ministry of Science.

problem where the input is revealed to the algorithm a little at a time. Decisions made by the algorithm are not allowed to depend upon the unrevealed portion of the input. Since an online algorithm must make decisions without complete knowledge of the input, it is not generally possible for the online algorithm to find the optimal solution. Typically, the quality of an online algorithm is judged using competitive analysis. An online scheduling algorithm is C-competitive if the algorithm cost is never more than C times the optimal cost.

Probably the most fundamental example of an online machine scheduling problem is $P|online|C_{max}$. In this problem we have a fixed number m of identical machines. The jobs and their processing times are revealed to the online algorithm one by one. When a job is revealed the online algorithm must irrevocably assign the job to a machine. By the load of a machine, we will mean the sum of the processing times of all jobs assigned to the machine. The objective is to minimize the maximum load, often called the makespan. There has been a great deal of work on this problem including [1–3,5], but the best possible competitive ratio for this problem is still unknown for $m > 3$. The result most relevant to our work is due to Graham [6]. Although the terminology of competitive analysis was not used, it was shown that a simple algorithm, List Scheduling, is $(2 - 1/m)$-competitive.

The problem $P|online, r_j|C_{max}$ is another standard online machine scheduling problem. Once again there are a fixed number of machines. Each job has a processing time and a release time. A job is revealed to the online algorithm at its release time. For each job the online algorithm must choose which machine it will run on and assign a start time. No machine may simultaneously run two jobs. Note that the algorithm is not required to immediately assign a job at its release time. However, if the online algorithm assigns a job at time t then it cannot use information about jobs released after time t and it cannot start the job before time t. The objective is to minimize the makespan. For results on this model see [4,7,8].

The first problem we will consider in this paper is a variant of $P|online|C_{max}$. The differences are that 1) no machines are initially provided, 2) when a job is revealed the algorithm has the option to purchase new machines, and 3) the objective is to minimize the sum of the makespan and cost of the machines. We will refer to this problem as the **List Model** for scheduling with machine cost. We will show that no online algorithm can be better than 4/3-competitive and provide a $\varphi = (1 + \sqrt{5})/2$-competitive ($\approx 1.618$) algorithm for this model.

The second problem we will consider is a variant of $P|online, r_j|C_{max}$. The differences are that 1) no machines are initially provided and 2) the algorithm may purchase machines at any time, and 3) the objective is to minimize the sum of the makespan and the cost of the machines. We will refer to this problem as the **Time Model** for scheduling with machine cost. We will show that no online algorithm can be better than $(\sqrt{33} - 1)/4$-competitive (≈ 1.186) and provide a $(6 + \sqrt{205})/12$-competitive (≈ 1.693) algorithm for this model.

2 Preliminaries

Throughout the remainder of the paper we will use the following notations. The jobs will be labeled $j_1, \ldots, j_n$ and presented to the online algorithm in this order. We denote the processing time of job j_i by p_i and the largest processing time by $L = \max\{p_i\}$. For a fixed algorithm the start time of job j_i is s_i and its completion time $c_i = s_i + p_i$. The total amount of processing needed by the first ℓ jobs is $P_\ell = \sum_{i=1}^{\ell} p_i$. In the Time Model the release time of job j_i is r_i. We will assume that the cost of purchasing a machine is 1. Since we could simply rescale the machine costs and job sizes, any other constant cost function is equivalent. Under either model, we will use $A(\sigma)$ to denote the cost of an algorithm A on a given sequence of jobs σ. Similarly, we represent the optimal offline cost on a sequence σ by $OPT(\sigma)$.

If $A(\sigma) \leq C \cdot OPT(\sigma)$ for every sequence of jobs σ then A is called C-competitive. The competitive ratio of A is the infimum of all values for which A is C-competitive.

We now propose a class of online algorithms for the List Model. Our upper bound on the optimal competitive ratio will follow from analyzing a particular algorithm from this class. For an increasing sequence $\varrho = (0 = \varrho_1, \varrho_2 \ldots \varrho_i \ldots)$ we will define an online algorithm A_ϱ. When job j_ℓ is revealed A_ϱ purchases machines (if necessary) so that the current number of machines i satisfies $\varrho_i \leq P_\ell < \varrho_{i+1}$. A_ϱ then assigns job j_ℓ to the least loaded machine.

For the Time Model we define a very similar class of online algorithms. For an increasing sequence $\varrho = (0 = \varrho_1, \varrho_2 \ldots \varrho_i \ldots)$ we will define an online algorithm B_ϱ. When job j_ℓ is revealed B_ϱ purchases machines (if necessary) so that the current number of machines i satisfies $\varrho_i \leq P_\ell < \varrho_{i+1}$. Whenever there is at least one machine that is not processing a job and at least one job that has been released but not started, A_ϱ assigns the job with the largest processing time to an idle machine.

The offline version of machine scheduling under both the List Model and Time Model can easily be seen to be NP-complete by simple transformations from PARTITION. Since finding the exact optimal offline solution is typically infeasible, in our upper bound proofs we will use the following lower bound on the optimal offline solution.

Lemma 1. *For both the List Model and Time Model the optimal offline cost is at least $2\sqrt{P}$. Further, if $L \geq \sqrt{P}$ then the optimal offline cost is at least $L + P/L$.*

Proof. Let m^* be the number of machines and M^* be the makespan of the optimal solution. Since the largest job must be placed on some machine, $L \leq M^*$. Since the total load on any machine is no more than M^*, the maximum amount of processing which can be accomplished is $m^* M^*$. So, $P \leq m^* M^*$. Therefore, the optimal offline cost must be greater than the solution to the following optimization problem: minimize $m^* + M^*$ subject to $P \leq m^* M^*$ and $L \leq M^*$. It is easy to see that this value is the one described.

We note that by checking each value of m from 1 to n, any offline α-approximation algorithm for either $P|online|C_{max}$ or $P|online, r_j|C_{max}$ yields an offline α-approximation algorithm for the modified version.

3 List Model

3.1 Lower Bound

Theorem 1. *For the List Model no online algorithm has a competitive ratio smaller than $\frac{4}{3}$.*

Proof. Consider a very long sequence of jobs with each job having a very small processing time, $p_i = \epsilon$ for all i. It is easy to see that any algorithm which never purchases a second machine is not C-competitive for any C. So assume the algorithm purchases a second machine when job j_ℓ is released. If $P_\ell \leq 2$ then the offline algorithm can serve all jobs with one machine and the competitive ratio can be no smaller than

$$\frac{P_\ell - \epsilon + 2}{P_\ell + 1} \geq \frac{4 - \epsilon}{3}.$$

If $P_\ell > 2$ then the offline algorithm can split the jobs nearly evenly between two machines and the competitive ratio can be no smaller than

$$\frac{P_\ell - \epsilon + 2}{P_\ell/2 + \epsilon + 2} \geq \frac{4}{3 + \epsilon}.$$

Since we can choose ϵ to be arbitrarily small, we obtain the result.

3.2 Upper Bound

Throughout this section we will consider A an online algorithm from the class described in the previous section. Let $A = A_\varrho$ for $\varrho = (0, 4, 9, 16, \ldots, i^2, \ldots)$. The basic intuition for selecting ϱ comes from Lemma 1. If the optimal cost is close to $2\sqrt{P}$ then the optimal algorithm uses approximately $\sqrt{P}$ machines. If $P \geq 4$ then A tries to mimic this behavior by purchasing at most $\sqrt{P}$ machines.

Theorem 2. *The competitive ratio of A is φ.*

Proof. We will first prove that A is φ-competitive.

Consider an arbitrary sequence of jobs $\sigma = j_1, \ldots, j_n$ and fix an optimal schedule. Let m be the number of machines used by A, j_ℓ be the last job that A completes and k be the number of machines that A owns immediately after j_ℓ is released.

Case A: $m = 1$.
If A purchases only one machine then the cost of the algorithm is $1 + P$ and $P < 4$. If the optimal offline schedule also uses one machine then the ratio of

costs is 1. If the optimal offline schedule uses two or more machines then the optimal cost is at least $2 + \frac{P}{2}$. Since $P < 4$, the cost ratio is no more than $5/4$.

In the remaining cases we will repeatedly use several simple inequalities. By our choice of $\mathcal{A}$, we know $m \leq \sqrt{P} < m + 1$ and $k \leq \sqrt{P_\ell} < k + 1$. We use Lemma 1 to estimate the $OPT(\sigma)$. Since $\mathcal{A}$ always assigns a job to the machine with the lightest load,

$$A(\sigma) \leq m + \frac{P_\ell - p_\ell}{k} + p_\ell = m + \frac{P_\ell}{k} + \frac{k-1}{k}p_\ell.$$

Case B: $m > k$ and $p_\ell \leq \sqrt{P}$.
Using the inequalities given above:

$$\frac{A(\sigma)}{OPT(\sigma)} \leq \frac{m + (k+1)^2/k + (k-1)\sqrt{P}/k}{2\sqrt{P}}$$
$$\leq \frac{m + k + 2 + 1/k + \sqrt{P} - \sqrt{P}/k}{2\sqrt{P}}$$
$$\leq \frac{3\sqrt{P} + (m - \sqrt{P})/(m-1)}{2\sqrt{P}}$$
$$\leq 3/2 \leq \varphi.$$

Case C: $m = k > 1$ and $p_\ell \leq \sqrt{P}$.
Since $(m + P/m + \sqrt{P})/2\sqrt{P}$ is increasing in P,

$$\frac{A(\sigma)}{OPT(\sigma)} \leq \frac{m + P/m + \sqrt{P}}{2\sqrt{P}} \leq \frac{3m + 3 + 1/m}{2(m+1)} \leq \frac{19}{12} \leq \varphi.$$

Case D: $m > k$ and $p_\ell > \sqrt{P}$.
Using the inequalities given above:

$$\frac{A(\sigma)}{OPT(\sigma)} \leq \frac{m + (k+1)^2/k + p_\ell(k-1)/k}{p_\ell + P/p_\ell}$$
$$\leq \frac{2\sqrt{P} + p_\ell}{p_\ell + P/p_\ell} = \frac{2 + p_\ell/\sqrt{P}}{p_\ell/\sqrt{P} + \sqrt{P}/p_\ell}$$
$$\leq \varphi.$$

The last inequality follows from the fact that $f(x) = (2 + x)/(x + 1/x)$ is never greater than φ.

Case E: $m = k > 1$ and $p_\ell > \sqrt{P}$.
Since $(m + P/m - p_\ell/m)/2\sqrt{P}$ is increasing in P,

$$\frac{m + P/m - p_\ell/m}{2\sqrt{P}} \leq \frac{m + (m+1)^2/m - p_\ell/m}{2(m+1)}$$
$$= 1 + \frac{1 - p_\ell}{2m(m+1)} \leq 1.$$

Therefore, $m + P/m - p_\ell/m \leq 2\sqrt{P}$ and

$$\frac{A(\sigma)}{OPT(\sigma)} \leq \frac{2\sqrt{P} + p_\ell}{p_\ell + P/p_\ell} = \frac{2 + p_\ell/\sqrt{P}}{p_\ell/\sqrt{P} + \sqrt{P}/p_\ell} \leq \varphi,$$

where the final inequality once again follows from the fact that the maximum of $f(x) = (2 + x)/(x + 1/x)$ is φ.

We now wish to show that A is not C-competitive for any $C < \varphi$. Consider a sequence of N^3 jobs of size $1/N$ followed by one job of size φN. A will schedule the first N^3 by purchasing N machines and putting N^2 jobs on each machine. The final job will be placed on an arbitrary machine. Therefore, A's cost will be $N + N + \varphi N$. The optimal cost is no more than $\varphi N + \lceil (N + \varphi)/\varphi \rceil$. So, the competitive ratio of A is at least

$$\frac{(2 + \varphi)N}{\varphi N + \lceil (N + \varphi)/\varphi \rceil} \xrightarrow{N \to \infty} \frac{2 + \varphi}{\varphi + 1/\varphi} = \varphi$$

Therefore the competitive ratio of A is φ.

4   Time Model

The Time Model differs from the List Model in two respects. The online algorithm has the advantage of not having to immediately assign jobs. However, the online algorithm has the disadvantage that if a machine is purchased at time t then it cannot be used before time t. For these reasons neither our upper nor lower bounds from the List Model directly apply to the Time Model.

4.1   Lower Bound

Theorem 3. *For the Time Model no online algorithm has a competitive ratio smaller than*

$$C = \frac{\sqrt{33} - 1}{4} \approx 1.186.$$

Proof. Fix an online algorithm and let $S = C + \frac{1}{2}$. Consider two jobs $(p_1, r_1) = (p_2, r_2) = (S, 0)$. Let $t = \max\{s_1, s_2\}$. If $t < S - 1$ then we will present a third job $(p_3, r_3) = (2S - t - \epsilon, t + \epsilon)$. The optimal offline cost is $S + 2$ for the first two jobs and $2S + 2$ if all three are given. If $t \geq S$ then the algorithm's makespan is at least $2S$. So, the cost ratio can be no smaller than $(2S + 1)/(S + 2) = C$. If $S - 1 \leq t < S$ then the algorithm must run the two jobs on different machines and have makespan at least $2S - 1$. So, the cost ratio can be no smaller than $(2S + 1)/(S + 2) = C$. If $t < S - 1$ then the third job is presented. Once again the algorithm must run the first two jobs on different machines. If it purchases exactly two machines then the makespan is at least $3S - t - \epsilon$. If it purchases at least three machines then the makespan is at least $2S$. So, the cost ratio can be no smaller than $\min\{3S - t - \epsilon + 2, 2S - \epsilon + 3\}/(2S + 2)$. As ϵ tends to zero this value tends to C.

Regardless of how the online algorithm schedules the first two jobs in this sequence the cost ratio can be made arbitrarily close to C. Therefore the competitive ratio must be at least C.

4.2 Upper Bound

Throughout this section we will consider $\mathcal{B}$, an algorithm from the class described in section 2. Let $\mathcal{B} = \mathcal{B}_\varrho$ for $\varrho = (0, 4, 9, 16, \ldots, i^2, \ldots)$. Once again we attempt to mimic the behavior of an offline algorithm which achieves a cost near $2\sqrt{P}$.

Theorem 4. *Algorithm $\mathcal{B}$ is $\frac{6+\sqrt{205}}{12} \approx 1.693$-competitive.*

Proof. Consider an arbitrary sequence of jobs $\sigma = j_1, \ldots, j_n$ and fix an optimal schedule. Let M^* be the optimal makespan, m^* be the optimal number of machines, M be the makespan of $\mathcal{B}$, and m be the number of machines used by $\mathcal{B}$.

Case A: $m = 1$.
If $\mathcal{B}$'s machine is never idle then let $t=0$. Otherwise, let t be the latest time that $\mathcal{B}$'s machine is idle. Let $W \le P < 4$ be the total processing time of all jobs released at or after time t. The cost of $\mathcal{B}$ is $1 + t + W$. The optimal cost is at least $m^* + t + W/m^*$. For this case:

$$\frac{\mathcal{B}(\sigma)}{OPT(\sigma)} \le \frac{1+t+W}{m^*+t+W/m^*} \le 5/4.$$

Case B: $m > 1$.
We claim that $M - M^* \le \frac{P}{m}$. Suppose that this was not true. Let j_ℓ be the last job to finish in $\mathcal{B}$'s schedule and k be the number of machines owned by $\mathcal{B}$ at time $M^* - p_\ell$. Note that none of $\mathcal{B}$'s machines can be idle during the time period $I = [M^* - p_\ell, M - p_\ell]$. So at least $(M - M^*)k$ processing is completed during I. Further, if $p_\ell \le M - M^*$ then an additional $(M - M^* - p_\ell)(m - k)$ processing is completed during $[M^*, M - p_\ell]$ by the remaining $m - k$ machines that $\mathcal{B}$ purchases.

If $k = m$ then more than P processing is completed, which is a contradiction. If $k < m$ then $P_\ell < (k+1)^2$. Since the processing time of any job released after $M - p_\ell$ must be less than p_ℓ, job j_ℓ will start before any job released after $M - p_\ell$. Therefore,

$$\begin{aligned} P_\ell &\ge (M - M^*)k + p_\ell + \max\{(M^* - M - p_\ell)(m - k), 0\} \\ &\ge (M - M^*)(k+1) > P/m(k+1) \\ &\ge (k+1)^2, \end{aligned}$$

which is also a contradiction. So the claim must be true.
Since $m \ge 2$, $P < (m+1)^2$, and $(P/m+m)/\sqrt{P}$ is increasing in P, it is easy to verify that $P/m + m \le 13\sqrt{P}/6$.

Putting these facts together we get:

$$B(\sigma) \leq M^* + P/m + m$$
$$\leq M^* + 2\sqrt{169P/144}$$
$$\leq M^* + \frac{(-6+\sqrt{205})M^*}{12} + \frac{(6+\sqrt{205})P}{12M^*}$$
$$= \frac{6+\sqrt{205}}{12}(M^* + P/M^*)$$
$$\leq \frac{6+\sqrt{205}}{12}OPT(\sigma).$$

The third inequality is an application of the arithmetic-geometric mean inequality. So we have shown that $B(\sigma)$ is $(6+\sqrt{205})/12$-competitive.

5 Final Comments

We have proposed adding the concept of machine cost to scheduling problems. When machine cost is added to list scheduling (List Model) we show a lower bound of $4/3$ on the competitive ratio of any online algorithm and give an algorithm with competitive ratio φ. When machine cost is added to scheduling over time (Time Model) we show a lower bound of $(\sqrt{33}-1)/4$ on the competitive ratio of any online algorithm and present a $(6+\sqrt{205})/12$-competitive algorithm. It would be nice to find the precise competitive ratios for these two problems.

Since almost any scheduling problem can be modified to include machine cost, there are nearly an infinite number of related open questions. We list several directions we feel could be interesting. Is there a collection of problems for which the competitive ratio always decreases (increases) when machine cost is considered? Can we place good bounds on the competitive ratios of problems for naturally arising classes of cost functions? Using computers as an example, as time passes the cost of a given machine is typically decreasing and bounded away from 0.

6 Acknowledgements

The authors would like to thank János Csirik for suggesting the problem. The first author completed some of this work while visiting Gerhard Woeginger at the TU Graz and is grateful for the kind hospitality.

References

1. Susanne Albers. Better bounds for online scheduling. In *Proc. 29th Symp. Theory of Computing*, pages 130–139, 1997.
2. Chandra Chekuri, Rajeev Motwani, Balas Natarajan, and Cliff Stein. Approximation techniques for average completion time scheduling. In *Proc. 8th Symp. on Discrete Algorithms*, pages 609–618, 1997.

3. Bo Chen, Andre van Vliet, and Gerhard J. Woeginger. New lower and upper bounds for on-line scheduling. *Operations Research Letters*, 16:221–230, 1994.
4. Bo Chen and Arjen P. A. Vestjens. Scheduling on identical machines: How good is LPT in an on-line setting? *Operations Research Letters*, 21:165–169, 1998.
5. Gábor Galambos and Gerhard J. Woeginger. An on-line scheduling heuristic with better worst case ratio than Graham's list scheduling. *SIAM Journal on Computing*, 22:349–355, 1993.
6. Ronald L. Graham. Bounds for certain multiprocessing anomalies. *Bell System Technical Journal*, 45:1563–1581, 1966.
7. John Noga and Steven Seiden. Scheduling two machines with release times. To appear at Integer Programming and Combinatorial Optimization, 1999.
8. David B. Shmoys, Joel Wein, and David P. Williamson. Scheduling parallel machines on-line. *SIAM Journal on Computing*, 24:1313–1331, 1995.

A Linear Time Approximation Scheme for the Job Shop Scheduling Problem [*]

Klaus Jansen[1], Roberto Solis-Oba[2], and Maxim Sviridenko[3]

[1] Instituto Dalle Molle di Studi sull'Intelligenza Artificiale, Lugano, Switzerland, **klaus@idsia.ch**
[2] Max Planck Institut für Informatik, Saarbrücken, Germany, **solis@mpi-sb.mpg.de**
[3] Sobolev Institute of Mathematics, Novosibirsk, Russia, **svir@math.nsc.ru**

Abstract. We study the preemptive and non-preemptive versions of the job shop scheduling problem when the number of machines and the number of operations per job are fixed. We present linear time approximation schemes for both problems. These algorithms are the best possible for such problems in two regards: they achieve the best possible performance ratio since both problems are known to be strongly NP-hard; and they have optimum asymptotic time complexity.

1 Introduction

In the job shop scheduling problem, there is a set $\mathcal{J} = \{J_1, \ldots, J_n\}$ of n jobs that must be processed on a group $M = \{1, \ldots, m\}$ of m machines. Each job J_j consists of a sequence of operations $O_{1j}, O_{2j}, \ldots, O_{\mu j}$, where O_{ij} must be processed on machine $m_{ij} \in \{1, \ldots, m\}$ during p_{ij} time units. The operations $O_{1j}, O_{2j}, \ldots, O_{\mu j}$ must be processed one after another in the given order and each machine can process at most one operation at a time. The problem is to schedule the jobs so as to minimize the overall *makespan*, or schedule *length*, which is the time by which all jobs are completed.

Two widely studied variants of the job shop scheduling problem are the open shop and the flow shop problems. In these problems each job has exactly one operation per machine. In the open shop problem there is no specified ordering for the processing of the operations of a job, so they can be processed in any order. In the flow shop problem the ordering of the operations is the same for all jobs.

All above problems are strongly NP-hard when the number of machines is part of the input [14]. The job shop problem is strongly NP-hard even if each job has at most three operations and there are only two machines [10]. The flow shop problem is strongly NP-hard as soon as $m = 3$ [1]. The open shop problem with a constant number of machines $m \geq 3$ is known to be weakly NP-hard [2], and the question of whether the problem is NP-hard in the strong sense is still

[*] This work was supported in part by EU ESPRIT LTR Project 20244 (ALCOM-IT) and by the Swiss Office Fédéral de l'éducation et de la Science Project 97.0315 titled "Platform".

open. Williamson et al. [14] proved that when the number of machines, jobs, and operations per job are part of the input there does not exist a polynomial time approximation algorithm with worst case bound smaller than $\frac{5}{4}$ for any of the above three problems unless $P = NP$. On the other hand the preemptive version of the job shop scheduling problem is NP-complete in the strong sense even when $m = 3$ and $\mu = 3$ [3].

Consider an instance I of one of the above shop scheduling problems. Let $OPT(I)$ be the length of an optimum solution, and let $\mathcal{A}(I)$ be the length of the schedule obtained by some algorithm $\mathcal{A}$. A polynomial-time approximation scheme (PTAS) for the problem is an algorithm $\mathcal{A}$ that, for any (constant) $\epsilon > 0$ and input I, outputs in time polynomial in the length of I a feasible schedule of length $R_{\mathcal{A}}(I, \epsilon) = \frac{A(I)}{OPT(I)} \leq 1 + \epsilon$. We can view such an algorithm $\mathcal{A}$ as a family of algorithms $\{\mathcal{A}_\epsilon | \epsilon > 0\}$ such that $\mathcal{A}_\epsilon(I) \leq (1 + \epsilon)OPT(I)$. A fully polynomial time approximation scheme is an approximation scheme $\{\mathcal{A}_\epsilon\}$ in which each algorithm $\mathcal{A}_\epsilon$ runs in time polynomial in the length of I and $\frac{1}{\epsilon}$.

When the number m of machines is fixed, Hall [6] developed a polynomial time approximation scheme for the flow shop problem, while Sevastianov and Woeginger [13] designed a polynomial time approximation scheme for the open shop problem. For the case of job shops when m and μ are fixed, we [9] designed a polynomial time approximation scheme. When m and μ are part of the input the best known result [4] for the job shop problem is an approximation algorithm with worst case bound $O((\log(m\mu)\log(min(m\mu, p_{max}))/\log\log(m\mu))^2)$, where p_{max} is the largest processing time among all operations.

In this paper we study the preemptive and non-preemptive versions of the job shop scheduling problem when the number of machines m and the number of operations per job μ are fixed. In this work we improve on the polynomial time approximation scheme described in [9] in two regards. First, we present a PTAS for the non-preemptive version of the problem that runs in linear time. This is a major improvement on the previous PTAS which is only guaranteed to run in polynomial time. Second, we show how to extend our PTAS to the preemptive case maintaining linear time complexity.

We briefly describe the algorithm for the non-preemptive version of the problem. The idea is to divide the set of jobs $\mathcal{J}$ into two groups $\mathcal{L}$ and $\mathcal{S}$ formed by jobs with "large" and "small" total processing time, respectively. We fix a relative ordering for the long jobs and find a (infeasible) schedule for the small jobs using a linear program. Then we round the solution for the linear program so that only a constant number of small jobs are preempted. These jobs are scheduled sequentially at the end of the solution. The linear program induces a partition on the set of small jobs, and the rounded solution might be infeasible in each group of this partition. We use Sevastianov's [12] algorithm independently on each group to find a feasible schedule for the whole set of jobs. Choosing the set $\mathcal{L}$ appropriately, but interestingly of constant size, we can show that the length of the schedule is at most $1 + \epsilon$ times the length of an optimum schedule.

There are two main contributions in this algorithm. The first is a novel rounding procedure that allows us to bring down to a constant the number of fractional assignments in any solution of the linear program described above. This rounding procedure exploits the sparsity and special structure of the constraint matrix B for the linear program. We show that it is possible to find in constant time certain singular submatrix B' of B. This submatrix defines a perturbation $\delta_{B'}$ for some variables in the linear program. When $\delta_{B'}$ is added to these variables we show that at least one of them is rounded up to 1 or down to 0. Hence the new solution for the linear program has at least one fewer fractional assignment. By finding a linear number of these submatrices B' and rounding the values of the variables as described above we can prove that in the resulting solution the number of variables with fractional assignments is constant.

This rounding procedure has a nice interpretation. Each submatrix B' defines a set of operations $O_{B'}$ that receive fractional assignments. By changing the values of the variables by $\delta_{B'}$ we are effectively shifting around pieces of size $\delta_{B'}$ from each operation in $O_{B'}$. The path that these shifted pieces follow forms a cycle, hence the total length of the schedule does not change. The shifting process is such that pieces belonging to the same operation are placed together, and thus the number of variables with fractional assignments decreases.

The second main contribution of the algorithm is a procedure for reducing the time complexity of Sevastianov's algorithm [12] from $O(n^2)$ to $O(n)$. To do this we exploit a remarkable feature of this algorithm: the quality of the solution that it produces does not depend on the number of jobs, only on the length of the largest operation. Hence by carefully "glueing" jobs (to form larger composed jobs) we can decrease the number of jobs, and thus reduce the complexity from $O(n^2)$ down to $O(n)$. We show that we can glue the jobs so that the processing time of the largest operation does not increase by too much, and hence the length of the schedule produced by the algorithm is only $1 + \epsilon$ times the length of an optimum solution.

The final ingredient to get a linear time approximation scheme for the problem is an algorithm for solving approximately the above linear program in $O(n)$ time. We show that our linear program can be rewritten so that the logarithmic potential price directive decomposition method of Grigoriadis and Khachiyan [5] can be used to find an approximate solution in linear time.

We show that the approximation scheme can be generalized also to the preemptive version of the job shop scheduling problem. The approach is similar to the non-preemptive case, but we have to handle carefully the set of long jobs $\mathcal{L}$ to ensure that we find a feasible solution and that the algorithm still runs in linear time.

2 Linear Program

In this section, we formulate the fractional job shop scheduling problem as a linear program. Let $\mathcal{L} \subset \mathcal{J}$ be a set formed by a constant k number of jobs with large total processing time. We call $\mathcal{L}$ the set of *long* jobs. We show later how to select the set of long jobs, the only thing that we need to keep in mind is that $\mathcal{L}$ contains a constant number of jobs. Set $\mathcal{S} = \mathcal{J} \setminus \mathcal{L}$ is the set of *short* jobs. Two operations O_{ij} and $O_{i'j'}$, $J_j, J_{j'} \in \mathcal{L}$ and $j \neq j'$, are *compatible* if they must be processed on different machines, i.e., $m_{ij} \neq m_{i'j'}$. A *snapshot* of $\mathcal{L}$ is a subset of compatible operations. A *relative schedule* of $\mathcal{L}$ is a sequence $M(1), \ldots, M(g)$ of snapshots of $\mathcal{L}$ such that

- $M(1) = \emptyset$ and $M(g) = \emptyset$,
- each operation O_{ij} for a job $J_j \in \mathcal{L}$ occurs in a subsequence of consecutive snapshots $M(\alpha_{ij}), \ldots, M(\beta_{ij})$, $2 \leq \alpha_{ij} \leq \beta_{ij} < g$, where $M(\alpha_{ij})$ is the first and $M(\beta_{ij})$ is the last snapshot that contain operation O_{ij},
- for two operations O_{ij} and $O_{i+1,j}$ of the same job $J_j \in \mathcal{L}$, $\beta_{ij} < \alpha_{i+1,j}$, and
- consecutive snapshots are different.

We observe that g can be bounded by $2k\mu+1$. A relative schedule corresponds to an order of execution for the jobs in $\mathcal{L}$. One can associate a relative schedule to each non-preemptive schedule of $\mathcal{L}$ by looking at every time in the schedule when a task of $\mathcal{L}$ starts or ends and creating a snapshot right after that time. Notice, that we have included one snapshot $M(1) = \emptyset$ without operations of jobs in $\mathcal{L}$. This corresponds to the case when the schedule for $\mathcal{J}$ does not contain any operation O_{ij} of a job $J_j \in \mathcal{L}$ at the beginning. For any snapshot $M(\ell)$ let $P(\ell) = \bigcup_{O_{ij} \in M(\ell), J_j \in \mathcal{L}} \{m_{ij}\}$, be the set of machines that process operations from long jobs during snapshot $M(\ell)$.

For each relative schedule $R = (M(1), \ldots, M(g))$ of $\mathcal{L}$, we formulate a linear program $LP(R)$ as follows. For each job $J_j \in \mathcal{S}$ we use a set of decision variables $x_{j,(i_1,\ldots,i_\mu)} \in [0, 1]$ for tuples $(i_1, \ldots, i_\mu) \in A$, where $A = \{(i_1, \ldots, i_\mu) | 1 \leq i_1 \leq i_2 \leq \cdots \leq i_\mu \leq g\}$.

The meaning of these variables is that $x_{j,(i_1,\ldots,i_\mu)} = 1$ if and only if each operation O_{kj} of job J_j is scheduled in snapshot i_k for each $1 \leq k \leq \mu$. Note that by the way in which we numbered the operations, any tuple $(i_1, \ldots, i_\mu) \in A$ represents a valid ordering for the operations. Let the load $L_{\ell,h}$ on machine h in snapshot $M(\ell)$ be defined as the total processing time of operations from short jobs that are executed by machine h during snapshot ℓ, i.e., $L_{\ell,h} = \sum_{J_j \in \mathcal{S}} \sum_{(i_1,\ldots,i_\mu) \in A} \sum_{k=1,\ldots,\mu | i_k=\ell, m_{kj}=h} x_{j,(i_1,\ldots,i_\mu)} p_{kj}$. Each variable t_ℓ, $1 \leq \ell \leq g$, denotes the length of snapshot $M(\ell)$. The linear program for a given relative schedule R is the following.

Minimize $\sum_{\ell=1}^{g} t_\ell$
s.t. (1) $t_\ell \geq 0$, for all $1 \leq \ell \leq g$
(2) $\sum_{\ell=\alpha_{ij}}^{\beta_{ij}} t_\ell = p_{ij}$ for every operation O_{ij}, $1 \leq i \leq \mu$, $J_j \in \mathcal{L}$,
(3) $x_{j,(i_1,\ldots,i_\mu)} \geq 0$ for each $J_j \in \mathcal{S}$, $(i_1, \ldots, i_\mu) \in A$,
(4) $\sum_{(i_1,\ldots,i_\mu) \in A} x_{j,(i_1,\ldots,i_\mu)} = 1$ for every $J_j \in \mathcal{S}$,
(5) $L_{\ell,h} \leq t_\ell$, for every $1 \leq \ell \leq g$, $1 \leq h \leq m$, and $h \notin P(\ell)$,
(6) $x_{j,(i_1,\ldots,i_\mu)} = 0$ for $J_j \in \mathcal{S}$ and $m_{kj} \in P(i_k)$.

To avoid that some operation O_{kj} of a short job $J_j \in \mathcal{S}$ is scheduled on a machine $m_{kj} \in P(i_k)$ occupied by a long job, we set $x_{j,(i_1,\ldots,i_\mu)} = 0$ (constraint (6)). Note that if a variable $x_{j,(i_1,\ldots,i_\mu)}$ has value one, then job $j \in \mathcal{S}$ has a unique snapshot assignment.

Lemma 1. *An optimum solution of $LP(R)$ has value no larger than the makespan of an optimum schedule for $\mathcal{J}$ that respects the relative schedule R.*

Proof. Consider an optimum schedule S^* for $\mathcal{J}$ that respects the relative schedule R. We only need to show that for any job $J_j \in S$ there is a feasible solution of $LP(R)$ that schedules all operations of J_j in the same positions as S^*.

Assume that S^* schedules each operation O_{kj} on machine m_{kj} for a time interval p_{kj} that spans consecutive snapshots $M(s_{kj}), \ldots, M(e_{kj})$, where s_{kj} might be equal to e_{kj} (corresponding to the case when the operation is completely scheduled in a single snapshot). Let $f_{kj}(i)$ be the fraction of operation O_{kj} that is scheduled in snapshot $M(i)$.

We assign values to the variables $x_{j,(i_1,\ldots,i_\mu)}$ as follows. Set $x_{j,(s_{1j},s_{2j},\ldots,s_{\mu j})} = f$, where $f = \min\{f_{kj}(s_{kj}) \mid 1 \le k \le \mu\}$. Then we have to assign values to the other variables $x_{j,(i_1,\ldots,i_\mu)}$ to cover the remaining $1-f$ fraction of each operation. To do this, for every operation O_{kj}, we make $f_{kj}(s_{kj}) = f_{kj}(s_{kj}) - f$. Clearly for at least one operation O_{kj} the new value of $f_{kj}(s_{kj})$ will be set to zero; for those operations with $f_{kj}(s_{kj}) = 0$ we make $s_{kj} = s_{kj} + 1$ since the beginning snapshot for the rest of the operation O_{kj} is snapshot $M(s_{kj} + 1)$. Then we assign value to the new $x_{j,(s_{1j},s_{2j},\ldots,s_{\mu j})}$ as above and repeat the process until $f = 0$.

Note that each iteration of this process assigns a value to a different variable $x_{j,(i_1,\ldots,i_\mu)}$ since from one iteration to the next at least one snapshot s_{kj} is changed. This assignment of values to variables $x_{j,(i_1,\ldots,i_\mu)}$ satisfies the constraints of the linear program and assigns the same time intervals than S^*. Thus the claim follows. □

Let $OPT(I)$ be the makespan of an optimum schedule, let $d_j = \sum_{i=1}^{\mu} p_{ij}$ be the total processing time of job $J_j \in \mathcal{J}$, and let $D = \sum_{J_j \in \mathcal{J}} d_j$.

Lemma 2.
$$\frac{D}{m} \le OPT(I) \le D.$$

By dividing all execution times p_{ij} by D, we may assume that $D = 1$ and $\frac{1}{m} \le OPT(I) \le 1$. Given a relative schedule R, the minimum makespan $OPT_R(I)$ of a schedule respecting R satisfies also $\frac{1}{m} \le OPT_R(I) \le 1$.

3 The Algorithm

For each relative schedule R we approximately solve the linear program $LP(R)$. Then, we round the solution to get only few fractional variables. Next, we transform the solution of the linear program into a feasible schedule. Finally, we select one schedule that has the smallest makespan.

3.1 Approximate Solution of the Linear Program

We guess the length s of an optimum schedule and add the constraint $\sum_{\ell=1}^{g} t_\ell \le s$ to $LP(R)$. Then we replace constraint (5) by constraint (5'), where λ is a nonnegative value:

(5') $\quad L_{\ell,h} - t_\ell + 1 \leq \lambda, \quad$ for $h \notin P(\ell)$.

This new linear program will be denoted as $LP(R, s, \lambda)$. The linear program $LP(R, s, \lambda)$ has a special *block angular* structure (for a survey see [5, 11]) that we describe below. For each job $J_j \in S$ we define a set $B_j = \{(x_{j,(1,...,1)}, x_{j,(1,...,1,2)}, ..., x_{j(g,g,...,g)}) \mid$ conditions (3), (4), and (6) are satisfied $\}$. Each set B_j is a g^μ-dimensional *simplex*. We also define a set $B_{|S|+1} = \{(t_1, t_2, ..., t_g) \mid \sum_{\ell=1}^{g} t_\ell \leq s$ and conditions (1)-(2) are satisfied $\}$. Set $B_{|S|+1}$ is a convex compact set. Linear inequalities (5') form a set of so called *coupling constraints*. Let $f_{\ell,h} = L_{\ell,h} - t_\ell + 1$. Since $t_\ell \leq s \leq 1$, these functions are non-negative.

The logarithmic potential price directive decomposition method [5] developed by Grigoriadis and Khachiyan [5] for a large class of problems with block angular structure provides a ρ-relaxed decision procedure for $LP(R, s, 1)$. This procedure either determines that $LP(R, s, 1)$ is infeasible, or computes (a solution that is nearly feasible in the sense that it is) a feasible solution of $LP(R, s, (1+\rho))$. The overall running time of the procedure for $LP(R, s, 1)$ is $O(n)$.

Recall that $OPT(I)$ is the optimum makespan for a given instance I of the job shop problem. For brevity in the following, we use OPT instead of $OPT(I)$ and OPT_R instead of $OPT_R(I)$.

Lemma 3. *The following assertions are true:*

(0) $LP(R, 1, 1)$ *is feasible for each relative schedule* R,
(1) *if* $LP(R, s, 1)$ *is feasible and* $s \leq s'$, *then* $LP(R, s', 1)$ *is feasible,*
(2) *if* $LP(R, s, 1)$ *is infeasible, then there exists no schedule (that respects the relative schedule* R *for the long jobs in* $\mathcal{L}$*) with makespan at most* s,
(3) $LP(R, OPT_R, 1)$ *is feasible for each relative schedule* R,

Let R be a relative schedule for $\mathcal{L}$ in an optimum schedule. This Lemma implies that we can use binary search on the interval $[\frac{1}{m}, 1]$ to find a value $s \leq (1+\frac{\epsilon}{4})OPT$ such that $LP(R, s, (1+\rho))$ is feasible for $\rho = \frac{\epsilon}{4mg}$. This search can be performed in $O(\log m/\epsilon)$ iterations. For this value s a solution to $LP(R, s, 1+\rho)$ has value $\sum_{\ell=1}^{g} t_\ell \leq OPT(1+\frac{\epsilon}{4})$, and machine loads $L_{\ell,h} \leq t_\ell + \rho$. Since $\rho = \frac{\epsilon}{4mg}$ we obtain a fractional solution of value $\sum_{\ell=1}^{g}(t_\ell + \rho) \leq (1 + \frac{\epsilon}{2})OPT$.

Lemma 4. *A solution for* $LP(R, s, 1+\rho)$, *with* $s \leq (1 + \frac{\epsilon}{4})OPT$ *and* $\rho = \frac{\epsilon}{4mg}$, *of value at most* $(1 + \frac{\epsilon}{2})OPT$ *can be found in linear time.*

3.2 Rounding Step

In this section we show how we can modify any feasible solution for $LP(R, s, 1+\rho)$ to get a new feasible solution in which all but a constant number of variables $x_{j,(i_1,...,i_\mu)}$ have value 0 or 1. Moreover we can do this rounding step in linear time. Let $n' = |S|$.

Let us write the linear program $LP(R, s, 1 + \rho)$ in matrix form as $Bx = b$, $x \geq 0$. We arrange the columns of B so that the last g columns contain the coefficients for variables t_ℓ and the first $n'g^\mu$ columns have the coefficients for variables $x_{j,(i_1,\ldots,i_\mu)}$. The key observation that allows us to perform the rounding in linear time is to note that matrix B is sparse. In particular, each one of the first $n'g^\mu$ columns has at most $\mu + 1$ non-zero entries, and it is not difficult to show that there exists a constant size subset B' of these columns in which the number of non-zero rows is smaller than the number of columns. The non-zero entries of B' induce a singular matrix of constant size, so we can find a non-zero vector y in the null space of this matrix, i.e., $B'y = 0$.

Let $\delta > 0$ be the smallest value such that some component of the vector $x+\delta y$ is either zero or one (if the dimension of y is smaller than the dimension of x we augment it by adding an appropriate number of zero entries). Note that the vector $x+\delta y$ is a feasible solution of $LP(R, s, 1+\rho)$. Let x^0 and x^1 be respectively the zero and one components of vector $x + \delta y$. We update the linear program by making $x = x + \delta y$ and then removing from x all variables in x^0 and x^1 and all columns of B corresponding to such variables. If $x^1 \neq \emptyset$ then vector b is set to $b - \sum_{i \in x^1} B[*, i]$, where $B[*, i]$ is the column of B corresponding to variable i.

This process rounds the value of at least one variable $x_{j,(i_1,\ldots,i_\mu)}$ to either 0 or 1 and it can be carried out in constant time since the sizes of the submatrices involved are constant. We note that the value of δ can be found in constant time also since y has constant size. We can repeat this process until only a constant number of variables $x_{j,(i_1,\ldots,i_\mu)}$ have fractional values. Since there is a linear number of these variables then the overall time is linear.

Now we describe the rounding algorithm in more detail. Let us assume that the first $n'g^\mu$ columns of B are indexed so that the columns corresponding to variables $x_{j,(i_1,\ldots,i_\mu)}$ for each job J_j appear in adjacent positions. We might assume that at all times during the rounding procedure each job J_j has associated at least two columns in B. This assumption can be made since if job J_j has only one associated column, then the corresponding variable $x_{j,(i_1,\ldots,i_\mu)}$ must have value 1. Let B' be the set formed by the first $2mg + 2$ columns of B. Note that at most $2mg+1$ rows of B' have non-zero entries. To see this observe that at most $mg + 1$ of these entries come from constraint (4) and by the above assumption on the number of columns for each job, while at most mg non-zero entries come from constraint (5').

To avoid introducing more notation let B' be the matrix induced by the non-zero rows. Since B' has at most $2mg+1$ rows and exactly $2mg+2$ columns then B' is singular and hence its null space has at least one non-zero vector y such that $B'y = 0$. Since the size of B' is constant, vector y can be found in constant time by using simple linear algebra.

After updating x, B, and b as described above, the procedure is repeated. This is done until there are at most $2mg + 1$ columns in B corresponding to variables $x_{j,(i_1,\ldots,i_\mu)}$. Hence the total number of iterations is at most $n'g^\mu - 2mg - 1$ and each iteration can be done in constant time.

Lemma 5. *A solution for $LP(R, s, 1+\rho)$ can be transformed in linear time into another feasible solution for $LP(R, s, 1+\rho)$ in which the set of jobs $\mathcal{F}$ that have fractional assignments in more than one snapshot after the rounding procedure has size $|\mathcal{F}| \leq mg \leq m(2k\mu + 1)$.*

Proof. By the above argument at most $2mg + 1$ variables $x_{j,(i_1,...,i_\mu)}$ might have value different from 0 and 1. Since at the end of the rounding procedure each job has either 0 or at least 2 columns in B, then at most mg jobs receive fractional assignments. □

3.3 Generating a Schedule

After rounding the solution for $LP(R, s, 1 + \rho)$ as described above, we remove all jobs $\mathcal{F}$ that received fractional assignment. These jobs will be placed at the end of the schedule. Using the approximate solution for the LP, our snapshots are enlarged from t_ℓ to at most $t'_\ell = t_\ell + \rho$.

Consider an snapshot $M(\ell)$, $\ell \in \{1, \ldots, g\}$. In this snapshot the set $P(\ell)$ contains machines that are reserved for the long jobs. The set of operations $O(\ell)$ from small jobs assigned by the solution of the linear program must be scheduled in machines $M \setminus P(\ell)$. Find in each machine $i \in M \setminus P(\ell)$ the set $O_\ell(i)$ formed by the $\mu^3 m$ largest operations assigned to it in $M(\ell)$. Let $V_\ell(i)$ be the set of jobs corresponding to operations $O_\ell(i)$ and let $V_\ell = \bigcup_{i \notin P(\ell)} V_\ell(i)$. Remove from $M(\ell)$ all jobs in V_ℓ (these jobs will be scheduled later).

Let $p_{max}(\ell)$ be the maximum processing time among the remaining operations from $O(\ell)$. Without loss of generality we may assume that the largest remaining operation is assigned to machine M_ℓ. Let $V = \bigcup_{\ell=1}^{g} V_\ell$.

Lemma 6.
$$|V| \leq \mu^3 m^2 g.$$

Let $S(O') = \sum_{O_{kj} \in O'} p_{kj}$ be the total processing time for some set of operations O'; and let $S(J')$ be the total processing time of all jobs in some set $J' \subset \mathcal{J}$. The load of machine M_ℓ is at most $t'_\ell - \mu^3 m p_{max}(\ell)$ and all operations have processing time no larger than $p_{max}(\ell)$. We notice that the load of the other machines $i \in M \setminus P(\ell)$, $i \neq M_\ell$, could be close to t'_ℓ. Since we have deleted the largest $\mu^3 m$ operations on machine M_ℓ (all with length at least $p_{max}(\ell)$), we get the following bound:

Lemma 7.
$$\mu^3 m p_{max}(\ell) \leq S(O_\ell(M_\ell)).$$

On the modified snapshot $M'(\ell)$ the remaining operations from the small jobs form an instance of the job shop problem with maximum machine load t'_ℓ and maximum operation length $p_{max}(\ell)$. We use the algorithm of Sevastianov [12] to find in $O(n^2 \mu^2 m^2)$ time a schedule for the operations in $M'(\ell)$ of length at most $\bar{t}_\ell = t_\ell + \rho + \mu^3 m p_{max}(\ell)$. Hence we get an enlarged snapshot $M'(\ell)$ of length at most $\bar{t}_\ell$. Summing these enlargements among all snapshots, we get:

Lemma 8.
$$\sum_{\ell=1}^{g} \mu^3 m p_{max}(\ell) \leq \sum_{\ell=1}^{g} S(O_\ell(M_\ell)) \leq S(V).$$

The total length of the snapshots $M'(\alpha_{ij}), \ldots, M'(\beta_{ij})$ that contain an operation O_{ij} of a long job J_j might be larger than p_{ij}. This creates some idle times on machine m_{ij}. We start operations O_{ij} for long jobs $\mathcal{L}$ at the beginning of the enlarged snapshot $M'(\alpha_{ij})$. The resulting schedule is clearly feasible.

Lemma 9. *A feasible schedule for the jobs $\mathcal{J}$ of length at most $(1+\frac{\epsilon}{2})OPT_R(I)+ S(\mathcal{V}) + S(\mathcal{F} \cup \mathcal{V})$ can be found in $O(n^2)$ time.*

Proof. The small jobs $\mathcal{F}$ that received fractional assignment and the jobs in $\mathcal{V}$ are scheduled sequentially at the end of the schedule produced by the above algorithm. Thus, by Lemmas 4 and 8 the claim follows. $\quad\square$

4 Analysis of the Algorithm

Lemma 10. *Let $k = |\mathcal{L}|$. The total number of relative schedules is at most $(2k^2\mu^2)^{k\mu}$.*

Since k, m and μ are constant, the number of relative schedules is also constant.

Lemma 11. *[7] Let $d_1 \geq d_2 \geq \ldots d_n > 0$ be a sequence of real numbers and $1 = \sum_{j=1}^{n} d_j$. Let q be a nonnegative integer, α be a positive value, and assume that $n \geq (q + 1)^{\lceil\frac{1}{\alpha}\rceil}$. There exists an integer k such that $d_k + \ldots + d_{k+qk-1} \leq \alpha$ and $k \leq (q + 1)^{\lceil\frac{1}{\alpha}\rceil - 1}$.*

Let $\alpha = \frac{\epsilon}{4m}$ and $qk = 5\mu^4 m^2 k$, i.e. $q = 5\mu^4 m^2$. By this lemma, $k \leq (5\mu^4 m^2 + 1)^{\lceil 4m/\epsilon\rceil - 1}$. We choose $\mathcal{L}$ as the set of $k - 1$ jobs with the largest values $d_j = \sum_{i=1}^{\mu} p_{ij}$.

Lemma 12.
$$S(\mathcal{F} \cup \mathcal{V}) \leq \frac{\epsilon}{4}OPT_R(I).$$

Proof. By Lemmas 5 and 6, $|\mathcal{F} \cup \mathcal{V}| \leq \mu^3 m^2 g + mg \leq 2\mu^4 m^2 k + \mu^3 m^2 + 2mk\mu + m \leq 5\mu^4 m^2 k$. By the preceding discussion on the choice of $\mathcal{L}$ and by Lemma 11, the total processing time for the jobs in $\mathcal{F} \cup \mathcal{V}$ is at most $\frac{\epsilon}{4m} \leq \frac{\epsilon}{4}OPT_R(I)$. $\quad\square$

Theorem 1. *For any fixed m and μ, there is a polynomial-time approximation scheme for the job shop scheduling problem that computes for any fixed $\epsilon > 0$ a feasible schedule of makespan at most $(1 + \epsilon)OPT$ in $O(n^2)$ time.*

Proof. By considering all relative schedules R for $\mathcal{L}$ the claim on the length of the makespan follows from Lemmas 9 and 12. For every fixed m and ϵ, all computations can be carried out in $O(n)$ time, with exception of the algorithm of Sevastianov that runs in $O(n^2)$ time. $\quad\square$

4.1 Speed Up to Linear Time

We show how to speed up the algorithm of Sevastianov [12] to get linear time in Theorem 1. Let $p_{max} = max_{ij} p_{ij}$ denote the maximum processing time of any operation. Let $\Pi_{max} = max_{1 \le i \le m} \sum_{j,k|m_{kj}=i} p_{kj}$ be the maximum load or total processing time assigned to any machine. Given an instance of the job shop problem with m machines and μ operations per job, Sevastianov's algorithm finds in $O(n^2 \mu^2 m^2)$ time a schedule of length at most $\Pi_{max} + \mu^3 m p_{max}$.

We may assume that $\Pi_{max} \le 1$ (by normalization). Let $d_j = \sum_{\ell=1}^{\mu} p_{\ell j}$. For a job J_j let $m_j = (m_{1j}, m_{2j}, \ldots, m_{\mu j})$ be a vector that describes the machines on which its operations must be performed. Let us partition the set of jobs $\mathcal{J}$ into m^{μ} groups $\mathcal{J}_1, \mathcal{J}_2, \ldots, \mathcal{J}_{m^{\mu}}$ such that all jobs J_j in some group $\mathcal{J}_i$ have the same machine vector m_j, and jobs from different groups have different machine vectors. Consider the jobs in one of the groups $\mathcal{J}_i$. Let γ be a constant to be defined later. Let J_j and J_h be two jobs from $\mathcal{J}_i$ such that for each one of them the total execution time of its operations is smaller than γ. We "glue" together these two jobs to form a composed job in which the processing time of the i-th operation is equal to the sum of the processing times of the i-th operations of J_j and J_h. We repeat this process until at most one job from $\mathcal{J}_i$ has processing time smaller than γ. At the end, all jobs in group $\mathcal{J}_i$ have processing times smaller than $max(p_{max}, 2\gamma)$.

The same procedure is performed in all other groups $\mathcal{J}_j$. At the end of this process, each one of the composed jobs has at most μ operations, and the total number of composed jobs is no larger than $\frac{m}{\gamma} + m^{\mu}$. Note that the procedure runs in linear time and a feasible schedule for the original jobs can be easily obtained from a feasible schedule for the composed jobs.

We run Sevastianov's algorithm on the set of composed jobs to get a schedule of length $\Pi_{max} + \mu^3 m r$, where $r = max\{p_{max}, 2\gamma\}$. The time needed to get this schedule is $O((\frac{m}{\gamma} + m^{\mu})^2 \mu^2 m^2)$. If we choose $\gamma = \frac{\epsilon}{8m^2 \mu^3 g}$ we get a total lengthening for the snapshots of at most $\frac{\epsilon}{4} OPT$. By this choice of γ Sevastianov's algorithm needs only constant time plus linear pre-processing time.

Theorem 2. *For any fixed m and μ, there is a polynomial-time approximation scheme for the job shop scheduling problem that computes for any fixed $\epsilon > 0$ in $O(n)$ time a feasible schedule whose makespan is at most $(1 + \epsilon)OPT$.*

5 Preemptive Job Shop Problem

As in the non-preemptive case we divide the set of jobs $\mathcal{J}$ into long jobs $\mathcal{L}$ and short jobs $\mathcal{S}$. Set $\mathcal{L}$ has a constant number of jobs, as in the non-preemptive case. Consider a preemptive schedule for $\mathcal{J}$. Look at the time at which any operation from a long job starts or ends. These times define a set of time intervals, similar to those defined in the non-preemptive case by the snapshots. For convenience we also call these time intervals snapshots.

Since $\mathcal{L}$ has a constant number of jobs (and hence there is a constant number of snapshots), we can consider all relative orderings of the long jobs in the snapshots. An operation of a long job is scheduled in consecutive snapshots $i, i+1, \ldots, i+t$, *but* only a fraction (possible equal to zero) of the operation might be scheduled in any one of these snapshots. However, and this is crucial for the analysis, in every snapshot there can be at most one operation from any given long job.

Now we define a linear program as in the case of the non-preemptive job shop. For each long job J_j we define variables $x_{j,(i_1,\ldots,i_\mu)}$ for those tuples $(i_1,\ldots,i_\mu)$ corresponding to snapshots where the operations of J_j might be scheduled, as described above. These variables indicate which fraction of each operation is scheduled in each snapshot. Let g be the number of snapshots, t_ℓ be the length of the ℓ snapshot, and $L_{\ell,h}$ be as before. The new linear program is the following.

Minimize $\sum_{\ell=1}^{g} t_\ell$
 s.t. **(1)** $t_\ell \geq 0$, for all $1 \leq \ell \leq g$
 (3) $x_{j,(i_1,\ldots,i_\mu)} \geq 0$ for each $J_j \in \mathcal{S}$, $(i_1,\ldots,i_\mu) \in A$,
 (3') $x_{j,(i_1,\ldots,i_\mu)} \geq 0$ for each $J_j \in \mathcal{L}$, and $(i_1,\ldots,i_\mu)$ as above,
 (4') $\sum_{(i_1,\ldots,i_\mu) \in A} x_{j,(i_1,\ldots,i_\mu)} = 1$ for every $J_j \in \mathcal{J}$,
 (5) $L_{\ell,h} \leq t_\ell$, for every $1 \leq \ell \leq g$, $1 \leq h \leq m$,

Note that in any solution of this linear program the schedule for the long jobs is always feasible, since there is at most one operation of a given job in any snapshot. We find an approximate solution for the linear program using the algorithm of Section 3.1, and then we apply our rounding procedure to this solution. After rounding there are at most mg small jobs that receive fractional assignments (see Section 3.2). These jobs are scheduled at the end, as before.

Next we find a feasible schedule for every snapshot as follows. Let us consider a snapshot. Remove from the snapshot the operations belonging to long jobs. These operations will be reintroduced to the schedule later. Then use Sevastianov's algorithm as described in Section 3.3 to find a feasible schedule for the (fractions of) small jobs in the snapshot. Finally we put back the operations from the long jobs, scheduling them in the empty gaps left by the small jobs. Note that it might be necessary to split an operation of a long job in order to make it fit in the empty gaps. At the end we have a feasible schedule because there is at most one operation of each long job in the snapshot. In this schedule the number of preemptions is at most $n\mu + mg$. By using more elaborated ideas we can further reduce the number of preemption to a constant.

Choosing the size of $\mathcal{L}$ as we did for the non-preemptive case we ensure that the length of the schedule is at most $1 + \epsilon$ times the length of an optimum schedule. This algorithm runs in linear time.

6 Conclusions

We have designed a polynomial time approximation scheme for the preemptive and non-preemptive job shop problems when the number of machines and the number of operations per job are fixed. The running time for the algorithm is

$O(n)$. Using our approach we can design linear time approximation schemes for open shops and flow shops. Recently we have extended our techniques to design a linear time approximation scheme for the flexible job shop problem with fixed number of machines and number of operations per job. The flexible job shop problem is a generalization of the job shop problem in which an operation O_{ij} can be processed on any machine $k \in M$ in time $p_{ij}(k)$ [8].

References

1. M.R. Garey, D.S. Johnson and R. Sethi, The complexity of flowshop and jobshop scheduling, *Mathematics of Operations Research* 1 (1976), 117-129.
2. T. Gonzales and S. Sahni, Open shop scheduling to minimize finish time, *Journal of the ACM* 23 (1976), 665-679.
3. T. Gonzales and S. Sahni, Flowshop and jobshop schedules: complexity and approximation, *Operations Research* 26 (1978), 36-52.
4. L.A. Goldberg, M. Paterson, A. Srinivasan, and E. Sweedyk, Better approximation guarantees for job-shop scheduling, *Proceedings of the 8th Symposium on Discrete Algorithms* SODA 97, 599-608.
5. M.D. Grigoriadis and L.G. Khachiyan, Coordination complexity of parallel price-directive decomposition, *Mathematics of Operations Research* 21 (1996), 321-340.
6. L.A. Hall, Approximability of flow shop scheduling, *Mathematical Programming* 82 (1998), 175-190.
7. K. Jansen and L. Porkolab, Linear-time approximation schemes for scheduling malleable parallel tasks, *Proceedings of the 10th Annual ACM-SIAM Symposium on Discrete Algorithms*, 1999, 490-498.
8. K. Jansen, R. Solis-Oba, and M. Mastrolilli, Linear time approximation scheme for the flexible job shop problem, unpublished manuscript.
9. K. Jansen, R. Solis-Oba and M.I. Sviridenko, Makespan minimization in job shops: a polynomial time approximation scheme, *Proceedings of the 31th Annual ACM Symposium on Theory of Computing*, 1999, 394-399.
10. E.L. Lawler, J.K. Lenstra, A.H.G. Rinnooy Kan and D.B. Shmoys, Sequencing and scheduling: Algorithms and complexity, in: Handbook in Operations Research and Management Science, Vol. 4, North-Holland, 1993, 445-522.
11. S.A. Plotkin, D.B. Shmoys and E. Tardos, Fast approximation algorithms for fractional packing and covering problems, *Mathematics of Operations Research* 20 (1995), 257-301.
12. S.V. Sevastianov, Bounding algorithms for the routing problem with arbitrary paths and alternative servers, *Cybernetics* 22 (1986), 773-780.
13. S.V. Sevastianov and G.J. Woeginger, Makespan minimization in open shops: A polynomial time approximation scheme, *Mathematical Programming* 82 (1998), 191-198.
14. D.P. Williamson, L.A. Hall, J.A. Hoogeveen, C.A.J. Hurkens, J.K. Lenstra, S.V. Sevastianov and D.B. Shmoys, Short shop schedules, *Operations Research* 45 (1997), 288-294.

Randomized Rounding for Semidefinite Programs – Variations on the MAX CUT Example

Uriel Feige

Weizmann Institute, Rehovot 76100, Israel

Abstract. MAX CUT is the problem of partitioning the vertices of a graph into two sets, maximizing the number of edges joining these sets. Goemans and Williamson gave an algorithm that approximates MAX CUT within a ratio of 0.87856. Their algorithm first uses a semidefinite programming relaxation of MAX CUT that embeds the vertices of the graph on the surface of an n dimensional sphere, and then cuts the sphere in two at random.

In this survey we shall review several variations of this algorithm which offer improved approximation ratios for some special families of instances of MAX CUT, as well as for problems related to MAX CUT.

1   Introduction

This survey covers an area of current active research. Hence, there is danger, or rather hope, that the survey will become outdated in the near future. The level of presentation will be kept informal. More details can be found in the references. Results are presented in a logical order that does not always correspond to the historical order in which they were derived.

The scope of the survey is limited to MAX CUT and to strongly related other problems (such as MAX BISECTION). Many of the recent approximation algorithms based on semidefinite algorithms are not included (such as those for COLORING [10] and for 3SAT [12]). Results in which the author was involved are perhaps over-represented in this survey, but hopefully, not in bad taste.

2   The Algorithm of Goemans and Williamson

For a graph $G(V, E)$ with $|V| = n$ and $|E| = m$, MAX CUT is the problem of partitioning V into two sets, such that the number of edges connecting the two sets is maximized. This problem is NP-hard to approximate within ratios better than 16/17 [9]. Partitioning the vertices into two sets at random gives a cut whose expected number of edges is $m/2$, trivially giving an approximation algorithm with expected ratio at least $1/2$. For many years, nothing substantially better was known. In a major breakthrough, Goemans and Williamson [8] gave an algorithm with approximation ratio of 0.87856. For completeness, we review their well known algorithm, which we call algorithm GW.

MAX CUT can be formulated as an integer quadratic program. With each vertex i we associate a variable $x_i \in \{-1, +1\}$, where -1 and $+1$ can be viewed as the two sides of the cut. With an edge $(i, j) \in E$ we associate the expression $\frac{1-x_i x_j}{2}$ which evaluates to 0 if its endpoints are on the same side of the cut, and to 1 if its endpoints are on different sides of the cut.

The integer quadratic program for MAX CUT:

Maximize: $\sum_{(i,j) \in E} \frac{1-x_i x_j}{2}$

Subject to: $x_i \in \{-1, +1\}$, for every $1 \leq i \leq n$.

This integer quadratic program is relaxed by replacing the x_i by unit vectors v_i in an n-dimensional space (the x_i can be viewed as unit vectors in a 1-dimensional space). The product $x_i x_j$ is replaced by an inner product $v_i v_j$. Geometrically, this corresponds to embedding the vertices of G on a unit n-dimensional sphere S_n, while trying to keep the images of vertices that are adjacent in G far apart on the sphere.

The geometric program for MAX CUT:

Maximize: $\sum_{(i,j) \in E} \frac{1-v_i v_j}{2}$

Subject to: $v_i \in S_n$, for every $1 \leq i \leq n$.

The geometric program is equivalent to a semidefinite program in which the variables y_{ij} are the inner products $v_i v_j$, and the n by n matrix Y whose i, j entry is y_{ij} is constrained to be positive semidefinite (i.e., the matrix of inner products of n vectors). The constraint $v_i \in S_n$ is equivalent to $v_i v_i = 1$, which gives the constraint $y_{ii} = 1$.

The semidefinite program for MAX CUT:

Maximize: $\sum_{(i,j) \in E} \frac{1-y_{ij}}{2}$

Subject to: $y_{ii} = 1$, for every $1 \leq i \leq n$,

and the matrix $Y = (y_{ij})$ is positive semidefinite.

This semidefinite program can be solved up to arbitrary precision in polynomial time, and thereafter a set of vectors v_i maximizing the geometric program (up to arbitrary precision) can be extracted from the matrix Y (for more details, see [8]).

The value of the objective function of the geometric problem is at least that of the MAX CUT problem, as any ± 1 solution for the integer quadratic problem is also a solution of the geometric problem, with the same value for the objective function.

One approach to convert a solution of the geometric program to a feasible cut in the graph is to partition the sphere S_n into two halves by passing a hyperplane through the origin of the sphere, and labeling vertices on one half by -1 and on the other half by $+1$. The choice of hyperplane may affect the quality of solution that is obtained. Surprisingly, a random hyperplane is expected to give a cut that is not far from optimal.

Consider an arbitrary edge (i, j). Its contribution to the value of the objective function is $\frac{1-v_i v_j}{2}$. The probability that it is cut by a random hyperplane is directly proportional to the angle between v_i and v_j, and can be shown to be exactly $\frac{\cos^{-1}(v_i v_j)}{\pi}$. Hence the ratio between the expected contribution of the edge (i, j) to the final cut and its contribution to the objective function of the geometric program is $\frac{2\cos^{-1}(v_iv_j)}{\pi(1-v_iv_j)}$. This ratio is minimized when the angle between v_i and v_i is $\theta \simeq 2.33$, giving a ratio of $\alpha \simeq 0.87856$. By linearity of expectation, the expected number of edges cut by the random hyperplane is at least α times the value of geometric program, giving an α approximation for MAX CUT.

We remark that a random hyperplane can be chosen algorithmically by choosing a random unit vector r, which implicitly defines the hyperplane $\{x|xr = 0\}$. See details in [8].

2.1 Outline of Survey

The algorithm of Goemans and Williamson, and variations of it, were applied to many other problems, some well known examples being MAX 2SAT [8, 3], MAX 3SAT [12], MIN COLORING [10] and MAX CLIQUE [1].

The work reported in this survey is partly motivated by the belief that algorithm GW does not exploit the full power of semidefinite programming in the context of approximating MAX CUT.

Research goal: Improve the approximation ratio of MAX CUT beyond $\alpha \simeq 0.87856$.

In the following sections, we shall survey several approaches that try to improve over algorithm GW.

In Section 3 we add constraints to the semidefinite program so as to obtain better embeddings of the graph on the sphere. In Section 4 we describe limitations of the random hyperplane rounding technique, and suggest an alternative "best" hyperplane rounding technique. In Section 5 we investigate rounding techniques that rearrange the points on the sphere prior to cutting. In Section 6 we describe approaches that rearrange the vertices after cutting.

This survey is limited to approaches that remain within the general framework of the GW algorithm.

3 Improving the Embedding

The GW algorithm starts with an embedding of the graph on a sphere. The *value* of this embedding is the value of the objective function of the geometric program. The quality of the embedding can be measured in terms of the so called *integrality ratio*: the ratio between the size of the optimal cut in the graph and the value of the geometric embedding. (We define here the integrality ratio as a number smaller than 1. For this reason we avoid the more common name *integrality gap*, which is usually defined as the inverse of our integrality ratio.) This measure of quality takes the view that we are trying to estimate the size of the maximum cut in the graph, rather than actually find this cut. We may output the value of the embedding as our estimate, and then the error in the estimation is bounded by the integrality ratio.

Goemans and Williamson show that the integrality ratio for their embedding may be essentially as low as α. As a simple example, let G be a triangle (a 3-cycle). Arranging the vertices uniformly on a circle (with angle of $2\pi/3$ between every two vectors) gives an embedding with value 9/4, whereas the maximum cut size is 2. This gives an integrality ratio of $8/9 \simeq 0.888$. For tighter examples, see [8].

To improve the value of the embedding on the sphere, one may add more constraints to the semidefinite program. In doing so, one is guided by two requirements:

1. The constraints need to be satisfied by the true optimal solution (in which the y_{ij} correspond to products of ± 1 variables).
2. The resulting program needs to be solvable in polynomial time (up to arbitrary precision).

Feige and Goemans [3] analyse the effect of adding *triangle constraints* of the form $y_{ij} + y_{jk} + y_{ki} \geq -1$ and $y_{ij} - y_{jk} - y_{ki} \geq -1$, for every i, j, k. Geometrically, these constraints forbid some embeddings on the sphere. In particular, if three vectors v_i, v_j, v_k lie in the same plane (including the origin), it now must be the case that either two of them are identical, or antipodal. The 3-cycle graph no longer serves as an example for a graph with bad integrality ratio. Moreover, it can be shown that for every planar graph, the value of the geometric program is equal to that of the maximum cut.

Feige and Goemans were unable to show that the more constrained semidefinite relaxation leads to an approximation algorithm with improved approximation ratio for MAX CUT (though they were able to show this for related problems such as MAX 2SAT).

Open question: Does addition of the triangle constraints improve the integrality ratio of the geometric embedding for MAX CUT?

Feige and Goemans also discuss additional constraints that can be added. Lovasz and Schriver [13] describe a systematic way of adding constraints to semidefinite relaxations. The above open question extends to all such formulations.

4 Improving the Rounding Technique

Goemans and Williamson use the random hyperplane rounding technique. The analysis or the approximation ratio compares the expected number of edges in the final cut to the value of the geometric embedding. We remark that for most graphs that have maximum cuts well above $m/2$, the random hyperplane rounding technique will actually produce the optimal cut. This is implicit in [2, 5].

Karloff [11] studies the limitations of this approach. He considers a family of graphs in which individual graphs have the following properties:

1. The maximum cut in the graph is not unique. There are $k = \Omega(\log n)$ different maximum cuts in the graph. (The graph is very "symmetric" – it is both vertex transitive and edge transitive.)

2. The value of the geometric program (and the semidefinite program) for this graph is exactly equal to the size of the maximum cut. Hence the integrality ratio is 1.

3. The vertices can be embedded on a sphere as follows. Each vertex is a vector in $\{+1, -1\}^k$ (and normalized by $1/\sqrt{k}$), where coordinate j of vector i is ± 1 depending on the side on which vertex i lies in the jth optimal cut. It follows that the value of this embedding is equal to the size of the maximum cut.

4. The sides of each maximum cut are labeled ± 1 in such a way that for the above embedding, the angle between the vectors of any two adjacent vertices is (arbitrarily close to) θ, where θ is the worst angle for the GW rounding technique. Hence the analysis of the random hyperplane rounding technique only gives an approximation ratio of α for the above graph and associated embedding.

The embedding described above, derived as a combination of legal cuts, satisfies all constraints discussed in Section 3. Hence we are led to conclude that if one wishes to get an approximation ratio better than α, one needs a rounding technique different than that of Goemans and Williamson.

For Karloff's graph and embedding, each of the k maximum cuts can be derived by using a hyperplane whose normal is a vector in the direction of the respective coordinate. Hence a rounding technique that uses the *best* hyperplane (the one that cuts the most edges) rather than a random one would find the maximum cut.

Open question: Design examples that show a large gap between the value of the geometric embedding and the cut obtained by the best hyperplane.

The above open question can serve as an intermediate step towards analysing the integrality ratio.

Remark: The best hyperplane can be approximated in polynomial time in the following sense. The dimension of the embedding can be reduced by projecting the sphere on a random d-dimensional subspace. When d is a large enough constant, the vast majority of distances are preserved up to a small distortion, implying the same for angles. To avoid degeneracies, perturb the location of each point by a small random displacement. The value of the objective function hardly changes by this dimension reduction and perturbation (the change becomes negligible the larger d is). Now each relevant hyperplane is supported by d points, allowing us to enumerate all hyperplanes in time n^d.

5   Rotations

In some cases, it is possible to improve the results of the random hyperplane rounding technique by first rearranging the vectors v_i on the sphere. This modifies the geometric embedding, making it suboptimal with respect to the objective function. However, this suboptimal solution is easier to round with the random hyperplane technique.

Feige and Goemans [3] suggested to use rotations in cases where the sphere has a distinct "north pole" (and "south pole"). Rotating each vector somewhat towards its nearest pole prior to cutting the sphere with a random hyperplane can lead to improved solutions. The usefulness of this approach was demonstrated for problems such as MAX 2SAT, where there is a unique additional vector v_0 that is interpreted as the +1 direction and can serve as the north pole. It is not clear whether a similar idea can be applied for MAX CUT, due to a lack of a natural candidate direction that can serve as the north pole of the sphere.

Zwick [15] has used a notion of *outward rotations* for several problems. For MAX CUT, Zwick observes that there are two "bad" angles for which the random hyperplane fails to give an expectation above α. One is the angle θ mentioned in Section 2. The other is the trivial angle 0, for which the contribution to the value of the geometric program is 0, and so is the contribution to the cut produced by a hyperplane. Hence worst case instances for the GW algorithm may have an arbitrary mixture of both types of angles for pairs of vertices connected by edges. In the extreme case, where all angles are 0 (though this would never be the optimal geometric embedding) it is clear that a random hyperplane would not cut any edge, whereas ignoring the geometric embedding and giving the vertices ± 1 values independently at random is expected to cut roughly half the edges. This latter rounding technique is equivalent to first embedding the vertices as mutually orthogonal unit vectors, and then cutting with a random hyperplane. *Outward rotation* is a technique of averaging between the two embeddings: the optimal geometric embedding on one set of coordinates and the mutually orthogonal embedding on another set of coordinates. It can be used in order to obtain approximation ratios better than α whenever a substantial fraction of the edges have angles 0, showing that essentially the only case when the geometric embedding (perhaps) fails to have integrality ratio better than α is when all edges have angle θ.

6   Modifying the Cut

The random hyperplane rounding technique produces a cut in the graph. This cut may later be modified to produce the final cut. Below we give two representative examples.

Modifying the cut to get a feasible solution. MAX BISECTION is the problem of partitioning the vertices into two *equal* size sets while maximizing the number of edges in the cut. Rounding the geometric embedding via a random hyperplane produces a cut for which the expected number of vertices on each side is $n/2$, but the variance may be very large. Hence, this cut may not be a feasible solution to the problem. Frieze and Jerrum [7] analysed a greedy algorithm that modifies the initial cut by moving vertices from the larger side to the smaller one until both sides are of equal size. As moving vertices from one side to the other may decrease the number of edges in the cut, it is necessary to have an estimate of the expected number of vertices that need to be moved. Such an estimate can be derived if we add a constraint such as $\sum v_i v_j = 0$ to the geometric embedding, which is satisfied if exactly half the vectors are +1 and half of them −1. Frieze and Jerrum used this approach to obtain an approximation ratio of roughly 0.65. This was later improved by Ye [14], who used outward rotations prior to the random hyperplane cut. This has a negative effect of decreasing the expected number of edges in the initial cut, and a positive effect of decreasing the expected number of vertices that need to be moved (note that in the extreme case for outward rotation all vectors are orthogonal and then with high probability each side of the cut contains $n/2 \pm O(\sqrt{n})$ vertices). Trading off these two effects, Ye achieves an approximation ratio of 0.699.

Other problems in which a graph needs to be cut into two parts of prescribed sizes are studied in [6]. An interesting result there shows that when a graph has a *vertex cover* of size k, then one can find in polynomial time a set of k vertices that covers more than $0.8m$ edges. The analysis in [6] follows approaches similar to that of [7], and in some cases can be improved by using outward rotations.

Modifying the cut to get improved approximation ratios. Given a cut in a graph, a *misplaced* vertex is one that is on the same side as most of its neighbors. The number of edges cut can be increased by having misplaced vertices change sides. This local heuristic was used by Feige, Karpinski and Langberg [4] to obtain an approximation ratio significantly better than $\alpha \simeq 0.87856$ for MAX CUT on graphs with maximum degree 3 (the current version claims an approximation ratio of 0.914). Recall that the integrality ratio of the geometric embedding is as bad as α only if all edges have angle θ. Assume such a geometric embedding, and moreover, assume that the triangle constraints mentioned in Section 3 are satisfied. The basic observation in [4] is that in this case, if we consider an arbitrary vertex and two of its neighbors, there is constant probability that all three vertices end up on the same side of a random hyperplane. Such a vertex of degree at most 3 is necessarily misplaced. This gives $\Omega(n)$ expected misplaced vertices, and $\Omega(n)$ edges added to the cut by moving misplaced vertices. As the total number of edges is at most $3n/2$, this gives a significant improvement in the approximation ratio.

7 Conclusions

The algorithm of Goemans of Williamson for MAX CUT uses semidefinite programming to embed the vertices of the graph on a sphere, and then uses the geometry of the embedding to find a good cut in the graph. A similar approach was used for many other problems, some of which are mentioned in this survey. For almost all of these problems, the approximation ratio achieved by the rounding technique (e.g., via a random hyperplane) does not match the integrality ratio of the known negative examples. This indicates that there is still much room for research on the use of semidefinite programs in approximation algorithms.

Acknowledgements

Part of this work is supported by a Minerva grant, project number 8354 at the Weizmann Institute. This survey was written while the author was visiting Compaq Systems Research Center, Palo Alto, California.

References

1. Noga Alon and Nabil Kahale. "Approximating the independence number via the ϑ-function". *Math. Programming.*
2. Ravi Boppana. "Eigenvalues and graph bisection: an average-case analysis. In *Proceedings of the 28th Annual IEEE Symposium on Foundations of Computer Science*, 1997, 280–285.
3. Uriel Feige and Michel Goemans. "Approximating the value of two prover proof systems, with applications to MAX 2-SAT and MAX DICUT". In *Proceedings of third Israel Symposium on Theory of Computing and Systems*, 1995, 182–189.
4. Uriel Feige, Marek Karpinski and Michael Langberg. "MAX CUT on graphs of degree at most 3". *Manuscript*, 1999.
5. Uriel Feige and Joe Kilian. "Heuristics for semirandom graph models". *Manuscript*, May 1999. A preliminary version appeared in *Proceedings of the 39th Annual IEEE Symposium on Foundations of Computer Science*, 1998, 674–683.
6. Uriel Feige and Michael Langberg. "Approximation algorithms for maximization problems arising in graph partitioning". *Manuscript*, 1999.
7. Alan Frieze and Mark Jerrum. "Improved approximation algorithms for MAX k-CUT and MAX Bisection". *Algorithmica*, 18, 67–81, 1997.
8. Michel Goemans and David Williamson. "Improved approximation algorithms for maximum cut and satisfiability problems using semidefinite programming". *Journal of the ACM*, 42, 1115-1145, 1995.
9. Johan Hastad. "Some optimal inapproximability results". In *Proceedings of the 29th Annual ACM Symposium on Theory of Computing*, 1997, 1-10.
10. David Karger, Rajeev Motwani and Madhu Sudan. "Approximate graph coloring by semidefinite programming". *Journal of the ACM*, 45, 246–265, 1998.
11. Howard Karloff. "How good is the Goemans-Williamson MAX CUT algorithm?" In *Proceedings of the 28th Annual ACM Symposium on Theory of Computing*, 1996, 427–434.
12. Howard Karloff and Uri Zwick. "A 7/8 approximation algorithm for MAX 3SAT?" In *Proceedings of the 38th Annual IEEE Symposium on Foundations of Computer Science*, 1997, 406–415.
13. Laszlo Lovasz and Alexander Schrijver. "Cones of Matrices and set-functions and 0-1 optimization". *SIAM J. Optimization*, 1(2), 166-190, 1991.
14. Yinyu Ye. "A .699-approximation algorithm for Max-Bisection". *Manuscript*, March 1999.
15. Uri Zwick. "Outward rotations: a tool for rounding solutions of semidefinite programming relaxations, with applications to MAX CUT and other problems". In *Proceedings of the 31st Annual ACM Symposium on Theory of Computing*, 1999, 679–687.

Hardness Results for the Power Range Assignment Problem in Packet Radio Networks

(Extended Abstract)

Andrea E. F. Clementi[1], Paolo Penna[1], Riccardo Silvestri[2]

[1] Dipartimento di Matematica, Università di Roma "Tor Vergata",
{clementi,penna}@axp.mat.uniroma2.it
[2] Dipartimento di Matematica Pura e Applicata, Università de L'Aquila
silver@dsi.uniroma1.it

Abstract. The *minimum range assignment problem* consists of assigning transmission ranges to the stations of a multi-hop packet radio network so as to minimize the total power consumption provided that the transmission range assigned to the stations ensures the strong connectivity of the network (i.e. each station can communicate with any other station by multi-hop transmission). The complexity of this optimization problem was studied by Kirousis, Kranakis, Krizanc, and Pelc (1997). In particular, they proved that, when the stations are located in a 3-dimensional Euclidean space, the problem is NP-hard and admits a 2-approximation algorithm. On the other hand, they left the complexity of the 2-dimensional case as an open problem.

As for the 3-dimensional case, we strengthen their negative result by showing that the minimum range assignment problem is APX-complete, so, it does not admit a polynomial-time approximation scheme unless P = NP.

We also solve the open problem discussed by Kirousis *et al* by proving that the 2-dimensional case remains NP-hard.

1   Introduction

A *Multi-Hop Packet Radio Network* [10] is a set of radio stations located on a geographical region that are able to communicate by transmitting and receiving radio signals. A transmission range is assigned to each station s and any other station t within this range can directly (i.e. by one *hop*) receive messages from s. Communication between two stations that are not within their respective ranges can be achieved by *multi-hop* transmissions. In general, Multi-Hop Packet Radio Networks are adopted whenever the construction of more traditional networks is impossible or, simply, too expensive.

It is reasonably assumed [10] that the power P_t required by a station t to correctly transmit data to another station s must satisfy the inequality

$$\frac{P_t}{d(t, s)^\beta} > \gamma \tag{1}$$

where $d(t, s)$ is the distance between t and s, $\beta \geq 1$ is the *distance-power gradient*, and $\gamma \geq 1$ is the *transmission-quality* parameter. In an ideal environment (see [10]) $\beta = 2$ but it may vary from 1 to 6 depending on the environment conditions of the place the network is located. In the rest of the paper, we fix $\beta = 2$ and $\gamma = 1$, however, our results can be easily extended to any $\beta, \gamma > 1$.

Combinatorial optimization problems arising from the design of radio networks have been the subject of several papers over the last years (see [10] for a survey). In particular, NP-completeness results and approximation algorithm for scheduling communication and power range assignment problems in radio networks have been derived in [2, 6, 13, 14].

More recently, Kirousis *et al*, in [9], investigated the complexity of the MIN RANGE ASSIGNMENT problem that consists of minimizing the overall transmission power assigned to the stations of a radio network, provided that (multi-hop) communication is guaranteed for any pair of stations (for a formal definition see Section 2). It turns out that the complexity of this problem depends on the number of dimensions of the space the stations are located on. In the 1-dimensional case (i.e. when the stations are located along a line) they provide a polynomial-time algorithm that finds a range assignment of minimum cost. As for stations located in the 3-dimensional space, they instead derive a polynomial-time reduction from MIN VERTEX COVER restricted to planar cubic[1] graphs thus showing that MIN RANGE ASSIGNMENT is NP-hard. They also provide a polynomial-time 2-approximation algorithm that works for any dimension.

In this paper, we address the question whether the approximation algorithm given by Kirousis *et al* for the MIN RANGE ASSIGNMENT problem in three dimensions can be significantly improved. More precisely, we ask whether or not the problem does admit a *Polynomial-Time Approximation Scheme* (PTAS). We indeed demonstrate the APX-completeness of this problem thus implying that it does not admit PTAS unless P = NP (see [12] for a formal definition of these concepts).

The standard method to derive an APX-completeness result for a given optimization problem Π is: *i)* consider a problem Π' which is APX-hard and then *ii)* show an *approximation-preserving* reduction from Π' to Π [12]. We emphasize that Kirousis *et al*'s reduction does not satisfy any of these two requirements. In fact, as mentioned above, their reduction is from MIN VERTEX COVER restricted to planar cubic graphs which cannot be APX-hard (unless P = NP) since it admits a PTAS [3]. Furthermore, it is not hard to verify that their reduction is not approximation-preserving.

In order to achieve our hardness result, we instead consider the MIN VERTEX COVER problem restricted to cubic graphs which is known to be APX-complete [11, 1] and then we show an approximation-preserving reduction from this variant of MIN VERTEX COVER to MIN RANGE ASSIGNMENT in three dimensions. Furthermore, our reduction is "efficient", we obtain an interesting explicit relationship between the approximability behaviour of MIN VERTEX COVER and that of the 2-dimensional MIN RANGE ASSIGNMENT problem.

---

[1] A graph is *cubic* when every node has degree 3.

In fact, we can state that if MIN VERTEX COVER on cubic graphs is not $\overline{\rho}$-approximable then MIN RANGE ASSIGNMENT in three dimensions is not $\frac{\overline{\rho}+4}{5}$-approximable.

Kirousis *et al*'s reduction works only in the 3-dimensional case. In fact, the reduction starts from a planar orthogonal drawing of a (planar) cubic graph G and replace each edge by a *gadget* of stations drawn in the 3-dimensional space that "simulates" the connection between the two adjacent nodes. In order to preserve pairwise "independence" of the drawing of gadgets, their reduction strongly uses the third dimension left "free" by the planar drawing of G. The complexity of the MIN RANGE ASSIGNMENT problem in two dimensions is thus left as an open question: Kirousis *et al* in fact conjectured the NP-hardness of this restriction.

It turns out that the gadget construction used in our approximation-preserving reduction for the 3-dimensional case can be suitably adapted in order to derive a polynomial-time reduction from MIN VERTEX COVER on planar cubic graphs to the 2-dimensional MIN RANGE ASSIGNMENT problem thus proving their conjecture. The following table summarizes the results obtained in this paper.

Problem version	Previous results	Our results
1-Dim. Case	in P[9]	-
2-Dim. Case	in APX[9]	NP-complete
3-Dim. Case	NP-complete, in APX[9]	APX-complete

Organization of the Paper. In Section 2, we give the preliminary definitions. For the sake of convenience, we first provide the reduction proving the NP-completeness result for the 2-dimensional case in Section 3. Then, in Section 4, we show the APX-completeness of MIN RANGE ASSIGNMENT in the 3-dimensional case. Finally, some open problems are discussed in Section 5. The proofs of the technical lemmas will be given in the full version of the paper.

2 Preliminaries

Let $S = \{s_1, \ldots, s_n\}$ be a set of n points (representing stations) of an Euclidean space $\mathcal{E}$ with distance function $d : \mathcal{E}^2 \to \mathcal{R}^+$, where $\mathcal{R}^+$ denotes the set of non negative reals. A *range assignment* for S is a function $r : S \to \mathcal{R}^+$. The *cost* $\text{cost}(r)$ of r is defined as

$$\text{cost}(r) = \sum_{i=1}^{n} (r(s_i))^2 .$$

Observe that we have set the distance-power gradient β to 2 (see Eq. 1), however our results can be easily extended to any constant $\beta > 1$.

The *communication graph* of a range assignment r is the directed graph $G_r(S, E)$ where $(s_i, s_j) \in E$ if and only if $r(s_i) \geq d(s_i, s_j)$. We say that an assignment r for S is *feasible* if the corresponding communication graph is strongly connected. Given a set S of n points in an Euclidean space, the MIN RANGE ASSIGNMENT problem consists of finding a feasible range assignment r_{min} for S of minimum cost. With 2D MIN RANGE ASSIGNMENT (respectively, 3D MIN RANGE ASSIGNMENT) we denote the MIN RANGE ASSIGNMENT problem in which the points are placed on $\mathcal{R}^2$ (respectively, on $\mathcal{R}^3$).

The MIN VERTEX COVER problem is to find a subset K of the set of vertices of V of a graph $G(V, E)$ such that K contains at least one endpoint of any edge in E and $|K|$ is as small as possible. MIN VERTEX COVER is known to be NP-hard even when restricted to planar cubic graphs [7]. Moreover, it is known to be APX-complete when restricted to cubic graphs [11, 1]. It follows that a constant $\bar{\rho} > 1$ exists such that MIN VERTEX COVER restricted to cubic graphs is not $\bar{\rho}$-approximable unless P = NP.

3 2D MIN RANGE ASSIGNMENT is NP-hard

We will show a polynomial-time reduction from MIN VERTEX COVER restricted to planar, cubic graphs to 2D MIN RANGE ASSIGNMENT.

Given a planar, cubic graph $G(V, E)$, it is always possible to derive a planar orthogonal drawing of G in which each edge is represented by a polyline having only one bend [15, 8]. We can then replace every edge whose drawing has one bend with a chain of three edges (we add two new vertices) in such a way that all edges are represented by straightline segments. The obtained drawing will be denoted by $D(G)$. It is easy to verify that, if $2h$ is the number of vertices added by this operation, then G has a vertex cover of size k if and only if $D(G)$ has a vertex cover[2] of size $k + h$. As we will see in Subsection 3.2, further vertices will be added in $D(G)$ still preserving the above relationship between the vertex covers of G and those of $D(G)$.

Our goal is to replace each edge (and thus both of its vertices) of $D(G)$ with a gadget of points (stations) in the Euclidean space $\mathcal{R}^2$ in order to construct an instance of the 2D MIN RANGE ASSIGNMENT problem and then show that this construction is a polynomial-time reduction. In the next subsection we provide the key properties of these gadgets and the reduction to 2D MIN RANGE ASSIGNMENT that relies on such properties. The formal construction of the 2-dimensional gadgets is instead given in Subsection 3.2.

3.1 The Properties of the 2-Dimensional Gadgets and the Reduction

The type of gadget used to replace one edge of $D(G)$ depends on the local "situation" that occurs in the drawing (for example it depends on the degree of its endpoints). However, we can state the properties that characterize any of these gadgets.

---

[2] In what follows, we will improperly $D(G)$ to denote both the drawing and the graph it represents.

Definition 1 (Gadget Properties). *Let $\delta, \delta', \epsilon \geq 0$ such that $\delta + \epsilon > \delta'$ and $\alpha > 1$ (a suitable choice of such parameters will be given later). For any edge (a, b) the corresponding gadget g_{ab} contains the sets of points $X_{ab} = \{x_1, \ldots, x_{l_1}\}$, $Y_{ab} = \{y_{ab}, y_{ba}\}$, $Z_{ab} = \{z_1, \ldots, z_{l_2}\}$ and $V_{ab} = \{a, b\}$, where l_1 and l_2 depend on the length of the drawing of (a, b). These sets of points are drawn in $\mathcal{R}^2$ so that the following properties hold:*

1. $d(a, y_{ab}) = d(b, y_{ba}) = \delta + \epsilon$.
2. X_{ab} *is a chain of points drawn so that $d(a, x_1) = \delta$ and $d(b, x_{l_1}) = \delta$. Furthermore, for any $i = 1, \ldots, l_1 - 1$, $d(x_i, x_{i+1}) = \delta$ and, for any $i \neq j$, $d(x_i, x_j) \geq \delta$.*
3. Z_{ab} *is a chain of points drawn so that $d(y_{ab}, z_1) = d(y_{ba}, z_{l_2}) = \delta'$. Furthermore, for any $i = 1, \ldots, l_2 - 1$, $d(z_i, z_{i+1}) = \delta'$ and, for any $i \neq j$, $d(z_i, z_j) \geq \delta'$.*
4. *For any $x_i \in X_{ab}$ and $z_j \in Z_{ab}$, $d(x_i, z_j) > \delta + \epsilon$. Furthermore, for any $i = 1, \ldots, l_1$, $d(x_i, y_{ab}) \geq \delta + \epsilon$ and $d(x_i, y_{ba}) \geq \delta + \epsilon$.*
5. *Given any two different gadgets g_{ab} and g_{cd}, for any $v \in g_{ab} \setminus g_{cd}$ and $w \in g_{cd} \setminus g_{ab}$, we have that $d(v, w) \geq \delta$ and if $v \notin V_{ab} \cup X_{ab}$ or $w \notin V_{cd} \cup X_{cd}$ then $d(v, w) \geq \alpha\delta$.*

From the above definition, it turns out that the gadgets consist of two components whose relative distance is $\delta + \epsilon$: the VX-*component* consisting of the "chain" of points in $X_{ab} \cup V_{ab}$, and the YZ-*component* consisting of the chain of points in $Y_{ab} \cup Z_{ab}$.

Let $S(G)$ be the set of points obtained by replacing each edge of $D(G)$ by one gadget having the properties described above.

Note 1. Let r^{min} be the range assignment of $S(G)$ in which every point in VX and in YZ have range δ and δ', respectively (notice that this assignment is not feasible). The corresponding communication graph consists of $m + 1$ strongly connected components, where m is the number of edges: the YZ-components of the m gadgets and the union $\mathcal{U}$ of all the VX-components of the gadgets. It thus follows that, in order to achieve a feasible assignment, we must define the "bridge-point" between $\mathcal{U}$ and every YZ-component.

The above note leads us to define the following *canonical* (feasible) solutions for $S(G)$.

Definition 2 (Canonical Solutions for $S(G)$). *A range assignment r for $S(G)$ is canonical if, for every gadget g_{ab} of $S(G)$, the following properties hold.*

1. *Either $r(y_{ab}) = \delta + \epsilon$ and $r(y_{ba}) = \delta'$ (so, y_{ab} is a radio "bridge" from the YZ-component to the VX one) or vice versa.*
2. *For every $v \in \{a, b\}$, either $r(v) = \delta$ or $r(v) = \delta + \epsilon$. Furthermore, there exists $v \in \{a, b\}$ such that $r(v) = \delta + \epsilon$ (so, v is a radio "bridge" from the VX-component to the YZ one).*
3. *For every $x \in X_{ab}$, $r(x) = \delta$.*

4. *For every $z \in Z_{ab}$, $r(z) = \delta'$.*

We observe that any canonical assignment is feasible.

Lemma 1. *Let us consider the construction $S(G)$ in which α, δ and ϵ are three positive constants such that*

$$\alpha^2 \delta^2 > (m-1)[(\delta + \epsilon)^2 - \delta^2] + (\delta + \epsilon)^2. \tag{2}$$

Then, for any feasible range assignment r for $S(G)$, there is a canonical range assignment r^c such that $\mathrm{cost}(r^c) \leq \mathrm{cost}(r)$.

We now assume that $S(G)$ satisfies the hypothesis of Lemma 1.

Lemma 2. *Given any planar cubic graph $G(V,E)$, assume that it is possible to construct the set of points $S(G)$ in the plane in time polynomial in the size of G. Then* MIN VERTEX COVER *is polynomial-time reducible to* 2D MIN RANGE ASSIGNMENT.

3.2 The Construction of the 2-Dimensional Gadgets

This section is devoted to the construction of the 2-dimensional gadgets that allow us to obtain the point set $S(G)$ corresponding to a given planar cubic graph G.

Definition 3 (Construction of $S(G)$). *Let $G(V,E)$ be a planar cubic graph, then the set of points $S(G)$ is constructed as follows:*

1. *Construct a planar orthogonal grid drawing of G with at most one bend per edge.*
2. *For any edge represented by a polyline with one bend, add two new vertices so that any edge is represented with a straight line segment.*
3. *Starting from the obtained graph $D(G)$, replace its edges with the gadgets satisfying Definition 1 and Eq. 2. This step may require further vertices to be added to $D(G)$ while preserving the relationship between the vertex cover solutions.*

Let us first observe that G has a vertex cover of size k if and only if $D(G)$ has a vertex cover of size $k + h$, where $2h$ is the number of new vertices added in the last two steps. As we will see in the sequel h is polynomially bounded in the size of G. We can therefore consider the problem of finding a minimum vertex cover for $D(G)$.

During the third step of the construction, it is required to preserve Property 5 of Definition 1, i.e., points from different gadgets are required to be within distance at least $\alpha\delta$. Informally speaking, the main technical problem is drawing the Z-chains corresponding to incident edges so that the properties of Definition 1 hold. To this aim, we adopt a set of suitable construction rules that are described in the full version of the paper.

In the sequel the term $S(G)$ will denote the network drawn from $D(G)$ according to the construction rules mentioned above. Let L_{min} be the minimum distance between any two V-points in $D(G)$. Then, any two V-points of the obtained network $S(G)$ have distance not smaller than L_{min}.

Lemma 3. *Let* $\delta = L_{min}/6$. *Then, an* $\epsilon > 0$ *exists for which the corresponding network* $S(G)$ *satisfies Eq. 2, i.e.,*

$$\alpha^2\delta^2 > (m-1)[(\delta+\epsilon)^2 - \delta^2] + (\delta+\epsilon)^2$$

where

$$\alpha = \frac{1+\sqrt{2}}{2} .$$

Combining Lemma 2 with Lemma 3 we obtain the following result.

Theorem 1. 2D MIN RANGE ASSIGNMENT *is* NP-*hard.*

4   3D MIN RANGE ASSIGNMENT is APX-complete

The APX-completeness of 3D MIN RANGE ASSIGNMENT is achieved by showing an approximation-preserving reduction from MIN VERTEX COVER restricted to cubic graphs, a restriction of MIN VERTEX COVER which is known to be APX-complete [11, 1]. The approximation-preserving reduction follows the same idea of the reduction shown in the previous section and thus requires a suitable 3-dimensional drawing of a cubic graph.

Theorem 2. *[5] There is a polynomial-time algorithm that, given any cubic graph* $G(V, E)$, *returns a 3-dimensional orthogonal drawing* $D(G)$ *of* G *such that:*

- *Every edge is represented as a polyline with at most three bends.*
- *Vertices are represented as points with integer coordinates, thus the minimum distance* L_{min} *between two vertices is at least 1.*
- *The maximum length* L_{max} *of an edge in* $D(G)$ *is polynomially bounded in* $m = |E|$.

4.1   The 3-Dimensional Gadgets

In what follows, we assume to have at hand the 3-dimensional, orthogonal drawing $D(G)$ of a cubic graph G that satisfies the properties of Theorem 2. Then the approximation-preserving reduction replaces each edge of $D(G)$ with a 3-dimensional gadget of stations having the following properties.

Definition 4 (Properties of 3-Dimensional Gadgets).
    Let l *and* ϵ *be positive constants (a suitable choice of such parameters will be given later). For any edge* (a, b) *the corresponding gadget contains the sets of points* $X_{ab} = \{x_1, \ldots, x_{l_1}\}$, $Y_{ab} = \{y_{ab}, y_{ba}\}$, $Z_{ab} = \{z_1, \ldots, z_{l_2}\}$ *and* $V_{ab} = \{a, b\}$, *where* l_1 *and* l_2 *depend on the distance* $d(a, b)$ *and* $d(y_{ab}, y_{ba})$, *respectively. The above set of points is drawn in such a way that the following properties hold:*

1. $d(a, y_{ab}) = d(b, y_{ba}) = l$.
2. X_{ab} and Z_{ab} are two chains of points drawn so that $d(a, x_1) = d(b, x_l) = \epsilon$ and $d(y_{ab}, z_1) = d(y_{ba}, z_m) = \epsilon$, respectively. Furthermore, for any $i = 1, \ldots, l-1$, $d(x_i, x_{i+1}) = \epsilon$ and for any $j = 1, \ldots, m-1$ $d(z_j, z_{j+1}) = \epsilon$.
3. For any $x_i \in X_{ab}$ and $z_j \in Z_{ab}$, $d(x_i, z_j) > l$. Furthermore $d(x_i, y_a b) \geq l$ and $d(x_i, y_b a) \geq l$.
4. Given any two different gadgets g_1 and g_2, for any $v \in g_1$ and $w \in g_2$ with $u \neq w$ of different type (for example, if u is a X-point then w is either a Y-point or a Z-point), we have that $d(v, w) > l$. Moreover, the minimum distance between the YZ-component[3] of g_1 and the YZ-component of g_2 is $2l$.
5. Given any two non adjacent gadgets g_1 and g_2, for any $v \in g_1$ and $w \in g_2$, $d(v, w) \geq L_{min}/2$.

Let l and ϵ two positive reals such that $l \leq L_{min}$ (this assumption guarantees Properties 4 and 5 of Definition 4) and $\epsilon < l$. The construction of the 3-dimensional gadgets can be obtained by adopting the same method of the 2-dimensional case. The technical differences will be discussed in the full version of the paper.

We emphasize that the 3-dimensional gadgets have two further properties which will be strongly used to achieve an approximation-preserving reduction (see Theorem 3).

Lemma 4. 1). *The set of V-points of $S(G)$ is the set of vertices of G, i.e. no new vertices will be added with respect to those of $D(G)$.*

2). *It is possible to make the overall range cost of both X and Z points of any gadget arbitrarily small by augmenting the number of equally spaced stations in these two chains. More formally, if L is the length of the polyline representing an edge (a, b) in $D(G)$ and k is the number of points in the X (or Z) component then the overall power needed for the X component is*

$$(k + 2)\left(\frac{L}{k+1}\right)^2 \qquad (3)$$

So, by increasing k, we can make the above value smaller than any fixed positive constant.

4.2 The Approximation-Preserving Reduction

Definition 5 (Canonical Solutions for $S(G)$.). *A range assignment r for $S(G)$ is canonical if, for every gadget g_{ab} of $S(G)$, the following properties hold.*

1. *Either $r(y_{ab}) = l$ and $r(y_{ba}) = \epsilon$ (so, y_{ab} is the radio "bridge" from the YZ-component to the VX one) or vice versa.*

---

[3] Similarly to the 2-dimensional case, the sets of points $V_{ab} \cup X_{ab}$ and $Y_{ab} \cup Z_{ab}$ will be denoted as VX-component and YZ-component, respectively.

2. For every $v \in \{a, b\}$, either $r(v) = \epsilon$ or $r(v) = l$. Furthermore, there exists $v \in \{a, b\}$ such that $r(v) = l$ (so, v is a radio "bridge" from the VX-component to the YZ one).

3. For every $x \in X_{ab}$, $r(x) = \epsilon$.

4. For every $z \in Z_{ab}$, $r(z) = \epsilon$.

Lemma 5. For any graph G, let us consider the construction $S(G)$ in which l is a positive real that satisfies the following inequality

$$l^2 < \frac{L_{min}^2}{m} . \tag{4}$$

Then, for any feasible range assignment r of $S(G)$, there is a canonical range assignment r^c such that $\mathrm{cost}(r^c) \leq \mathrm{cost}(r)$.

Informally speaking, the presence of the third dimension in placing the gadgets allows to keep a polynomially large gap between the value of l (i.e. the minimum distance between the VX component and the YX one of a gadget) and that of ϵ (i.e. the minimum distance between points in the same chain component). This gap yields the significant weight of each bridge-point of type V in a canonical solution and it will be a key ingredient in proving the next theorem. Notice also that this gap cannot be smaller than a fixed positive constant in the 2-dimensional reduction shown in the previous section.

Theorem 3. 3D MIN RANGE ASSIGNMENT *is* APX-*complete.*

Proof. The outline of the proof is the following. We assume that we have at hand a polynomial-time ρ-approximation algorithm $\mathcal{A}$ for 3D MIN RANGE ASSIGNMENT. Then, we show a polynomial-time method that transforms $\mathcal{A}$ into a ρ'-approximation algorithm for MIN VERTEX COVER on cubic graphs with $\rho' \leq 5\rho - 4$. Since a constant $\overline{\rho} > 1$ exists such that MIN VERTEX COVER restricted to cubic graphs is not $\overline{\rho}$-approximable unless P = NP [11, 1], the theorem follows.

Assume that a 3-degree graph $G(V, E)$ is given. Then, from the 3-dimensional orthogonal drawing $D(G)$ of G, we construct the radio network $S(G)$ by replacing each edge in $D(G)$ with one 3-dimensional gadget whose properties are described in Definition 4. It is possible to prove (see the full version of the paper) that these gadgets can be constructed and correctly placed in the 3-dimensional space in polynomial time. We also assume that the parameter l of $S(G)$ satisfies Inequality 4. Using the same arguments in the proof of Lemma 2, we can show that any vertex cover $K \subseteq V$ of G yields a canonical assignment r_K whose cost is

$$\mathrm{cost}(r_K) = \kappa l^2 + m l^2 + \overline{\epsilon_K}, \tag{5}$$

where $\kappa = |K|$ and $\overline{\epsilon_K}$ is the overall cost due to all points v that have range ϵ. Since each gadget of $S(G)$ has at most $4L_{max}/\epsilon$ points, it holds that

$$\overline{\epsilon_K} \leq 4 m L_{max} \epsilon . \tag{6}$$

On the other hand, from Lemma 5, we can consider only canonical solutions of $S(G)$. Thus, given a canonical solution r^c, we can consider the subset K of V-points whose range is l. It is easy to verify that K is a vertex cover of G. Furthermore, the cost of r^c can be written as follows

$$\text{cost}(r^c) = |K|l^2 + ml^2 + \overline{\epsilon_K} \ .$$

Let K^{opt} be an optimum vertex cover for G, from the above equation we have that the optimum range assignment cost opt_r can be written as

$$\text{opt}_r = |K^{opt}|l^2 + ml^2 + \overline{\epsilon_{K^{opt}}} \tag{7}$$

Since G has maximum degree 3 then $|K^{opt}| \geq m/3$; so, the above equation implies that

$$\text{opt}_r \leq 4|K^{opt}|l^2 + \overline{\epsilon_{K^{opt}}}. \tag{8}$$

Let us now consider a ρ-approximation algorithm for 3D MIN RANGE ASSIGN-MENT such that given $S(G)$ in input it returns a solution r^{apx} whose cost is less than $\rho \cdot \text{opt}_r$. From Lemma 5, we can assume that r^{apx} is canonical. It thus follows that the cost $\text{cost}(r^{apx})$ can be written as

$$\text{cost}(r^{apx}) = |K^{apx}|l^2 + ml^2 + \overline{\epsilon_{K^{apx}}} \ .$$

From Eq.s 7 and 8 we obtain

$$\frac{\text{cost}(r^{apx})}{\text{opt}_r} = \frac{\text{cost}(r^{apx}) - \text{opt}_r}{\text{opt}_r} + 1 \tag{9}$$

$$= \frac{|K^{apx}|l^2 + ml^2 + \overline{\epsilon_{K^{apx}}} - |K^{opt}|l^2 - ml^2 - \overline{\epsilon_{K^{opt}}}}{\text{opt}_r} + 1 \tag{10}$$

$$\geq \frac{|K^{apx}|l^2 - |K^{opt}|l^2}{4|K^{opt}|l^2 + \overline{\epsilon_{K^{opt}}}} + 1 \tag{11}$$

Note that we can make $\overline{\epsilon_{K^{opt}}}$ arbitrarily small (independently from l) by reducing the parameter ϵ in the construction of $S(G)$: this is in turn obtained by increasing the number of X and Z points in the gadgets (see Lemma 4).

From Eq. 6, from the fact that L_{max} is polynomially bounded in the size of G and from the fact that l and L_{max} are polynomially related, we can ensure that $\overline{\epsilon_{K^{opt}}} \leq l^2$ by adding a polynomial number of points (see again Lemma 4). So, from Eq. 9 we obtain

$$\frac{\text{cost}(r^{apx})}{\text{opt}_r} \geq \frac{|K^{apx}|l^2 - |K^{opt}|l^2}{4|K^{opt}|l^2 + \overline{\epsilon_{K^{opt}}}} + 1 \geq \frac{|K^{apx}|}{5|K^{opt}|} + \frac{4}{5} \ .$$

Finally, it follows that the approximation ratio for MIN VERTEX COVER is bounded by

$$\frac{|K^{apx}|}{|K^{opt}|} \leq \frac{5\text{cost}(r^{apx})}{\text{opt}_r} - 4.$$

Q.E.D.

5 Open Problems

The first open problem is whether 2D MIN RANGE ASSIGNMENT is APX-complete or admits a PTAS. Notice that a possible APX-completeness reduction should be from a different problem, since MIN VERTEX COVER restricted to planar graphs is in PTAS. As regard the 3D MIN RANGE ASSIGNMENT problem it could be interesting to reduce the large gap between the factor 2 of the approximation algorithm and the inaproximability bound than can be derived by combining our reduction with the approximability lower bound of MIN VERTEX COVER on cubic graphs. As far as we know, there is no known significant explicit lower bound for the latter problem (an explicit 1.0029 lower bound for MIN VERTEX COVER on degree 5 graphs is given in [4] that – if it could be extended to cubic graphs and then combined with our reduction – would give a lower bound for 3D MIN RANGE ASSIGNMENT of 1.00059).

A crucial characteristic of the optimal solutions for the 3D MIN RANGE ASSIGNMENT instances given by our reduction is that stations that communicate directly have relative distance either l or ϵ, where $l >> \epsilon$. It could be interesting to consider instances in which the above situation does not occur. Notice that this is the case of the 2D MIN RANGE ASSIGNMENT instances of our reduction. Thus, the problem on such restricted instances remains NP-hard. However, it is an open problem whether a better approximation factor or even a PTAS can be obtained.

Another interesting aspect concerns the maximum number of hops required by any two stations to communicate. This corresponds to the diameter h of the communication graph. Our constructions yield solutions whose communication graph has unbounded (i.e. linear in the number of stations) diameter. So, the complexity of MIN RANGE ASSIGNMENT with bounded diameter remains open also in the 1-dimensional case. A special case where stations are placed at uniform distance on a line and either h is constant or $h \in O(\log n)$ has been solved in [9].

References

1. P. Alimonti and V. Kann. Hardness of approximating problems on cubic graphs. *Proc. 3rd Italian Conf. on Algorithms and Complexity*, LNCS, Springer-Verlag(1203):288–298, 1997.
2. E. Arikan. Some complexity results about packet radio networks. *IEEE Transactions on Information Theory*, IT-30:456–461, 1984.
3. B.S. Baker. Approximation algorithms for np-complete problems on planar graphs. *Journal of ACM*, 41:153–180, 1994.
4. P. Berman and M. Karpinski. On some tighter inapproximability results. *Electronic Colloquium on Computational Complexity*, (29), 1998.
5. P. Eades, A. Symvonis, and S. Whitesides. Two algorithms for three dimensional orthogonal graph drawing. *Graph Drawing'96, LNCS*, (1190):139–154, 1996.
6. A. Ephemides and T. Truong. Scheduling broadcast in multihop radio networks. *IEEE Transactions on Communications*, 30:456–461, 1990.
7. M.R. Garey and D.S. Johnson. *Computers and Intractability - A Guide to the Theory of NP-Completness*. Freeman and Co., New York, 1979.

8. G. Kant. Drawing planar graphs using the canonical ordering. *Algorithmica - Special Issue on Graph Drawing*, (16):4–32, 1996. (Extended Abstract in 33-th IEEE FOCS (1992)).

9. L. M. Kirousis, E. Kranakis, D. Krizanc, and A. Pelc. Power consumption in packet radio networks. *14th Annual Symposium on Theoretical Aspects of Computer Science (STACS 97), LNCS.*

10. K. Pahlavan and A. Levesque. *Wireless Information Networks.* Wiley-Interscince, New York, 1995.

11. C. H. Papadimitriou and M. Yannakakis. Optimization, approximation, and complexity classes. *J. Comput. System Science*, 43:425–440, 1991.

12. C.H. Papadimitriou. *Computational Complexity.* Addison-Wesley Publishing Company, Inc., 1994.

13. S. Ramanathan and E. Lloyd. Scheduling boradcasts in multi-hop radio networks. *IEEE/ACM Transactions on Networking*, 1:166–172, 1993.

14. R. Ramaswami and K. Parhi. Distributed scheduling of broadcasts in radio network. *INFOCOM*, pages 497–504, 1989.

15. L. Valiant. Universality considerations in vlsi circuits. *IEEE Transactions on Computers*, C-30:135–140, 1981.

A New Approximation Algorithm for the Demand Routing and Slotting Problem with Unit Demands on Rings

Christine T. Cheng

Department of Mathematical Sciences
Johns Hopkins University
Baltimore, MD 21218
cheng@mts.jhu.edu

Abstract. The *demand routing and slotting problem* on unit demands (unit-DRSP) arises from constructing a SONET ring to minimize cost. Given a set of unit demands on an n-node ring, each demand must be routed clockwise or counterclockwise and assigned a slot so that no two routes that overlap occupy the same slot. The objective is to minimize the total number of slots used.

It is well known that unit-DRSP is NP-complete. The best approximation algorithm guarantees a solution to within twice of optimality. In the special case when the optimal solution uses many colors, a recent algorithm by Kumar [12] beats the approximation factor of 2. A demand of unit-DRSP can be viewed as a chord on the ring whose endpoints correspond to the source and destination of the demand. Let w denote the size of the largest set of demand chords that pairwise intersect in the interior of the ring. We first present an algorithm that achieves an approximation factor of $2 - 2/(w+1)$ in an n-node network. We then show how to combine our algorithm with Kumar's to achieve a hybrid algorithm with an an approximation factor of $(2 - \max\{4/n, 1/(50 \log n)\})$.
Keywords: DRSP, slotting, SONET rings, bandwidth allocation problem, WDM networks

1 Introduction

A main issue that arose in optical networks because of the concentration of large volume of traffic into small amounts of nodes and links is the *survivability* of the networks. That is, when a link or node of a network breaks down, the network must be capable of rerouting all traffic as quickly as possible. This led to the development of the technology called *synchronous optical networks* (SONETs). At present, it is one of the dominant technological standards used in the United States [18].

Among the most popular configurations of SONETs are rings. In a SONET ring, nodes are connected by links made of optical fibers. Each link in the ring has the same capacity K and is divided into K *slots*, labeled from 1 to K, where each slot has size equal to a unit of ring capacity. To transmit a *unit* demand between two nodes, a route, whether clockwise or counterclockwise, is chosen and is assigned a slot number. The slot number indicates which slot of the links along the route are used to transmit the demand, as unit demands must stay in the same slot throughout transmission. To transmit a demand of size d units between two pairs of nodes, a route must again be chosen and d slot numbers assigned for the route. Once a slot in a link is assigned to a demand, it is occupied. All other demands that go through the link cannot use the same slot. We note that assigning slots to routes is equivalent to a assigning colors to the routes so that no two overlapping routes are assigned the same color(s).

The cost of constructing a SONET ring is an increasing function of the capacity of the ring. Thus, before purchasing or constructing a SONET ring, it is important to determine the minimum slots needed to transmit all the demands in the network. This is known as the *demand routing and slotting problem* (DRSP) in rings. (In practice, the goal is to satisfy the *expected* demands of the network).

In [4], Carpenter, et al. showed that DRSP is NP-complete. They presented several approximation algorithms whose solutions are within a factor of 2 of the optimal. An important subcase of the problem is when all the demands have unit size. Sometimes, demands *can* be split, but only at integral values, and can thus be regarded as a multiplicity of unit demands [18]. Surprisingly, the complexity of the problem remains the same [5]. The best approximation algorithm (given first by Raghavan and Upfal in [16]) is also not any better. We shall call this subcase of DRSP as *unit-DRSP*.

Improving the approximation factor of 2 for unit-DRSP has, until now, seemed difficult. In the special case that the optimal solution uses many slots, a recent algorithm by Kumar [12] beats the factor of 2. By formulating the problem as a $0-1$ integer program, solving the relaxation and doing randomized rounding, Kumar was able to achieve a solution that used at most $(1.68 + o(1) + 2\sqrt{2\log n/z^*})z^*$ slots where z^* is the objective value of the optimal fractional solution and n is the size of the ring network. When $z^* > 81\log n$, the algorithm is guaranteed to perform better than the 2-approximation algorithm of Raghavan and Upfal.

Our Result. A demand of unit-DRSP can be viewed as a chord on the ring whose endpoints correspond to the source and destination of the demand. Let w denote the size of the largest set of demand chords that pairwise intersect in the interior of the ring. In this paper, we first present an algorithm for unit-DRSP that achieves an approximation factor of $2-2/(w+1)$ in an n-node network. This immediately implies that we achieve an approximation factor of $2 - 4/(n + 1)$. We then show how to combine our algorithm with Kumar's to achieve a hybrid algorithm with an approximation factor of $2 - 1/(50\log n)$.

While the hybrid algorithm gives our best asymptotic bounds, we note that our original algorithm outperforms it when either n or the optimal solution for DRSP, OPT, is small since $w \le \min(n/2, OPT)$. For example, if the optimal solution for DRSP uses 4 colors, our original algorithm uses at most 6 colors, whereas the hybrid algorithm and the 2-approximation algorithm are guaranteed to use at most 8 colors. In addition, currently deployed SONET rings typically have at most 24 nodes [3]. In this case, our algorithm guarantees a solution that is at most $1.83 * OPT$.

Related Work. We remark that unit-DRSP has also been studied extensively in SONETs and WDM (wavelength division multiplexing) networks with different topologies, including trees, trees of rings and meshes [16, 5, 13–15, 2]. [1] surveys the most recent results in the area. While the earliest model of the unit-DRSP in WDM ring networks just reduces to the problem above [16, 12], another model assumes that the underlying ring network is *directed* and *symmetric*. That is, if edge $(i, i + 1) \in E(G)$, then so is $(i + 1, i) \in E(G)$. Thus, two demands that traverse a link of the network, but in opposite directions, can be assigned the same slot number since they use, technically, different edges of the network. Wilfong and Winkler considered this problem and showed that unit-DRSP remains NP-complete in this setting [19]. The best approximation algorithm known for this problem also has a factor of 2.

Content of the Paper. The remainder of the paper is organized as follows. In Section 2, we define the problem and give two lower bounds that are relevant to our method. We present a greedy algorithm in Section 3 and the main algorithm that will make use of this greedy algorithm repeatedly in Section 4.

2 Preliminaries

The unit-DRSP is defined on an n-node ring. We are given a set of demands I on a ring network. Each demand is a (*source*, *destination*) pair where the source and destination are nodes on the network. A *routing* for I is an assignment of either the clockwise (which, henceforth, we abbreviate as cw) or counterclockwise (cw) source-destination path to each demand. A *slotting* for a routing of I is equivalent to assigning colors to the paths so that no two overlapping paths are assigned the same color. (Paths P_1 and P_2 are said to *overlap* if they have at least one edge of the network in common.) Since a path on the ring network is also an arc on a circle, we shall use the terms "paths" and "arcs" interchangeably.

A fixed choice of one of the cw or ccw paths for each demand in I determines a set of circular arcs C; conversely, a set of circular arcs C is *derivable* from I if the arcs are obtained from routing all the demands in I. We let $\mathcal{D}(I)$ denote the collection of sets derivable from I. A solution of I consists of some $C \in \mathcal{D}(I)$ and a valid coloring of C. The optimal solution to unit-DRSP is the one that uses the fewest number of colors among all possible solutions.

Let C be a set of circular arcs. We say that a pair of paths in C forms a *conflicting pair* if the paths overlap and the union of the paths is the ring. If C is derivable from I then C is a *parallel routing* if C does not contain any conflicting pairs. We say that $A \subseteq C$ is an *independent set of arcs* if all the arcs in A do not overlap. Thus, arcs that are assigned the same color in a solution of unit-DRSP form an independent set.

Lower Bounds. We first establish some lower bounds on the number of colors required by an optimal solution in terms of properties of I and the routings in $\mathcal{D}(I)$. One way to view a demand in I is to consider it as a chord on the circle where the endpoints of the chord correspond to the source and destination of the demand. Suppose two demand chords intersect in the interior of the circle. Clearly, regardless of how the demands are routed, the paths of the two demands will overlap and must be assigned different colors. Thus, the size of the largest set of demand chords such that the demand chords pairwise intersect each other (in the interior of the circle) is a natural lower bound to the number of colors needed for an optimal solution of unit-DRSP.

Let G_I be the the graph whose vertices correspond to the demands in I and whose edges correspond to pairs of demand chords that overlap. The parameter of interest to us is the clique number of G_I, $\omega(G_I)$. By the above discussion, $\omega(G_I)$ is, clearly, a lower bound on the optimal number of colors for the unit-DRSP. Gavril showed how to find $\omega(G_I)$ in polynomial time [8].

Let e be an edge on the ring. The set of circular arcs C induces a load L_e on e where L_e is the number of arcs in C that contain e. The largest load induced by C on an edge is the *ringload* of the network, which we denote as L_C. It is not difficult to see that if we wish to assign colors to each arc in C so that overlapping arcs are assigned different colors, then at least L_C colors must be used.

Another lower bound for unit-DRSP can be obtained by modifying the objective function to minimizing the ringload of the network. The ringloading problem defined on I asks for C^* such that $L_{C^*} = \min_{C \in \mathcal{D}(I)} L_C$. Since an optimal solution to unit-DRSP must have a ringload of at least L_{C^*}, a coloring of the optimal solution uses at least L_{C^*} colors.

The ringloading problem can be solved in polynomial time [6]. In [18], Schrijver, et al., showed how to find an optimal solution that is also a parallel routing. We have, thus, shown the following:

Proposition 1. *The optimal solution to unit-DRSP uses at least* $\max\{\omega(G_I), L(C^*)\}$ *colors where* C^* *is an optimal solution to the ringloading problem on* I *and is also a parallel routing.*

Recall that a solution to unit-DRSP consists of a routing $C \in \mathcal{D}(I)$ and a valid coloring of C. In the 2-approximation algorithm of unit-DRSP, C^* is the chosen routing. The arcs in C^* are colored as follows. Let $S_e \subseteq C^*$ be the set of arcs that pass through edge e. All the arcs in S_e are assigned different colors since they overlap. On the other hand, the set of arcs that do not pass through e, $C^* \backslash S_e$ form what is called an *interval set* since the arcs can be considered as intervals on a line. It is known that such a set of arcs can be colored optimally in linear time where the number of colors used is $L_{C^* \backslash S_e}$ [9]. Thus, the total number of colors used is at most $|S_e| + L_{C^* \backslash S_e} \leq 2L_{C^*}$.

The 2-approximation algorithm above is an example of one basic approach to solving the unit-DRSP: first, route all demands and then color the arcs. Unfortunately, coloring arcs on a ring optimally is very hard. Our algorithm also initially routes all demands. What differentiates our algorithm from this basic approach is that whenever it detects sub-optimality as it colors, it reroutes some of the demands. This crucial step allows us to show that the approximation factor of our algorithm is better than 2.

3 Coloring Circular Arcs

We examine the problem of coloring a fixed $C \in D(I)$. Recall a valid coloring for C assigns a color to each arc in C so that no two overlapping arcs have the same color. The problem of finding a valid coloring for C that uses as few colors as possible has been studied extensively. Garey, et al. showed that the problem is NP-complete [7]. Tucker gave a simple 2-approximation that used the ringload of C as a lower bound [17]. Hsu and Shih was able to reduce the approximation ratio to 5/3 [10]. More recently, Kumar gave an algorithm that achieves an approximation ratio of $1 + 1/e + o(1)$ for instances where the optimal solution uses a relatively large number of colors [11].

We introduce the following natural greedy algorithm for coloring the arcs of C due to Tucker [17] that will be used repeatedly as a subroutine in our new algorithm. Let $cmax = 1$. Start with an arbitrary arc a in C and color it with $cmax$. We proceed to the next arc b whose ccw endpoint is closest to the cw endpoint of a and color it with $cmax$ unless a and b overlap, in which case we color b with $cmax + 1$. We continue this process where the next arc chosen is the uncolored arc whose ccw endpoint is closest to the cw endpoint of the current arc. We always color the next arc with $cmax$ whenever possible. Otherwise, we use $cmax + 1$ and increment $cmax$.

Let us denote the ccw endpoint of arc a as vertex p. We say that the algorithm has completed k *rounds* if the algorithm has traversed p $k + 1$ times. We let R_k be the set of arcs colored during the kth round. If an arc's ccw endpoint was traversed during the kth round but its cw endpoint during the $k + 1$ round, we consider the arc to be in R_k.

4 Description of Algorithm

From [18], we know that there exists C^*, an optimal solution to the ringloading problem such that C^* is also a parallel routing, i.e., C^* contains no conflicting pairs. Furthermore, C^* can be found in polynomial time. We use C^* as the initial routing for our algorithm. Some demands may be rerouted later; i.e. some demands routed in the cw direction may be rerouted in the ccw direction or vice versa. Arcs in C^* will be colored in phases. At each phase, the greedy algorithm in Section 3 will be used for an arbitrary number of rounds. A phase ends only when the number of colors used is less than twice the number of completed rounds of coloring. This requirement allows us to prove later that the ratio of the total number of colors used by our algorithm to the optimal solution is bounded away from 2.

Let us first establish some properties when we apply our greedy algorithm to a set of arcs with no conflicting pairs.

Suppose we color arcs in C for r rounds using the greedy algorithm. Without loss of generality, we label the nodes in the network clockwise from 1 to n such that vertex 1 corresponds to the ccw endpoint of the first colored arc. Let C_i denote the set of arcs colored i and $c_{i,j}$ denote the jth arc colored i. The last arc colored i will also be denoted c_{i,n_i}. Let $l(c_{i,j})$ and $r(c_{i,j})$ denote the ccw ("left") and cw ("right") endpoints of $c_{i,j}$ respectively. Notice that because of the way the greedy algorithm assigns colors, arc $c_{i,1}$ must always overlap with $c_{i-1,1}$. (See Figure 1.) Here are some important properties about the endpoints of $c_{i-1,1}$ and $c_{i,1}$ which we will make use repeatedly. Properties 2 and 3 can be easily deduced from Property 1.

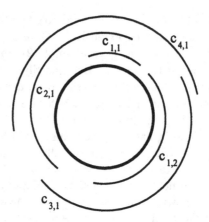

Fig. 1. This is an example of how the greedy algorithm labels the arcs if it started with $c_{1,1}$. Since the chords of $c_{1,1}$ and $c_{4,1}$ do not intersect, *COLOR_ARCS* will reroute $c_{4,1}$.

Property 1. If $c_{i-1,1}$ or $c_{i,1}$ does not contain vertex 1 except as an endpoint then $r(c_{i,1}) > l(c_{i-1,1})$.

Property 2. It is always the case that $r(c_{i-1,1}) \neq l(c_{i,1})$.

Property 3. The greedy algorithm uses at most $2r$ colors at the end of r rounds.

Surprisingly, if the greedy algorithm uses $2r$ colors in r rounds, we can deduce how the endpoints of some of the colored arcs are related to each other.

Lemma 1. *If $R_r \cap C_{2r} \neq \emptyset$, then the following set of inequalities are true.*

(i). $l(c_{1,*}) < r(c_{2,1}) < l(c_{3,1}) < \ldots < l(c_{2i-1,1}) < r(c_{2i,1}) < \ldots < r(c_{2r,1})$ *and*
(ii). $r(c_{1,*}) < l(c_{2,1}) < r(c_{3,1}) < \ldots < r(c_{2i-1,1}) < l(c_{2i,1}) < \ldots < l(c_{2r,1})$,

where $c_{1,*} \in C_1$ such that $l(c_{1,*}) < r(c_{2,1}) \leq l(c_{1,*+1})$.

Proof: First, suppose $r = 1$. If $R_1 \cap C_2 \neq \emptyset$ then from Property 3, $C_1 \cup \{c_{2,1}\} = R_1$. By definition, $1 \leq l(c_{1,*}) < r(c_{2,1})$ and since $r(c_{1,n_1}) \leq l(c_{2,1}) < n$, $r(c_{1,*}) \leq l(c_{2,1})$. Note, however, that $r(c_{1,*}) \neq l(c_{2,1})$ since equality would imply that $c_{1,*}$ and $c_{2,1}$ form a conflicting pair. Thus, the inequalities in (i) and (ii) are true when $r = 1$.

Suppose after i rounds $R_i \cap C_{2i} \neq \emptyset$ and our Lemma holds for $r \leq i - 1$. Clearly, all the arcs in $C_{2i-1} \cup \{c_{2i,1}\}$ were colored during the ith round. Hence,

$$1 < r(c_{2i-2,1}) < l(c_{2i-1,1}) < r(c_{2i-1,1}) < l(c_{2i,1}) < n. \tag{1}$$

From Property 1, we have

$$l(c_{2i-2,1}) < r(c_{2i-1,1}) \qquad l(c_{2i-1,1}) < r(c_{2i,1}). \tag{2}$$

Combining inequalities 1 and 2, these inequalities follow:

$$r(c_{2i-2,1}) < l(c_{2i-1,1}) < r(c_{2i,1}) \qquad l(c_{2i-2,1}) < r(c_{2i-1,1}) < l(c_{2i,1}).$$

The above inequalities together with our assumption that the inequalities in (i) and (ii) hold for $r \leq i-1$ show that our Lemma holds for $r \leq i$. By induction, the lemma holds in general. □

The next theorem tells us that whenever the greedy algorithm uses $2r$ colors in r rounds, the number of colors used is within 1 of optimal. Furthermore, if the number of colors used is not optimal, we can reroute some arc so that there is no need to use the extra color.

Theorem 1. *If $R_r \cap C_{2r} \neq \emptyset$ then the demand chords of $c_{2,1}, c_{3,1}, \ldots, c_{2r,1}$ form a K_{2r-1} in G_I. In addition, if the demand chords of $c_{1,*}, c_{2,1}, c_{3,1}, \ldots, c_{2r,1}$ do not form a K_{2r} in G_I then by rerouting $c_{2r,1}$ we can color the arcs with at most $2r-1$ colors.*

Proof: Suppose $R_r \cap C_{2r} \neq \emptyset$. From Lemma 1, we know that $c_{2,1}$ and $c_{2r,1}$ were the last arcs colored in their respective rounds and that $l(c_{2,1}) > r(c_{2,1})$, $l(c_{2r,1}) > r(c_{2r,1})$. If $r(c_{2r,1}) > l(c_{2,1})$, then $c_{2,1}$ and $c_{2r,1}$ form a conflicting pair. Thus, $r(c_{2r,1}) < l(c_{2,1})$. Combining this fact with the inequalities in (I) and (ii), we have

$$r(c_{2,1}) < \ldots < l(c_{2r-1,1}) < r(c_{2r,1}) < l(c_{2,1}) < \ldots < r(c_{2r-1,1}) < l(c_{2r,1}). \quad (3)$$

This implies that the demand chords of arcs $c_{2,1}, c_{3,1}, \ldots, c_{2r,1}$ pairwise intersect each other and consequently form a K_{2r-1} in G_I.

Suppose we also know that the demand chords associated with the arcs in the set $\{c_{1,*}, c_{2,1}, c_{3,1}, \ldots, c_{2r,1}\}$ do not form a K_{2r}, then clearly $r(c_{2r,1}) > r(c_{1,*})$. That is, the demand chords of $c_{1,*}$ and $c_{2r,1}$ do not intersect in the interior of the circle. Let us reroute $c_{2r,1}$ and denote the new arc as $\hat{c}_{2r,1}$. It is not difficult to see that $\{c_{1,1}, c_{1,2} \ldots c_{1,*}\} \cup \{\hat{c}_{2r,1}\}$, $C_1 \setminus \{c_{1,1}, c_{1,2} \ldots c_{1,*}\} \cup \{c_{2,1}\}$, $C_2 \setminus \{c_{2,1}\} \cup C_3$, $C_4, C_5, \ldots, C_{2r-1}$ are all independent sets. If we assign all arcs belonging to the same independent set the same colors then we have used exactly $2r-1$ colors. □

Theorem 2. *If the demand chords of $c_{1,*}, c_{2,1}, \ldots, c_{2r,1}$ form a K_{2r} in G_I but together with the demand chord of $c_{2r+1,1}$ do not form a K_{2r+1} in G_I then we can restart the greedy algorithm at $c_{1,*+1}$ so that after r rounds, only $2r-1$ colors are used.*

Proof: Suppose, instead of starting the coloring algorithm with $c_{1,1}$, we start with $c_{1,*+1}$ and continue for r rounds. Let us assume that $l(c_{2r+1,1}) \geq l(c_{1,*+1})$. It is not difficult to see that the algorithm will encounter the same set of arcs as if it started from $c_{1,1}$. After r rounds, since $l(c_{2r+1,1}) \geq l(c_{1,*+1})$, it would color the arcs in $C_1 \setminus \{c_{1,1}, c_{1,2} \ldots c_{1,*}\}$, $C_2, C_3, \ldots C_{2r-1}$ and some arcs of C_{2r} and none of the arcs in C_{2r+1}. Furthermore, it would assign arcs in $C_1 \setminus \{c_{1,1}, c_{1,2} \ldots c_{1,*}\}$ and C_2 the color 1 (since $l(c_{3,1}) < r(c_{1,*})$ which follows from the assumption that the chords of $c_{1,*}, c_{2,1}, \ldots, c_{2r,1}$ form a K_{2r}) and the arcs in C_i the color $i-1$ for $i = 3, \ldots, 2r$. Hence, at most $2r-1$ colors are used.

But suppose $l(c_{2r+1,1}) < l(c_{1,*+1})$. We note that $r(c_{2r,1}) < r(c_{1,*}) < l(c_{1,*+1})$ since the demand chords of $c_{1,*}, c_{2,1}, \ldots, c_{2r,1}$ form a K_{2r}. If $r(c_{2r,1}) < l(c_{2r+1,1}) <$

$r(c_{1,*})$ then $r(c_{2r+1,1})$ cannot lie between $l(c_{2r,1})$ and $l(c_{1,*})$ else the chords of $c_{1,*}, c_{2,1}, \ldots, c_{2r,1}$ together with $c_{2r+1,1}$ form a K_{2r+1} in G_I. Thus, $r(c_{2r+1,1}) \geq l(c_{1,*})$. This means, however, that $c_{1,*}$ and $c_{2r+1,1}$ form a conflicting pair, violating our original assumption that C^* is a parallel routing. If $r(c_{1,*}) \leq l(c_{2r+1,1}) < l(c_{1,*+1})$, the greedy algorithm would have chosen $c_{2r+1,1}$ as the next arc after $c_{1,*}$ to color 1. Since this was not the case, it must be that $l(c_{2r+1,1}) \geq l(c_{1,*+1})$.

$\square$

Unit-DRSP Algorithm(I**)**
Solve for C^* and $\omega(G_I)$.
$t \leftarrow 2\lceil \omega(G_I)/2 \rceil$;
$\mathcal{C} = \emptyset$.
while (no edge has load 0)
    $\mathcal{Q} = \emptyset$;
    $(\mathcal{Q}, C^*) \leftarrow COLOR\_ARCS(C^*, t)$;
    $\mathcal{C} \leftarrow \mathcal{C} \cup \mathcal{Q}$;
If some edge has load 0, color remaining arcs using algorithm for interval graphs.
return($\mathcal{C}$);

Fig. 2. Our new approximation algorithm for routing and slotting unit demands in rings.

From hereon, we shall let $t = 2\lceil \omega(G_I)/2 \rceil$.

Corollary 1. *If the demand chords of* $c_{1,*}, c_{2,1}, \ldots, c_{t,1}$ *form a* K_t *in* G_I, *then we can restart the greedy algorithm at* $c_{1,*+1}$ *so that after* $t/2$ *rounds, only* $t-1$ *colors are used.*

We are now ready to describe the algorithm. Our algorithm makes use of the proofs in Theorem 1 and Corollary 1. We know that whenever the greedy algorithm uses $2r$ colors after r rounds, it is coloring the arcs almost optimally. Thus, we let the greedy algorithm continue until some round k, when at most $2k - 1$ colors are used or it is possible to reroute arcs so that $2k - 1$ colors are used. We remove the colored arcs and start coloring again, this time using a new set of colors. Our algorithm eventually outputs $\mathcal{C}$, a collection of independent sets.

Our algorithm, *Unit-DRSP Algorithm*, appears in Figure 2. The main portion of the algorithm is a call to the greedy coloring and rerouting algorithm described above, and summarized in procedure $COLOR\_ARCS$ which appears in Figure 3.

5   Analysis of Algorithm

In the full version of the paper, we show the proof of the following lemma.

Lemma 2. *Let* $a \in C^*$. *If we apply the greedy algorithm starting from* a *for* r *rounds and* $R_r \neq \emptyset$, *then* $L_{C^* \setminus \bigcup_{i=1}^{r} R_i} \leq L_{C^*} - r$.

COLOR_ARCS(C^*, t)
$\mathcal{R} \leftarrow \emptyset$, $C_i \leftarrow \emptyset, i = 1, \ldots, t$;
$r = 1$;
Choose $a \in C^*$ and apply greedy algorithm for 1 round;
while $C_{2r} \neq \emptyset$
    if chords of $c_{1,*}, c_{2,1}, \ldots, c_{2r,1}$ do not form a K_{2r}
        then $c_{2r,1} \leftarrow$ rerouted $c_{2r,1}$
            $C_3 \leftarrow C_3 \cup C_2 \backslash \{c_{2,1}\}$
            $C_2 \leftarrow \{c_{2,1}\} \cup C_1 \backslash \{c_{1,1}, c_{1,2} \ldots c_{1,*}\}$
            $C_1 \leftarrow \{c_{1,1}, c_{1,2} \ldots c_{1,*}\} \cup \{c_{2r,1}\}$
            $C_{2r} \leftarrow \emptyset$
    else if $2r = t$
        then remove colors assigned to $c_{1,1}, c_{1,2} \ldots c_{1,*}$
                $C_1 \leftarrow C_1 \backslash \{c_{1,1}, c_{1,2} \ldots c_{1,*}\} \cup C_2$
                $C_i \leftarrow C_{i+1}$ for $i = 2, \ldots 2r - 1$
                $C_{2r} \leftarrow \emptyset$
        else continue applying the greedy algorithm for 1 round (from $c_{2r,1}$, the last
            arc colored in R_r), r++
$\mathcal{R} \leftarrow \bigcup_{i=2}^{2r-1} C_i$;
$C^* \leftarrow C^* \backslash \mathcal{R}$;
return$(\mathcal{R}, C^*)$;

Fig. 3. The procedure COLOR_ARCS.

Lemma 3. *Suppose in an iteration of* COLOR_ARCS, *k rounds of the greedy algorithm were completed. Then (a). $k \leq t/2$, (b). each C_i, $1 \leq i \leq 2k - 1$ is an independent set and $C_{2k} = \emptyset$, and (c). $L_{C^*_{NEW}} \leq L_{C^*} - k$ where we let C^*_{NEW} denote the set that contains the remaining uncolored arcs.*

Proof: If $k < t/2$, the chords of $c_{1,*}, c_{2,1}, \ldots, c_{2k,1}$ do not form a K_t. COLOR_ARCS ends by rerouting $c_{2k,1}$ so that $C_t = \emptyset$. If $k = t/2$ and the chords of $c_{1,*}, c_{2,1}, \ldots, c_{t,1}$ form a K_t, COLOR_ARCS recolors the arcs so that $C_t = \emptyset$. Claim (a) of the lemma follows.

If no arcs were rerouted in COLOR_ARCS, (b) is clearly true. Suppose, on the other hand, some arcs were rerouted and the chords of $c_{1,*}, c_{2,1}, \ldots, c_{2k,1}$ do not form a K_{2k}. From the proof of Theorem 1, we showed that the chords of $c_{1,*}$ and $c_{2r,1}$ are parallel and hence the rerouted $c_{2r,1}$ together with $c_{1,1}, c_{1,2} \ldots c_{1,*}$ form an independent set. Furthermore, since COLOR_ARCS terminated after k rounds of the greedy algorithm, the chords of $c_{1,*}, c_{2,1}, \ldots, c_{2(k-1),1}$ must form a $K_{2(k-1)}$. Thus, $\{c_{2,1}\} \cup C_1 \backslash \{c_{1,1}, c_{1,2} \ldots c_{1,*}\}$ and $C_3 \cup C_2 \backslash \{c_{2,1}\}$ are independent sets. Finally, none of the arcs in $C_i, i > 3$ were rerouted and $c_{2k,1}$ has been rerouted and recolored so (b) follows. If $k = t/2$ and the chords of $c_{1,*}, c_{2,1}, \ldots, c_{t,1}$ form a K_t, it is easy to check that $C_1 \backslash \{c_{1,1}, c_{1,2} \ldots c_{1,*}\} \cup C_2$

is an independent set. And since the other sets remained the same, claim *(b)* is again true for this case.

Claim *(c)* follows from Lemma 2 since the arcs added to $\mathcal{R}$ were obtained from traversing k rounds of the greedy algorithm starting from $c_{1,1}$ or $c_{1,*+1}$. $\quad\square$

Theorem 3. *There exists a solution in $\mathcal{D}(I)$ that can be colored with at most $(2 - 2/t)L_{C^*}$ colors where L_{C^*} is the optimal ringload of I.*

Proof: Consider $\mathcal{C}$, the collection of sets of arcs produced by the *unit-DRSP Algorithm*. By construction, each set added to $\mathcal{C}$ has non-overlapping arcs and all arcs in the same set are assigned the same color. It is sufficient to show that the number of sets added to $\mathcal{C}$ is at most $(2 - 2/t)L_{C^*}$.

Suppose our algorithm performed m iterations of *COLOR_ARCS* and then an edge has load 0. Let k_i be the number of rounds of coloring completed by *COLOR_ARCS* at iteration i. From Lemma 2, we know that the remaining set of uncolored arcs has load at most $L_{C^*} - \sum_{i=1}^{m} k_i$. Since this set is an interval set, that is, all the arcs do not pass through the edge with load 0, at most $L_{C^*} - \sum_{i=1}^{m} k_i$ colors are needed to finish coloring all the arcs. The total number of colors used is at most $\sum_{i=1}^{m}(2k_i - 1) + L_{C^*} - \sum_{i=1}^{m} k_i = L_{C^*} + \sum_{i=1}^{m} k_i - m$. But each $k_i \leq \lceil \omega(G_I)/2 \rceil \leq t/2$. Thus, $m \geq \sum_{i=1}^{m} k_i/(t/2)$. We have

$$\text{total \# of colors used} \leq L_{C^*} + \sum_{i=1}^{m} k_i - m$$

$$\leq L_{C^*} + \sum_{i=1}^{m} k_i - (2/t)\sum_{i=1}^{m} k_i$$

$$\leq (2 - 2/t)L_{C^*}.$$

Consider the case when the last iteration of *COLOR_ARCS* finishes the coloring. Then the number of colors used is at most $\sum_{i=1}^{m}(2k_i - 1)$. Again, using the fact that $m \geq \sum_{i=1}^{m} k_i/(t/2)$, we have

$$\text{total \# of colors used} \leq 2\sum_{i=1}^{m} k_i - m$$

$$\leq 2\sum_{i=1}^{m} k_i - (2/t)\sum_{i=1}^{m} k_i$$

$$\leq (2 - 2/t)\sum_{i=1}^{m} k_i$$

$$\leq (2 - 2/t)L_{C^*}.\square$$

Since $\omega(G_I) \leq \lfloor n/2 \rfloor$, C^* and $\omega(G_I)$ can be determined in polynomial time [18, 8], and the greedy algorithm for coloring circular arcs runs in polynomial time, the following corollary is immediate.

Corollary 2. *We have a polynomial-time algorithm for the unit-DRSP in rings that produces a $(2 - 4/n)$-approximation to the optimal solution.*

Finally, we show how we can improve our approximation factor by taking advantage of Kumar's result in [12]. Kumar presented an algorithm based on an integer program formulation of the unit-DRSP problem. He proved that the algorithm will use at most $(1.68 + o(1) + 2\sqrt{2\log n/z^*})z^*$ colors where z^* is the optimal value of the fractional solution and n is the size of the network. As mentioned earlier, when $z* > 81\log n$, the approximation factor of the algorithm is guaranteed to be better than 2. Furthermore, the approximation factor becomes smaller as z^* gets larger. We combine Kumar's algorithm and the *Unit-DRSP Algorithm* so that whether z^* is "small" or "large", our solution is guaranteed to be at most $(2 - 1/\theta(\log n)) * OPT$.

It is not difficult to show that $\omega(G_I)$ is a lower bound for z^*. When $z^* \geq 100\log n$, the approximation factor for Kumar's algorithm is at most 1.96. When $z^* < 100\log n$, from Theorem 3, unit-DRSP has an approximation factor of at most $2 - 1/(50\log n)$. Thus, if we route and color I using the best solution from Kumar's algorithm and the *Unit-DRSP Algorithm*, the number of colors used is at most $(2 - 1/(50\log n)) * OPT$. In cases when $4/n > 1/(50\log n)$, then the $(2 - 4/n) * OPT$ is a better guarantee.

Theorem 4. *The exists an algorithm for the unit-DRSP in rings that produces a $(2 - \max\{4/n, 1/(50\log n)\})$-approximation to the optimal solution.*

6 Discussion and Open Questions

Our algorithm colors within 1 of optimality each time procedure $COLOR\_ARCS$ is called. However, we are unable to make the same conclusion after $COLOR\_ARCS$ is called twice. That is, if Z_1 and Z_2 are the sets of demands routed and colored in two iterations of $COLOR\_ARCS$, the algorithm uses $\omega(G_{Z_1}) + \omega(G_{Z_2})$ colors but the best lower bound we have is $\max\{\omega(G_{Z_1}), \omega(G_{Z_2})\}$.

Question 1. Can we find a better lower bound for the unit-DRSP on $Z_1 \cup Z_2$?

In Section 3, we mentioned several ways of coloring circular arcs. Let $C \in \mathcal{D}(I)$. If we use any of these coloring algorithms, we have a solution to our problem. How close is the number of colors used in the solution to optimality?

Tucker's coloring scheme gives a 2-approximation based on the ringload of C. Thus, if this coloring was applied to $C^* \in \mathcal{D}(I)$ where C^* is a routing of I that minimizes ringload then the number of colors used is at most $2L_{C^*}$.

Let G_C be the graph whose vertices correspond to the arcs in C and the edges correspond the the arcs that intersect in C. Hsu and Shih's coloring scheme gives a 5/3-approximation based on $\omega(G_C)$, the largest clique in G_C. This suggests a very interesting problem.

Question 2. How do we route I so that the largest set of arcs that pairwise overlap is as small as possible?

Like his approach for the unit-DRSP, Kumar's coloring scheme for circular arc graphs is based on an integer program formulation of the problem. The number of colors used in the approximation algorithm is then compared to the optimal fractional solution. Again, we ask,

Question 3. How do we route I so that the optimal fractional solution for coloring the routing using Kumar's formulation is as small as possible?

7 Acknowledgments

I am very grateful for Lenore Cowen's advice and encouragement throughout the course of this work. I would also like to thank Leslie Hall and Tami Carpenter for many helpful discussions on topics related to the paper, Dany Kleitman and Tom Leighton for introducing me to DRSP, and Vijay Kumar for preprints of his two papers. I was supported in part by ONR grant N00014-96-1-0829.

References

1. B. Beauquier, J.-C Bermond, L. Gargano, P. Hell, S. Perennes and U. Vaccaro, Graph Problems Arising from Wavelength-Routing in All-Optical Networks, *In Proc. 2nd Workshop on Optics and Computer Science, part of IPPS* (1997).
2. J.-C Bermond, L. Gargano, S. Perennes, A. A. Rescigno, and U. Vaccaro, Efficient Collective Communication in Optical Networks, *Lecture Notes on Computer Science* 1099 (1996) 574–585.
3. T. Carpenter, *personal communication.*
4. T. Carpenter, S. Cosares, and I. Saniee, Demand Routing and Slotting on Ring Networks, *Technical Report 97-02, Bellcore* 1997.
5. T. Erlebach and K. Jansen, Call Scheduling in Trees, Rings and Meshes, *In Proc. 30th Hawaii Int'l Conf. on System Sciencs* 1 (1997) 221–222.
6. A. Frank, Edge-disjoint Paths in Planar Graphs, *J. Comb. Theory Series B* 39 (1985) 164–178.
7. M. Garey, D. Johnson, G. Miller, and C. Papadimitriou, The Complexity of Coloring Circular Arcs and Chords, *SIAM J. Alg. Disc. Methods* 1(2) (1980) 216–227.
8. F. Gavril, Algorithms for a Maximum Clique and a Maximum Independent Set of Circle Graphs, *Networks* 3 (1973) 261–273.
9. M. Golumbic, *Algorithmic Graph Theory and Perfect Graphs*, Academic Press Inc., (1980).
10. W.L. Hsu, W.K. Shih, An Approximation Algorithm for Coloring Circular Arc Graphs. *SIAM Conf. on Disc. Math.*, (1990).
11. V. Kumar, An Approximation Algorithm for Circular Arc Coloring, *preprint* (1998).
12. V. Kumar, Efficient Bandwidth Allocation in Ring Networks, *preprint* (1998) (Note: An early version of this work was presented at APPROX '98.)
13. V. Kumar and E.J. Schwabe, Improved Access to Optical Bandwidth in Trees, *In Proc. 8th Annual ACM-SIAM Symp. on Disc. Algs.*, (1997) 437–444.
14. A. Litman and A.L. Rosenberg, Balancing Communication in Ring-Structured Networks, *Technical Report 93-80, University of Massachusetts*, (1993).
15. M. Mihail, C. Kaklamanis, and S. Rao, Efficient Access to Optical Bandwidth, *In Proc. 36th IEEE Symp. on Foundations of Computer Science*, (1995) 548–557.
16. P. Raghavan and E. Upfal, Efficient Routing in All-optical Networks, *In Proc. 26th ACM Symp. on Theory of Comput.*, (1994) 134–143.
17. A. Tucker, Coloring a Family of Circular Arcs, *SIAM J. Appl. Math.*, 229(3) (1975) 493–502.
18. A. Schrijver, P. Seymour, and P. Winkler, The Ringloading Problem, *SIAM J. Disc. Math.*, 11(1) (1998) 1–14.
19. G. Wilfong, P. Winkler. Ring Routing and Wavelength Translation, *In Proc. 9th Annu. ACM-SIAM Symp. on Disc. Algs.*, (1998) 333–341.

Algorithms for Graph Partitioning on the Planted Partition Model

Anne Condon*
condon@cs.wisc.edu
Computer Sciences Department
University of Wisconsin
1210 West Dayton St.
Madison, WI 53706

Richard M. Karp**
karp@cs.washington.edu
Department of Computer Science and Engineering
University of Washington
Seattle, WA 98195

Abstract. The NP-hard graph bisection problem is to partition the nodes of an undirected graph into two equal-sized groups so as to minimize the number of edges that cross the partition. The more general graph l-partition problem is to partition the nodes of an undirected graph into l equal-sized groups so as to minimize the total number of edges that cross between groups.

We present a simple, linear-time algorithm for the graph l-partition problem and analyze it on a random "planted l-partition" model. In this model, the n nodes of a graph are partitioned into l groups, each of size n/l; two nodes in the same group are connected by an edge with some probability p, and two nodes in different groups are connected by an edge with some probability $r < p$. We show that if $p - r \geq n^{-1/2+\epsilon}$ for some constant ϵ, then the algorithm finds the optimal partition with probability $1 - \exp(-n^{\Theta(\epsilon)})$.

1 Introduction

The graph l-partition problem is to partition the nodes of an undirected graph into l equal-sized groups so as to minimize the *cut size*, namely the total number of edges that cross between groups. There is an extensive literature on algorithms for this problem because of its many applications, which include VLSI circuit placement, parallel task scheduling, and sparse matrix factorization. Unfortunately even the special case of this problem when $l = 2$, which is the well-known graph bisection problem, is NP-hard [9]. In light of this, much of the literature on

* Condon's research supported by NSF grants HRD-627241 and CCR-9257241.
** Karp's research supported by NSF grant DB1-9601046.

algorithms for graph bisection reports on average-case performance of algorithms on random graphs.

A popular random graph model is the $G(n, m)$ model in which a graph is selected randomly and uniformly from the set of all graphs with n nodes and m edges. A closely related model is the $G(n, p)$ model in which each pair of nodes is connected by an edge independently with probability p. No polynomial time algorithm is known that provably finds the minimum bisection with high probability on either of these models for general m or p. The lack of such an analysis may stem from the fact that, if $m/n \to \infty$, then for almost all graphs with n nodes and m edges the cut sizes of the best and worst bisections differ by only a low order term (see Bui et al. [3]).

Instead, some researchers have worked with random graph models in which the cut size of the best bisection is much smaller than the expected cut size. The earliest results, due to Bui et al. [2, 3] concerned the $G(n, m, b)$ model, in which a graph is chosen randomly and uniformly from the set of graphs that have n nodes, m edges, and minimum cut size b. Bui et al. describe an algorithm, based on network flow techniques, that with probability $1 - o(1)$ finds an optimal bisection on this model with the additional constraints that the graph is regular, say with degree d and $b = o(n^{1-1/\lfloor (d+1)/2 \rfloor})$. Since every graph with dn edges has average cut size $dn/2$, the minimum bisection for the Bui et al. graphs is asymptotically smaller than the average bisection. Dyer and Frieze [5] analyze an algorithm for (not necessarily regular) graphs with $\Omega(n^2)$ edges and $b \leq (1 - \epsilon)m/2$ for a fixed $\epsilon > 0$. The Dyer-Frieze algorithm is based on comparison of vertex degrees; it finds the minimum bisection in polynomial expected time. Boppana [4] presents a graph bisection algorithm based on eigenvector methods. He shows that if m is $\Omega(n \log n)$ and $b \leq (m - 5\sqrt{mn \log n})/2$, then his algorithm finds the minimum bisection with probability $1 - O(1/n)$. Thus, Boppana's analysis applies to a larger class of graphs than the analysis of either Bui et al. or Dyer and Frieze. However, the running time of Boppana's algorithm is high since the algorithm uses the ellipsoid method for finding the maximum of a concave function.

Jerrum and Sorkin [10] analyzed a constant-temperature simulated annealing algorithm for graph bisection on a slightly different random graph model, known as the planted bisection model. In this model, for n even, $n/2$ of the nodes of the graph are assigned one color and the remaining nodes are assigned a different color. The probability of an edge between like-colored nodes is p and the probability of an edge between differently-colored nodes is $r < p$. (The planted bisection model is roughly equivalent to the $G(n, m, b)$ model with $b = rn^2/4$ and $m = (p + r)n^2/4$.) If $p - r > n^{-1/2+\epsilon}$, for any fixed $\epsilon > 0$, then with probability $1 - \exp(-n^{\Omega(\epsilon)})$ the planted bisection is the unique bisection with minimum cut size (see [3, 10]). Boppana's analysis of his eigenvector algorithm applies also to the planted bisection model, with $p - r = \Omega(\log n/n)$.

Roughly, the algorithm of Jerrum and Sorkin proceeds from an initial bisection by repeatedly employing the following procedure. A pair of nodes, one on each side of the current bisection, is chosen. These are swapped with some probability that depends on the change in cut size that results if the nodes are swapped and on the "temperature" parameter of simulated annealing. Jerrum and Sorkin show that there is a choice of the temperature parameter for which, in $\Theta(n^2)$ iterations of the node swap procedure, the algorithm finds the minimum bisection with probability $1 - \exp(-n^{\Omega(\epsilon)})$ if $p - r \geq n^{-1/6+\epsilon}$. Their analysis centers on the evolution of the *maximum imbalance* of a color in a bisection (L, R), where the imbalance of a color is the difference between the number of nodes of that color in L and the number in R, all divided by 2. They show that the maximum imbalance behaves roughly as a random walk in which the bias for increase grows over time. In related work, Juels [12] analyzes a simple hill-climbing algorithm on the planted bisection model and shows that within $\Theta(n^2)$ iterations, this algorithm succeeds in finding the minimum bisection with probability $\Omega(1)$ if $p - r = \Omega(1)$. (Here and in what follows, the notation $\Omega(1)$ means some constant > 0 that is independent of the graph.)

In this paper, we analyze a linear-time algorithm, based on successive augmentation, for the graph l-partition problem. Our random graph model, the planted l-partition model, generalizes the planted bisection model: each node is assigned one of l colors, with n/l nodes of each color, and the probability of an edge between nodes is just as for the planted bisection model. We show that, for the planted l-partition model with $p - r \geq n^{-1/2+\epsilon}$, our algorithm outputs the minimum partition with probability $1 - \exp(-n^{\Omega(\epsilon)})$. For the graph bisection problem, our algorithm is faster, and our analysis holds for a larger range of $p - r$, than the algorithm of Jerrum and Sorkin.

Our linear-time algorithm is presented in Section 2. The heart of our algorithm is a procedure that builds up a 2-partition with high imbalance from a 2-partition that is initially empty, by repeatedly selecting a new pair of nodes and adding one node to each side in a greedy fashion. Our analysis in Section 2.1 describes the evolution of the maximum imbalance in the partition, taken over all colors. Our algorithm analysis is similar to that of Jerrum and Sorkin for the Metropolis algorithm, but is simpler because of the independence between the relevant edge probabilities at different iterations.

One phenomenon that we observed in experimental tests of our algorithm (see Section 3) is that our greedy partition-building procedure not only produces a partition in which some color has high imbalance (as predicted by our analysis), but in fact *all* of the imbalances in the partition follow a simple pattern. For example, with two colors the partition appears to evolve towards one in which, if there are k nodes on each side of the partition, then the imbalances are $\approx k/4$ and $-k/4$. With three colors the imbalances are $\approx 2k/9, 0$, and $-2k/9$. More generally, with l colors, the max imbalance is $\approx (l-1)k/l^2$ and the gap between successive imbalances is $2k/l^2$. In Section 3 we show that indeed $\Theta(n)$ iterations of our procedure produce a partition in which the difference between any pair of imbalances is $\Theta(n)$. Using this insight, we obtain a somewhat simpler algorithm for the l-partition problem in Section 3.

1.1   Related Work

Perhaps the best-known algorithm for graph bisection is the Kernighan-Lin (K-L) heuristic [13] and its modification by Fiduccia and Mattheyses [7]. Johnson et al. [11] experimentally compared the performance of the K-L and simulated annealing algorithms on several random graph models.

Regarding approximation algorithms for the graph bisection problems on general graphs, no polynomial-time algorithm is known that is guaranteed to output a bisection with cut size that is bounded by a constant times the minimum cut size. For dense graphs, i.e. graphs in which the minimum node degree is $\Omega(n)$, two polynomial time approximation schemes (PTAS) for the graph bisection problem were recently proposed [1, 8]: given a graph and a constant $\epsilon > 0$, these algorithms output a bisection with cut size at most $(1 + \epsilon)$ times the minimum cut size. (The running time of these algorithms is exponential in $1/\epsilon$.)

2   Algorithm 1 and its Analysis

Our algorithm consists of four phases. Briefly, the purpose of the first two phases is to build up a partition (L_2, R_2) with $|L_2| = |R_2| = \Theta(n)$ in which some color has an imbalance of $\Theta(n)$. By the imbalance of a color in a partition (L, R), we mean the number of nodes of that color in L less the number of nodes of that color in R, all divided by 2. In the third phase, partition (L_2, R_2) is used to partition the remaining unexamined nodes into two non-empty groups L and R such that no node in L is the same color as a node in R. In the fourth phase, all remaining unexamined nodes plus all nodes examined in phases 1 and 2 are added to the "correct" side of the partition (L, R). The problem can now be solved recursively on L and R.

Phase 1: L_1 and R_1 are initially empty. In each of $n_1 = \lceil n^{1-\epsilon/2} \rceil$ steps, choose a pair of nodes $(1, 2)$ randomly and uniformly from the unexamined nodes. Let $l_1(i)$ and $r_1(i)$ be the number of edges from node $i, i \in \{1, 2\}$ to nodes in L_1 and R_1 respectively, and let $X = l_1(1) - r_1(1) - l_1(2) + r_1(2)$. If $X > 0$, place nodes 1 and 2 in L_1 and R_1 respectively and if $X < 0$, place nodes 2 and 1 in L_1 and R_1 respectively. If $X = 0$ then draw i at random from the uniform distribution over $\{1, 2\}$, place node i in L_1 and node $3 - i$ in R_1.

Phase 2: L_2 and R_2 are initially empty. Let $n_2 = \lceil n/4 \rceil$. Choose n_2 new pairs of nodes randomly and uniformly from the unexamined nodes. As in phase 1, greedily assign one node from each pair $(1, 2)$ to each of L_2 and R_2, depending on the value of $X = l_1(1) - r_1(1) - l_1(2) + r_1(2)$. Note that all pairs may be assigned concurrently to (L_2, R_2).

Phase 3: For each remaining unexamined node v, let $l_2(v)$ denote the number of edges from node v to nodes in L_2. Let $o_0 < o_1 < \ldots < o_j$ be the ordered set of values $l_2(v)$ and let $o_a - o_{a-1}$ be the maximum difference between consecutive numbers in this ordered list. If $l_2(v) \geq o_a$, put node v in L and if $l_2(v) < o_a$ put node v in R.

Phase 4: In parallel for each node v examined in phases 1 and 2, assign v greedily to L if the fraction of nodes in L that have edges to v is greater than the fraction of nodes in R that have edges to v, and assign v to R otherwise.

We note that phase 2 could be removed if instead phase 1 were repeated for $\lceil n/4 \rceil$ steps and (L_1, R_1) were used in phase 3 instead of (L_2, R_2). The reason for including phase 2 is that it could be implemented in parallel, as could phases 3 and 4, in which case the parallel running time of the algorithm would be sublinear.

2.1 Analysis

We assume that $p - r = \Delta = n^{-1/2+\epsilon}$. The analysis shows that the following facts are true with probability $1 - \exp(-n^{\Theta(\epsilon)})$, referred to as "high probability" throughout. At the end of phase 1, some color in (L_1, R_1) has an imbalance $\geq n^{1-\epsilon}$. At the end of phase 2, some color in (L_2, R_2) has imbalance $\Theta(n)$. At the end of phase 3, no node in L is the same color as a node in R and both L and R are of size $\Theta(n)$. Finally, at the end of phase 4, by which time all nodes are assigned either to L or to R, no node in L is the same color as a node in R.

Phase 1 The following claim is key to the analysis of Phase 1. It shows (part (ii)) that at every iteration of the greedy partition-building algorithm of phase 1, maximum imbalance is at least as likely to increase as to decrease. Moreover (part (iii)), the higher the maximum imbalance, the more likely it is to increase.

Claim 1 Let $x(= x(T))$ be the maximum imbalance in partition (L_1, R_1) at time step T of phase 1. Then at each step $T + 1$ of phase 1, for any execution of the algorithm up to step T,
  (i) Prob[x increases] $= \Omega(1)$,
  (ii) Prob[x increases] $-$ Prob[x decreases] ≥ 0, and
  (iii) if $x = \Omega(n^{1/2-\epsilon/2})$ then

$$\text{Prob}[x \text{ increases}] - \text{Prob}[x \text{ decreases}] = \Omega(\min\{x\Delta/\sqrt{T}, 1\}).$$

Proof. : Since x cannot decrease if $x = 0$, we need only consider the case $x > 0$. Also, if there are at least two colors with imbalance x at the end of step T, then x cannot decrease at step $T + 1$ and can increase with probability $\Omega(1)$ (namely if nodes in the chosen pair have distinct colors, both of which have imbalance x). Hence, in the rest of the proof we assume that there is exactly one color with imbalance x.

Let $[x, x']$ denote the event that the colors of nodes 1 and 2 chosen at step $T + 1$ have imbalance x and x', respectively. Note that in the event $[x, x']$, x increases at step $T + 1$ if and only if node 1 is placed in L_1 and decreases otherwise. Let $X = l_1(1) - r_1(1) - l_1(2) + r_1(2)$, where $l_1(1), r_1(1), l_1(2)$, and

$r_1(2)$ are as defined in phase 1 of the algorithm, assuming the event $[x, x']$. We have that

$$\text{Prob}[x \text{ increases}| \text{ event } [x, x']] = \text{Prob}[X > 0] + \frac{1}{2}\text{Prob}[X = 0] \quad \text{and}$$

$$\text{Prob}[x \text{ decreases}| \text{ event } [x, x']] = \text{Prob}[X < 0] + \frac{1}{2}\text{Prob}[X = 0].$$

Therefore, to prove part (ii) of the claim it is sufficient to show that $\text{Prob}[X > 0] \geq \text{Prob}[X < 0]$. Averaging over all events $[x, x']$ then gives the result. Part (i) of the claim follows from this and the additional fact that at each step of phase 1, the probability of each event $[x, x']$ such that there are colors with imbalances x and x' is $\Omega(1)$.

To analyze $\text{Prob}[X \geq 0]$ we use the fact that each term $l_1(i)$ and $r_1(i)$ is binomially distributed. Let $B(n, p)$ denote the number of successes in n independent Bernoulli trials, each with probability p of success. Let $t + x$ and $t' + x'$ be the number of nodes in L_1 which have the same color as nodes 1 and 2, respectively, at the end of step T. Then X is the sum

$$B(t+t'+x-x',p)+B(2T-t-t'-x+x',r)-B(t+t'-x+x',p)-B(2T-t-t'+x-x',r).$$

Since $x - x' > 0$, X dominates the symmetric random variable X' with mean value 0, defined by

$$X' = B(\lceil t + t' \rceil, p) + B(2T - \lceil t - t' \rceil, r) - B(\lceil t + t' \rceil, p) - B(2T - \lceil t - t' \rceil, r).$$

Thus, $\text{Prob}[X > 0] \geq \text{Prob}[X' > 0] = \text{Prob}[X' < 0] \geq \text{Prob}[X < 0]$. This completes the proof of parts (i) and (ii) of the claim.

To prove part (iii), define X as before except that X is now conditioned on the following event: if 1 and 2 are the two nodes chosen at step $T + 1$, then the color of node 1 has imbalance x and the color of node 2 has imbalance at most 0. Call this event $[x, \leq 0]$. At every step of phase 1, the probability that event $[x, \leq 0]$ occurs is $\Theta(1)$ because some color must have imbalance at most 0. Hence, to prove part (iii) of the claim, it is sufficient to show that $\text{Prob}[X > 0] = 1/2 + \Omega(\min\{(x\Delta)/\sqrt{T}, 1\})$.

First note that since the color of node 2 has imbalance at most 0, the expected value of $l_1(2) - r_1(2)$ is at most 0. Hence EX, the expected value of X, is at least $2x\Delta$. To bound $\text{Prob}[X > 0]$ we use Esseen's Inequality:

Theorem 1. Esseen's Inequality *(Petrov [14, Theorem 3, p.111]) Let $X_1, \ldots,$ X_n be independent random variables such that $EX_j = 0$ and $E|X_j|^3 < \infty$, $j = 1, \ldots, n$. Let $\sigma_j^2 = EX_j^2$,*

$$B_n = \sum_{j=1}^{n} \sigma_j^2, \quad F_n(x) = \text{Prob}[B_n^{-1/2} \sum_{j=1}^{n} X_j < x], \quad \text{and} \quad L_n = B_n^{-3/2} \sum_{j=1}^{n} E|X_j|^3.$$

Then $\sup_x |F_n(x) - \Phi(x)| \leq AL_n$,
where A is an absolute constant and $\Phi(x)$ denotes the normal $(0, 1)$ distribution function.

Note that $X = \sum_{j=1}^{4T} X_j + EX$, where each random variable X_j is one of the following: (i) $Y - p$ where Y is 1 with probability p and 0 with probability $1-p$, (ii) $Y - r$ where Y is 1 with probability r and 0 with probability $1-r$, (iii) $Y + p$ where Y is -1 with probability p and 0 with probability $1-p$, or (iv) $Y + r$ where Y is -1 with probability r and 0 with probability $1-r$. Therefore $EX_j = 0$ and $E|X_j|^3 = O(1)$ and so the random variables X_j satisfy the conditions of Esseen's Inequality. Using the notation in Theorem 1 with $n = 4T$, we have that

$$\text{Prob}[X > 0] = \text{Prob}[B_{4T}^{-1/2} \sum_{j=1}^{4T} (-X_j) < B_{4T}^{-1/2} EX]$$

$$\geq \Phi(B_{4T}^{-1/2} EX) - AL_{4T} = 1/2 + \Omega(\min\{x\Delta/\sqrt{B_{4T}}, 1\}) - O(L_{4T}).$$

If $x = \Omega(n^{1/2 - \epsilon/2})$, then $L_{4T} = o(\min\{x\Delta/\sqrt{B_{4T}}, 1\})$ (details omitted). Hence, if $x = \Omega(n^{1/2 - \epsilon/2})$,

$$\text{Prob}[X > 0] = 1/2 + \Omega(\min\{\frac{x\Delta}{\sqrt{B_{4T}}}, 1\}).$$

Part (iii) of the claim follows from the observation that $\sqrt{B_{4T}} = O(\sqrt{T})$.

The analysis of the evolution of imbalance x is somewhat complicated by the fact that steps in the random process $x(t)$ are not independent and the transition probabilities vary depending on the history of the algorithm. It is convenient to relate the behavior of x to a (simpler) random walk with identical independent increments. The next lemma does this (proof omitted). Throughout, when we refer to the probability that $x(t+1)$ takes some value, we mean that probability given the history of the algorithm up to step t.

Lemma 2. *Let $\epsilon : \mathbf{N} \to \mathbf{R}$ be a function such that at step $t+1$ of any execution of the algorithm,*

$$\text{Prob}[x \text{ decreases}] - \text{Prob}[x \text{ increases}] \leq \epsilon(x(t)).$$

There exist constants c and d such that the following holds. Let a, b be nonnegative integers, with $a \leq b$ and let $Y(t), t = 0, 1, \ldots$ be a random walk with the following properties:

$$\text{Prob}[Y(t+1) = 1] = 1 \qquad\qquad \text{if } Y(t) = 0,$$
$$\text{Prob}[Y(t+1) = Y(t) - 1] = 1/2 + c \max_{j \in [a,b]}\{\epsilon(j)\} \quad \text{if } Y(t) > 0, \text{ and}$$
$$\text{Prob}[Y(t+1) = Y(t) + 1] = 1/2 - c \max_{j \in [a,b]}\{\epsilon(j)\} \quad \text{if } Y(t) > 0.$$

Then for any nonnegative integer i, $a \leq i \leq b$, the probability that, starting at $i/2$, x leaves $[a/2, b/2]$ at the right end within k steps is at least the probability that, starting at i, Y leaves $[a, b]$ at the right end within dk steps $- \exp(-\Theta(k))$.

Theorem 3. *In partition (L_1, R_1) at the end of phase 1, some color has imbalance at least $n^{1-\epsilon}$ with high probability.*

Proof. : We partition phase 1 into subphases, based on the value of the maximum imbalance x. The first subphase starts at time 0 and continues until $x \geq n^{1/2-\epsilon/2}/2$. By Claim 1, at every step of this subphase, the probability that x increases is at least the probability that x decreases. If Y is the random walk of Lemma 2 with $\epsilon() = 0$, then the probability that, starting at 0, x reaches $n^{1/2-\epsilon/2}/2$ within $n_1/2$ steps is at least the probability that, starting at 0, Y reaches $n^{1/2-\epsilon/2}$ within dn_1 steps $- \exp(-n^{\Theta(n_1)})$, where $d > 0$ is a constant. From Feller [6, XIV.3], the expected time for the unbiased random walk Y to reach $n^{1/2-\epsilon/2}$, starting at 0, is $n^{1-\epsilon}$. Applying Markov's inequality, it follows that the probability that Y reaches $n^{1/2-\epsilon/2}$ within dn_1 steps is $1 - \exp(-\Omega(n^{\epsilon/2}))$.

The tth subphase starts when the $(t-1)$st subphase ends. If $i/2 > 0$ is the value of x at the start of a subphase, then that subphase ends when $x = i$ or when $x = \lfloor i/4 \rfloor$ (or when Phase 1 ends).

Let Y be a random walk with no loop probabilities in which the difference between the probability of an increase and a decrease is $\delta = \Omega(\min\{i\Delta/\sqrt{T}, 1\}) = \Omega(\min\{i\Delta/\sqrt{n_1}, 1\})$. By Lemma 2 and Claim 1, the probability that, starting at $i/2$, x leaves $[\lfloor i/4 \rfloor, i]$ at the right end within $n^{1-3\epsilon/4}$ steps is at least the probability that, starting at i, Y leaves $[\lfloor i/2 \rfloor, 2i]$ at the right end within $dn^{1-3\epsilon/4}$ steps $- \exp(-n^{\Theta(n_1)})$, for some sufficiently small constant d (independent of i).

We first bound Prob[starting at i, Y leaves $[\lfloor i/2 \rfloor, 2i]$ at the right end]. Let s be the ratio of the probability of a decrease over the probability of an increase, that is, $s = \frac{1/2-\delta}{1/2+\delta} = 1 - \Theta(\delta)$. Using Feller [6, XIV.2.4], the probability that, starting at i, Y reaches $2i$ before $\lfloor i/2 \rfloor$ is at least

$$1 - \frac{s^{\lceil i/2 \rceil} - s^{2i-\lfloor i/2 \rfloor}}{1 - s^{2i-\lfloor i/2 \rfloor}} \geq 1 - \frac{s^{\lfloor i/2 \rfloor}}{1 - s^{3\lceil i/2 \rceil}}.$$

If $i = \Omega(n^{1/2-\epsilon/2})$ then $s^{i/2} = \exp(-\Omega(n^{1-\epsilon}\Delta/\sqrt{n_1})) = \exp(-\Omega(n^{\epsilon/4}))$. Hence as long as $i = \Omega(n^{1/2-\epsilon/2})$, the probability that, starting at i, Y reaches $2i$ before $\lfloor i/2 \rfloor$ is $1 - \exp(-\Omega(n^{\epsilon/4}))$.

It is also that case that, starting at $i = \Omega(n^{1/2-\epsilon/2})$, Y leaves the interval $[\lfloor i/2 \rfloor, 2i]$ within $dn^{1-3\epsilon/4}$ steps with high probability, where $d > 0$ is a constant. Therefore, for $i = \Omega(n^{1/2-\epsilon/2})$, with high probability, starting at $i/2$, x leaves $[\lfloor i/4 \rfloor, i]$ at the right end within $n^{1-3\epsilon/4}$ steps. It follows that with high probability, within $n_1 = \lceil n^{1-\epsilon/2} \rceil$ steps, sufficiently many subphases of phase 1 are completed, all ending by leaving the corresponding interval at the right end so that the imbalance is at least $2n^{1-\epsilon}$. A similar analysis shows that once the imbalance is $2n^{1-\epsilon}$, then with high probability it remains at least $n^{1-\epsilon}$ for the rest of phase 1. This completes the proof of the theorem.

Phases 2, 3, and 4

Theorem 4. *At the end of phase 2, with high probability the imbalance of some color is $\Theta(n)$.*

The proof of Theorem 4 follows in a straightforward way from the following. Let C be a color of greatest imbalance in (L_1, R_1). Let y be the imbalance of color C in (L_2, R_2) at the end of phase 2. Suppose that the maximum imbalance x at the end of the first phase is at least $n^{1-\epsilon}$. Let $(1,2)$ be a pair of nodes examined in phase 2. Then Prob[$(1,2)$ contributes positively to y] $-$ Prob[$(1,2)$ contributes negatively to y] $= \Omega(1)$. This can be shown using a similar argument to that of Claim 1 part (iii): if exactly one node in the pair $(1,2)$ has color C, and the other has a color with imbalance at most 0, then

Prob[$(1,2)$ contributes positively to y] $-$ Prob[$(1,2)$ contributes negatively to y]

$$= \Omega(\min\{\frac{x\Delta}{\sqrt{n_1}}, 1\}) = \Omega(1),$$

where the last equality follows from the fact that $n_1 = \lceil n^{1-\epsilon/2} \rceil$ and $x\Delta \geq n^{1-\epsilon} n^{-1/2+\epsilon}$.

Next, consider Phase 3. For each node v that is examined in phase 3, let $l_2(v)$ be the number of edges from v to a node in L_2. For each color C, let $EL_2(C)$ be the expected number of edges of an unexamined node of color C to nodes in the set L_2. First, we show that with high probability the values $l_2(v)$ are distributed as follows: for all nodes v of color C, the values $l_2(v)$ are clustered in a short interval centered at $EL_2(C)$. More precisely, in Claim 2 it is shown that with high probability, $|l_2(v) - EL_2(C)| \leq n^{1/2+\epsilon/2}$. Second, the interval spanned by the values $EL_2(C)$ is relatively large, namely of length $\Omega(n^{1/2+\epsilon})$. This is shown in Claim 3. Simple algebra then shows (Theorem 5) that two adjacent "clusters" must be far apart, implying that the quantity $o_a - o_{a-1}$ used as the partitioning criterion in phase 3 is large. The analysis of phase 4 is very similar to that of phase 3. Proofs are omitted in this extended abstract.

Claim 2 Let v be a node of color C. Then w.h.p. $|l_2(v) - EL_2(C)| \leq n^{1/2+\epsilon/2}$.

Claim 3 Let C_{max} and C_{min} be the colors with the largest and smallest number of nodes, respectively, in L_2. W.h.p. $EL_2(C_{max}) - EL_2(C_{min}) = \Omega(n^{1/2+\epsilon})$.

Theorem 5. At the end of phase 3, with high probability no node in L is the same color as a node in R and moreover, both L and R are non-empty.

3   A Non-Recursive Algorithm

In simulations of phase 1 of Algorithm 1, the maximum imbalance tended to increase over time, as we expected. Moreover, based on our experiments we hypothesize the following: The partition evolves towards one with maximum imbalance $\approx (l-1)k/l^2$ and a gap of $2k/l^2$ between successive imbalances.

The following table presents some evidence that this indeed is the case, based on experiments on our algorithm. The values listed in the third row of the table are the expected imbalances in our experiments, divided by k. The numbers presented are averaged over 20 runs of phase 1 of our algorithm with $n = 256,000$,

$p = 1/2$, $\Delta = n^{-1/2+.2} = .0829$, and $k = 100,000$. The values listed in the second row of the table are the numbers towards which we believe the imbalances evolve in the limit. In each case, the variance is that for the maximum imbalance.

No. colors	2	3	4
Hypothesis	.250,-.250	.222,.000,-.222	.1875, .0625, -.0625, -.1875
E[Imbalances]	.248,-.248	.215,.000,-.215	.1753, .0619, -.0606, -.1766
Variance:	0.000004	0.000011	0.000041

A heuristic explanation of this hypothesis is as follows. First, in case of two colors, consider the evolution of phase 1 once the maximum imbalance is large. Let C be the color with maximum imbalance. Roughly, in $1/2$ of the steps, exactly one of the chosen pair of nodes has color C and this is likely to be put in L. In $1/4$ of the steps, the chosen pair of nodes are both of color C. In the remaining $1/4$ of the steps, neither of the chosen nodes are of color C and a node that is not of color C is placed in L. Since in $3/4$ of the steps, the node placed in L is of color C, roughly 75% of the nodes in L should be of color C and by symmetry, roughly 25% of the nodes in R should be of color C. This heuristic explanation can be generalized to three or more colors,

In light of the above observations, we should expect a gap of $\Theta(n)$ between any pair of imbalances at the end of phase 2. In this event, it should be possible to separate the nodes from phase 2 into l distinct color classes in phase 3, rather than simply grouping the nodes into two groups as is done in Algorithm 1. In this way, the recursive phase 4 can be avoided. The following algorithm partitions all l color classes directly from the partition of phase 1.

Phases 1,2: Construct L_1, R_1, L_2, R_2 as in Algorithm 1.

Phase 3: The remaining unexamined nodes are partitioned into l, rather than 2, groups as follows. For each remaining unexamined node v, let $l_2(v)$ denote the number of edges from node v to nodes in L_2. Let $o_0 < o_1 < \ldots < o_j$ be the ordered set of values $l_2(v)$. Let the $l-1$ largest differences between pairs of consecutive numbers in this ordered list be

$$o_{a_1} - o_{a_1-1}, o_{a_2} - o_{a_2-1}, \ldots, o_{a_{l-1}} - o_{a_{l-1}-1}.$$

If $l_2(v) < o_{a_1}$ then put v in S_1. For $2 \leq i \leq l-1$, if $o_{a_{i-1}} \leq l_2(v) < o_{a_i}$ then put node v in S_i. Finally, if $o_{a_{l-1}} < l_2(v)$ then put node v in S_l.

Phase 4: In parallel for each node v examined in phases 1 and 2, assign v greedily to S_i if the fraction of nodes in S_i that have edges to v is greater than the fraction of nodes in S_j that have edges to S_i for all $j \neq i$ (breaking ties arbitrarily).

We claim that the following facts are true of Algorithm 2 with high probability. At the end of phase 1, the difference between the imbalances of any two distinct colors is at least $n^{1-\epsilon}/l$. At the end of phase 2, the difference between the imbalances of any two distinct colors is $\Theta(n)$. At the end of phase 3, within each set S_i all nodes have the same color. Finally, each node v examined in phase 4 is assigned to the set S_i with nodes of the same color as v.

We next sketch the analysis of phase 1 of Algorithm 2. Phases 2, 3, and 4 of Algorithm 2 can be analyzed by similar extensions of the corresponding phases of Algorithm 1.

Theorem 6. *At the end of phase 1, with high probability, the difference between the imbalances of any two distinct colors is at least $n^{1-\epsilon}/l$.*

Let $x_1 \geq x_2 \geq \ldots \geq x_l$ be the ordered sequence of imbalances of the colors in partition (L_1, R_1), as a function of the number of steps of phase 1. From the analysis of Algorithm 1, we already know that with high probability, in $n^{1-\epsilon/2}$ steps, $x_1 - x_l \geq 2n^{1-\epsilon}$. Therefore, for some i, $x_i - x_{i+1} \geq 2n^{1-\epsilon}/l$. We say that *a good gap arises* between i and j at some step of phase 1 if after that step, for the first time $x_i - x_j \geq 2n^{1-\epsilon}/l$. We say that phase 1 is *well behaved* if, once a good gap arises between a pair of contiguous imbalances, a gap of at least $n^{1-\epsilon}/l$ remains in all further steps of phase 1. We say that phase 1 is *normal* if for every color class C and for all k, the number of nodes of color C that have not been examined after k steps of Phase 1 lies between $\frac{n-2k}{l} - n^{1/2+\frac{\epsilon}{4}}$ and $\frac{n-2k}{l} + n^{1/2+\frac{\epsilon}{4}}$.

It is not hard to show that Phase 1 is normal with high probability. The following lemmas state that with high probability phase 1 is well behaved and, given that phase 1 is well behaved and normal, a good gap arises between each pair of contiguous imbalances during phase 1. The proof of Theorem 6 then follows directly from Lemmas 7 and 8.

Lemma 7. *Phase 1 is well behaved with high probability.*

Lemma 8. *Suppose that phase 1 is well behaved and normal. Suppose also that at some step of phase 1, a good gap has not arisen between x_i and x_j, where $i < j$, but that (i) either $i = 1$ or a good gap has arisen between between $i - 1$ and i, and (ii) either $j = l$ or a good gap has arisen between j and $j + 1$. Then, within $n^{1-\epsilon/2}/(2l)$ more steps, for some $k, i \leq k < j$, a good gap will arise between x_k and x_{k+1}, with high probability.*

4 Conclusions

A variant of our algorithm considers only one node per step, rather than a pair of nodes, and places it on the side of the partition to which it has the greatest fraction of edges. We observed that this variant performs better experimentally than our two-node algorithm but do not currently have an analysis of this variant. It would also be interesting to extend our results to the case where the number of color classes is unknown and where the color classes are of unequal size. Such cases arise in certain clustering applications.

The following related problem may also be relevant to data clustering applications. Consider a set of data samples, each of which has some attributes from a given set. Let M be a boolean matrix with entry $[i, j]$ having value 1 if and only if sample i has attribute j. The simplest version of the problem is to bisect both the samples (rows of the matrix) and the attributes (columns of the matrix) into two equal-sized groups, say $R1, R2$ and $C1, C2$, respectively, so as to minimize the number of 1-entries in the submatrices $R1 \times C2$ and $R2 \times C1$.

Acknowledgements

We thank the reviewers for their valuable comments.

References

1. S. Arora, D. Karger and M. Karpinski. "Polynomial time approximation schemes for dense instances of NP-hard problems," in *Proc. 27th Annual ACM Symp. on Theory of Computing*, 1995, 284–293.
2. T. Bui. "On bisecting random graphs," Report Number MIT/LCS/TR-287, Laboratory for Computer Science, Massachusetts Institute of Technology, Cambridge, MA, 1983.
3. T. Bui, S. Chaudhuri, T. Leighton, and M. Sipser. "Graph bisection algorithms with good average case behavior," *Combinatorica*, 7:2 (1987), 171–191.
4. R. B. Boppana. "Eigenvalues and graph bisection: an average-case analysis," in *Proceedings of the 28th Annual IEEE Symposium on Foundations of Computer Science*, 1987, 280–285.
5. M. E. Dyer and A. M. Frieze. "Fast solution of some random NP-hard problems," in *Proceedings of the 27th Annual IEEE Symposium on Foundations of Computer Science*, 1986, 331–336.
6. W. Feller. *An introduction to probability theory and its applications*, Volume 1, Third Edition, John Wiley and Sons, New York, 1968.
7. C. M. Fiduccia and R. M. Mattheyses. "A linear-time heuristic for improving network partitions," in *Proceedings of the ACM IEEE Nineteenth Design Automation Conference*, 1982, 174–181.
8. A. Frieze and R. Kannan. "The regularity lemma and approximation schemes for dense problems," in *Proceedings of the 37th Annual IEEE Symposium on Foundations of Computer Science*, 1996, 12–20.
9. M. R. Garey, D. S. Johnson, and L. Stockmeyer. "Some simplified NP-complete graph problems," *Theoretical Computer Science*, 1 (1976), 237–267.
10. M. Jerrum and G. B. Sorkin. "Simulated Annealing for graph bisection," in *Proceedings of the 34th Annual IEEE Symposium on the Foundations of Computer Science*, 1993, 94–103.
11. D. S. Johnson, C. R. Aragon, L. A. McGeoch, and C. Schevon. "Optimization by simulated annealing: an experimental evaluation; part 1, graph partitioning," *Operations Research*, 37:6 (Nov-Dec 1989), 865–892.
12. A. Juels. *Topics in Black Box Optimization*, Ph.D. Thesis, EECS Dept., U. California at Berkeley, 1996.
13. B. W. Kernighan and S. Lin. "An efficient heuristic procedure for partitioning graphs," Bell. Syst. Tech. J. 49, 291–307.
14. V. V. Petrov. *Sums of independent random variables*, Springer-Verlag, New York, 1975.

A Randomized Time-Work Optimal Parallel Algorithm for Finding a Minimum Spanning Forest

Seth Pettie and Vijaya Ramachandran

Department of Computer Sciences
The University of Texas at Austin
Austin, TX 78712
{seth,vlr}@cs.utexas.edu

Abstract. We present a randomized parallel algorithm to find a minimum spanning forest (MSF) in a weighted, undirected graph. On an EREW PRAM our algorithm runs in logarithmic time and linear work w.h.p. This is both time and work optimal and is the first provably optimal parallel algorithm under both measures.

1 Introduction

In this paper we present a randomized parallel minimum spanning forest (MSF) algorithm that is optimal with respect to both time and work. Finding an MSF is an important problem and there has been considerable prior work on parallel algorithms for the MSF problem. Following the linear-time sequential MSF algorithm of Karger, Klein and Tarjan [KKT95] came linear-work parallel MSF algorithms for the CRCW PRAM [CKT94,CKT96] and the EREW PRAM [PR97]. The best CRCW PRAM algorithm known to date [CKT96] runs in logarithmic time and linear work, but the time bound is not known to be optimal. The best EREW PRAM algorithm known prior to our work is the result of Poon and Ramachandran which runs in $O(\log n \log \log n \, 2^{\log^* n})$ time and linear work. All of these algorithms are randomized. Recently a deterministic EREW PRAM algorithm for MSF was given in [CHL99], which runs in logarithmic time with a linear number of processors, and hence with work $O((m+n)\log n)$, where n and m are the number of vertices and edges in the input graph. It was observed by Poon and Ramachandran [PR98] that the algorithm in [PR97] could be speeded up to run in $O(\log n \cdot 2^{\log^* n})$ time and linear work by using the algorithm in [CHL99] as a subroutine (and by modifying the 'Contract' subroutine in [PR97]).

In this paper we improve on the running time of the algorithm in [PR97,PR98] to $O(\log n)$, which is asymptotically optimal and we improve on the algorithm in [CKT96] by matching the logarithmic time bound on the weaker EREW PRAM.

The structure of our algorithm is fairly simple. The most complex portion turns out to be the subroutine calls made to the 'CHL algorithm' [CHL99] (which we use as a black-box). As a result our algorithm can be used as a simpler alternative to several other parallel algorithms.

1. For the CRCW PRAM we can replace the calls to the CHL algorithm by calls to the simple logarithmic time, linear-processor CRCW algorithm in [AS87]. The resulting algorithm runs in logarithmic time and linear work and is considerably simpler than the MSF algorithm in [CKT96].

2. As modified for the CRCW PRAM, our algorithm is simpler than the linear-work logarithmic-time CRCW algorithm for connected components given in [Gaz91].

3. Our algorithm improves on the EREW connectivity and spanning tree algorithms in [HZ94,HZ96] since we compute a *minimum* spanning tree within the same time and work bounds. Our algorithm is arguably simpler than the algorithms in [HZ94,HZ96].

In the following we say that a result holds *with high probability (or w.h.p.) in* n if the probability that it fails to hold is less than $1/n^c$, for any constant $c > 0$.

2   The High-Level Algorithm

Our algorithm is divided into two phases along the lines of the CRCW PRAM algorithm of [CKT96]. In Phase 1, the algorithm reduces the number of vertices in the graph from n to n/k vertices, where n is the number of vertices in the input graph, and $k = (\log^{(2)} n)^2$. [†] To perform this reduction the algorithm uses the familiar recursion tree of depth $\log^* n$ [CKT94,CKT96,PR97], which gives rise to $O(2^{\log^* n})$ recursive calls, but the time needed per invocation in our algorithm is well below $O(\log n/2^{\log^* n})$. Thus the total time for Phase 1 is $O(\log n)$. We accomplish this by requiring Phase 1 to find only a subset of the MSF. By contracting this subset of the MSF we obtain a graph with $O(n/k)$ vertices. Phase 2 then uses an algorithm similar to the one in [PR97], but needs no recursion due to the reduced number of vertices in the graph. Thus Phase 2 is able to find the MSF of the contracted graph in $O(\log n)$ time and linear work.

High-Level(G)
    (Phase 1) $G_t :=$ For all $v \in G$, keep lightest k edges of edge-list(v)
                 $M :=$ Find-k-Min$(G_t, \log^* n)$
                 $G':=$Contract all edges in G appearing in M
    (Phase 2) $G_s:=$Sample edges of G' with prob. $1/\sqrt{k} = 1/\log^{(2)} n$
                 $F_s :=$Find-MSF(G_s)
                 $G_f :=$ Filter(G', F_s)
                 $F :=$Find-MSF(G_f)
                 Return$(M \cup F)$

Theorem 1. *With high probability, High-Level(G) returns the MSF of G in* $O(\log n)$ *time using* $(m + n)/\log n$ *processors.*

[†] We use $\log^{(r)} n$ to denote the log function iterated r times, and $\log^* n$ to denote the minimum r s.t. $\log^{(r)} n \leq 1$.

In the following sections we describe and analyze the algorithms for Phases 1 and 2, then present the proof of the Theorem 1 for the expected running time. We then obtain a high probability bound for the running time and work. When analyzing the performance of our algorithm we assume perfect processor allocation. This can be achieved w.h.p to within a constant factor, using the load-balancing scheme in [HZ94], which requires superlinear space, or the linear-space scheme claimed in [HZ96]. In [PR99], we describe a simple scheme which uses linear space running on the $QRQW$ PRAM [GMR94], which is slightly stronger than the EREW PRAM. The usefulness of the QRQW PRAM lies in the fact that algorithms designed on that model map on to general-purpose models such as QSM [GMR97] and BSP [Val90] just as well as the EREW PRAM.

3 Phase 1

In Phase 1, we reduce the number of vertices in the input graph to $< n/k$ by identifying and contracting a large subset of the MSF. The challenge here is to identify these edges in logarithmic time and linear work.

Phase 1 achieves the desired reduction in vertices by identifying a *k-Min forest* (defined below). This is similar to the algorithm in [CKT96]. However, our algorithm is considerably simpler. We design a procedure *Borůvka-A*, which performs only the requisite work to identify edges in a *k*-Min forest. The result is that edge lists grow at a manageable rate, allowing us to process them in time $o(\log n / 2^{\log^* n})$. Phase 1 also needs a *Filter* routine, which removes '*k*-Min heavy' edges. For this we show that an MSF verification algorithm, slightly modified, is sufficient. The overall algorithm for Phase 1, Find-*k*-Min uses these two subroutines to achieve the stated reduction in the number of vertices.

3.1 k-Min Forest

Phase 1 resembles the earlier MSF algorithms of [KKT95,CKT94,CKT96,PR97]. However, instead of computing the MSF, we construct the *k-Min tree* [CKT96] of each vertex, where $k = (\log^{(2)} n)^2$. Contracting the edges in these *k*-Min trees will produce a graph with $O(n/k)$ vertices. To understand what a *k*-Min tree is, consider the Dijkstra-Jarnik-Prim minimum spanning tree algorithm:

Dijkstra-Jarnik-Prim(G)
    $S := \{v\}$ *(choose an arbitrary starting vertex v)*
    $T := \emptyset$
    Repeat until T contains the MST of G
        Choose minimum weight edge (a, b) s.t $a \in S, b \notin S$
        $T := T \cup \{(a, b)\}$
        $S := S \cup \{b\}$

The edge set k-Min(v) consists of the first k edges chosen by this algorithm, when started at vertex v. A forest F is a *k-Min forest* of G if $F \subseteq$ MSF(G) and for all $v \in G$, k-Min$(v) \subseteq F$.

Let $P_T(x, y)$ be the set of edges on the path from x to y in tree T, and let $maxweight\{A\}$ be the maximum weight in a set of edges A.

For any forest F in G, define an edge (a, b) in G to be F-heavy if $weight(a, b) > maxweight\{P_F(a, b)\}$ and to be F-light otherwise. If a and b are not in the same tree in F then (a, b) is F-light.

Let M be k-Min(v). We define $weight_v(w)$ to be $maxweight\{P_M(v, w)\}$ if w appears in k-Min(v), otherwise $weight_v(w) = maxweight\{k\text{-Min}(v)\}$. Define an edge (a, b) to be k-Min-heavy if $weight(a, b) > \max\{weight_a(b), weight_b(a)\}$, and to be k-Min-light otherwise.

Claim 1 *Let the measure $weight_v(w)$ be defined with respect to any k in the range $[1..n]$. Then $weight_v(w) \leq maxweight\{P_{MSF}(v, w)\}$.*

Proof. There are two cases, either w is inside k-Min(v), or it is outside. If w is inside k-Min(v), then $weight_v(w)$ is the same as $maxweight\{P_{MSF}(v, w)\}$ since k-Min(v) $\subseteq MSF$. Now suppose that w falls outside k-Min(v) and $weight_v(w) > maxweight\{P_{MSF}(v, w)\}$. There must be a path from v to w in the MSF consisting of edges lighter than $maxweight\{k\text{-Min}(v)\}$. However, at each step in the Dijkstra-Jarnik-Prim algorithm, at least one edge in P_{MSF} is eligible to be chosen in that step. Since $w \notin k$-Min(v), the edge with weight $maxweight\{k\text{-Min}(v)\}$ is never chosen. Contradiction.

Let K be a vector of n values, each in the range $[1..n]$. Each vertex u is associated with a value of K, denoted k_u. Define an edge (u, v) to be K-Min-light if $weight(u, v) < \max\{weight_u(v), weight_v(u)\}$, where $weight_u(v)$ and $weight_v(u)$ are defined with respect to k_u and k_v respectively.

Lemma 1. *Let H be a graph formed by sampling each edge in graph G with probability p. The expected number of edges in G that are K-Min-light in H is less than n/p, for any K.*

Proof. We show that any edge that is K-Min-light in G is also F-light where F is the MSF of H. The lemma then follows from the sampling lemma of [KKT95] which states that the expected number of F-light edges in G is less than n/p. Let us look at any K-Min-light edge (v, w). By Claim 1, $weight_v(w) \leq maxweight\{P_{MSF}(v, w)\}$, the measure used to determine F-lightness. Thus the criterion for K-Min-lightness, $\max\{weight_v(w), weight_w(v)\}$, must also be less than or equal to $maxweight\{P_{MSF}(v, w)\}$.

3.2 Borůvka-A Steps

In a basic Borůvka step [Bor26], each vertex chooses its minimum weight incident edge, inducing a number of disjoint trees. All such trees are then contracted into single vertices, and useless edges discarded. Call edges connecting two vertices in the same tree *internal* and all others *external*. All internal edges are useless, and if multiple external edges join the same two trees, all but the lightest are useless.

Phase 1 uses a modified Borůvka step to reduce the time bound to $o(\log n)$ per step. All vertices are classified as *live* or *dead*. After a modified Borůvka step, vertex v's *parent pointer* is $p(v) = w$, where (v, w) is v's min-weight edge. Each vertex has a *threshold* which keeps the weight of the lightest discarded edge adjacent to v. The algorithm discards edges known not to be in the k-Min tree of any vertex. The threshold variable guards against vertices choosing edges which may not be in the MSF. A dead vertex v has the useful property that for any edge e in k-Min(v), weight$(e) \leq$ weight$(v, p(v))$, thus dead vertices *need not participate* in any more Borůvka steps.

It is well-known that a Borůvka step generates a forest of *pseudo-trees*, where each pseudo-tree is a tree together with an extra edge forming a cycle of length 2. We assume that a Borůvka step also removes one of the edges in the cycle so that it generates a collection of rooted trees.

The following four claims refer to any tree resulting from a modified Borůvka step. Their proofs are straightforward and are omitted.

Claim 2 *The sequence of edge weights encountered on a path from v to root(v) is monotonically decreasing.*

Claim 3 *If depth$(v) = d$ then d-Min(v) consists of the edges in the path from v to root(v). Furthermore, $(v, p(v))$ is the heaviest edge in d-Min(v).*

Claim 4 *If u's min-weight edge is (u, v), then k-Min$(u) \subseteq k$-Min$(v) \cup \{(u, v)\}$.*

Claim 5 *Let T be a tree induced by a Borůvka step, T' a subtree of T, and e the min-weight edge on T, then the min-weight edge on T' is e or an edge of T.*

The procedure Borůvka-A(H, l, F) returns a contracted version of H with $\leq n/l$ live vertices. F contains edges identified as being in the MSF of H.

Borůvka-A(H, l, F)
    Repeat $\log l$ times: *($\log l$ modified Borůvka steps)*
        $F' := \emptyset$
        For each live vertex v
            Choose min. weight edge (v, w)
(1)   If weight$(v, w) >$ threshold(v), v becomes dead, stop else
                $p(v) := w$
                $F' := F' \cup \{(v, p(v))\}$
        Each tree T induced by edges of F' is one of two types:
            If root of T is dead, then
(2)   Every vertex in T becomes dead
            If T contains only live vertices
(3)   If depth$(v) \geq k$, v becomes dead
                Contract the subtree of T made up of live vertices.
                The resulting vertex is live, has no parent pointer, and
                keeps the smallest threshold of its constituent vertices.
    $F := F \cup F'$

Lemma 2. *If a vertex is pronounced dead, its k-Min tree has already been found.*

Proof. Vertices become dead only at the lines indicated by a number. Since we only discard edges that are in no vertex's k-Min tree, if the min-weight edge on any vertex is discarded, we know its k-Min tree has been found. This covers line (1). The correctness of line (2) follows from Claim 4. Since $(v, p(v))$ is the lightest incident edge on v, k-Min$(v) \subseteq k$-Min$(p(v)) \cup \{(v, p(v))\}$. If $p(v)$ is dead, then v can also be called dead. Since the root of a tree is dead, vertices at depth one are dead, implying vertices at depth two are dead, and so on. The validity of line (3) follows directly from Claim 3. If a vertex finds itself at depth $\geq k$, its k-Min tree lies along the path from the vertex to its root.

Lemma 3. *After a call to Borůvka-A$(H, k+1, F)$, F is a k-Min forest.*

Proof. By Lemma 2, dead vertices already satisfy the lemma. After a single Borůvka step, the set of parent pointers associated with live vertices induce a number of trees. Let $T(v)$ be the tree containing v. We assume inductively that after $\lceil \log i \rceil$ Borůvka steps, the $(i-1)$-Min tree of each vertex in the original graph has been found (this is clearly true for $i = 1$). For any live vertex v let (x, y) be the min-weight edge s.t. $x \in T(v), y \notin T(v)$. By the inductive hypothesis, the $(i-1)$-Min trees of v and y are subsets of $T(v)$ and $T(y)$ respectively. By Claim 5, (x, y) is the first external edge of $T(v)$ chosen by the Dijkstra-Jarnik-Prim algorithm, starting at v. Thus $(2(i-1)+1)$-Min(v) is a subset of $T(v) \cup \{(x, y)\} \cup T(y)$. Since edge (x, y) is chosen in the next Borůvka step, $(2i-1)$-Min(v) is a subset of $T(v)$ after $\lceil \log i \rceil + 1 = \lceil \log 2i \rceil$ Borůvka steps. Thus after $\log(k+1)$ steps, the k-Min tree of each vertex has been found.

Lemma 4. *After b modified Borůvka steps, all edge lists have length $\leq k^{k^b}$.*

Proof. This is true for $b = 0$. Inductively assume edge lists have length $\leq k^{k^{b-1}}$ after $b - 1$ steps. Since we only contract trees of height $< k$, the length of any edge list after b steps is $< (k^{k^{b-1}})^k = k^{k^b}$.

Later we show that our algorithm only deals with graphs resulting from $O(\log k)$ modified Borůvka steps. Hence the longest edge list is $k^{k^{O(\log k)}}$.

The costliest step in Borůvka-A is calculating the depth of each vertex. After the edge selection process, the root of each induced tree broadcasts its depth to all depth 1 vertices, which in turn broadcast to depth 2 vertices, etc. Vertices at depth $k - 1$ stop, letting their descendents infer that they are at depth $\geq k$.

Lemma 5. *Let G_1 have m_1 edges. Then a call to Borůvka-A(G_1, l, F) can be executed in time $O(k^{O(\log k)} + \log l \cdot \log n \cdot (m_1/m))$ with $(m+n)/\log n$ processors.*

Proof. (Sketch) The first term is a bound on the time needed to process edge lists. The second term represents the slowdown given $O(m/\log n)$ processors.

3.3 The Filtering Step

The Filter Forest

Concurrent with each modified Borůvka step, we will maintain a Filter forest, a structure that records which vertices merged together at what time, and the edge weights involved. (This structure appeared first in [King97]). If v is a vertex of the original graph, or a new vertex resulting from contracting a set of edges, there is a corresponding vertex $\phi(v)$ in the Filter forest. During a Borůvka step, if a vertex v becomes dead, a new vertex w is added to the Filter forest, as well as a directed edge $(\phi(v), w)$ having the same weight as $(v, p(v))$. If live vertices $v_1, \ldots, v_j$ are contracted into a live vertex v, a vertex $\phi(v)$ is added to the Filter forest in addition to directed edges $(\phi(v_1), \phi(v)), \ldots, (\phi(v_j), \phi(v))$, having the weights of edges $(v_1, p(v_1)), \ldots, (v_j, p(v_j))$, respectively.

It is shown in [King97] that the heaviest weight in the path from u to v in the MSF is the same as the heaviest weight in the path from $\phi(u)$ to $\phi(v)$ in the Filter forest (if there is such a path). Hence the measures $weight_v(w)$ can be easily computed in the following way. Let $P_f(x, y)$ be the path from x to y in the Filter forest. If $\phi(v)$ and $\phi(w)$ are not in the same Filter tree, then

$$weight_v(w) = maxweight\{P_f(\phi(v), root(\phi(v)))\} \text{ and}$$
$$weight_w(v) = maxweight\{P_f(\phi(w), root(\phi(w)))\}$$

If v and w are in the same Filter tree, then

$$weight_v(w) = weight_w(v) = maxweight\{P_f(\phi(v), \phi(w))\}$$

Claim 6 *The maximum weight on the path from $\phi(v)$ to $root(\phi(v))$ is the same as the maximum weight edge in r-Min(v), for some r.*

Proof. If $root(\phi(v))$ is at height h, then it is the result of h Borůvka steps. Assume that the claim holds for the first $i < h$ Borůvka steps. After a number of contractions, vertex v of the original graph is now represented in the current graph by v_c. Let T_{v_c} be the tree induced by the ith Borůvka step which contains v_c, and let e be the minimum weight incident edge on T_{v_c}. By the inductive hypothesis, $maxweight\{P_f(\phi(v), \phi(T_{v_c}))\} = maxweight\{r'\text{-Min}(v)\}$ for some r'. As was shown in the proof of Claim 5, all edges on the path from v_c to edge e have weight at most $\max\{weight(v_c, p(v_c)), weight(e)\}$. Each of the edges $(v_c, p(v_c))$ and e has a corresponding edge in the Filter forest, namely $(\phi(v_c), p(\phi(v_c)))$ and $(\phi(T_{v_c}), p(\phi(T_{v_c})))$. Since both these edges are on the path from $\phi(v)$ to $p(\phi(T_{v_c}))$, $maxweight\{P_f(\phi(v), p(\phi(T_{v_c})))\} = maxweight\{r\text{-Min}(v)\}$ for some $r \geq r'$. Thus the claim holds after $i + 1$ Borůvka steps.

The Filter Step

In a call to Filter(H, F) in Find-k-Min, we examine each edge $e = (v, w)$ in $H - F$, and delete e from H if $weight(e) > \max\{weight_v(w), weight_w(v)\}$ The Filter procedure is derived by modifying the $O(\log n)$ time, $O(m)$ work MSF verification algorithm of [KPRS97]. When v and w are not in the same tree, Filter tests the pairs $(\phi(v), root(\phi(v)))$ and $(\phi(w), root(\phi(w)))$. We delete e if both of these pairs are identified to be deleted. This computation will take time $O(\log r)$ where r is the size of the largest Filter tree.

Filter discards edges that cannot be in the k-Min tree of any vertex. When it discards an edge (a, b), it updates the *threshold* variables of both a and b, so that threshold(a) is the weight of the lightest discarded edge adjacent to a. If a's minimum weight edge is ever heavier than threshold(a), k-Min(a) has already been found, and a becomes dead.

The proof of the following Claim follows easily from Lemma 1 and Claim 6, and can be found in [PR99].

Claim 7 *Let H' be a graph formed by sampling each edge in H with probability p, and F be a k-Min forest of H'. The call to Filter(H, F) returns a graph containing a k-Min forest of H, whose expected number of edges is n/p.*

3.4 Finding a k-Min Forest

We are now ready to present the main procedure of Phase 1, Find-k-Min. (Recall that the initial call – given in Section 2 – is Find-k-Min$(G_t, \log^* n)$, where G_t is the graph obtained from G by removing all but the k lightest edges on each adjacency list.)

> **Find-k-Min(H, i)**
> $H_c := \text{Borůvka-A}(H, (\log^{(i-1)} n)^4, F)$
> if $i = 3$, return(F)
> $H_s := \text{sample edges of } H_c \text{ with prob. } 1/(\log^{(i-1)} n)^2$
> $F_s := \text{Find-}k\text{-Min}(H_s, i-1)$
> $H_f := \text{Filter}(H_c, F_s)$
> $F' := \text{Find-}k\text{-Min}(H_f, i-1)$
> Return$(F \cup F')$

H is a graph with some vertices possibly marked as dead; i is a parameter that indicates the level of recursion.

Lemma 6. *Find-k-Min$(G_t, \log^* n)$ returns a k-Min forest of G_t.*

Proof. The proof is by induction on i. For $i = 3$, Find-k-Min$(H, 3)$ returns F, which by Lemma 3 contains the k-Min tree of each vertex. Assume inductively that Find-k-Min$(H, i-1)$ returns the k-Min tree of H. Consider the call Find-k-Min(H, i). By the inductive assumption the call to Find-k-Min$(H_s, i-1)$ returns the k-Min tree of each vertex in H_s. By Claim 7 the call to Filter(H_c, F_s) returns in H_f a set of edges that contains the k-Min trees of all vertices in H_c. Finally, by the inductive assumption, the set of edges returned by the call to Find-k-min$(H_f, i-1)$ contains the k-Min trees of all vertices in H_f. Since F' contains the $(\log^{(i-1)} n)$-Min tree of each vertex in H, and Find-k-Min(H, i) returns $F \cup F'$, it returns the edges in the k-Min tree of each vertex in H.

Claim 8 *The number of live vertices in $H \leq n/(\log^{(i)} n)^4$, and the expected number of edges in $H \leq m/(\log^{(i)} n)^2$.*

Proof. These invariants hold initially, when $i = \log^* n$. By Lemma 3, the contracted graph H_c has no more than $n/(\log^{(i-1)} n)^4$ live vertices. Since H_s is derived by sampling edges with probability $1/(\log^{(i-1)} n)^2$, the expected number of edges in H_s is $\leq m/(\log^{(i-1)} n)^2$, maintaining the invariants for the first recursive call. By Lemma 1, the expected number of edges in $H_f \leq \frac{n(\log^{(i-1)} n)^2}{(\log^{(i-1)} n)^4} = n/(\log^{(i-1)} n)^2$. Since H_f has the same number of vertices as H_c, both invariants are maintained for the second recursive call.

3.5 Performance of Find-k-Min

Lemma 7. *Find-k-min$(G_t, \log^* n)$ runs in expected time $O(\log n)$ and work $O(m)$.*

Proof. The total running time is the sum of the local computation performed in each invocation. Aside from randomly sampling the edges, the local computation consists of calls to Filter and Borůvka-A.

In a given invocation of Find-k-min, the number of Borůvka steps performed on graph H is the sum of all Borůvka steps performed in all ancestral invocations of Find-k-min, i.e. $\leq \sum_{i=3}^{\log^* n} O(\log^{(i)} n) = O(\log^{(3)} n)$. From Lemma 4 we can infer that the size of any tree in the Filter forest is $k^{k^{O(\log^{(3)} n)}}$, thus the time needed for each modified Borůvka step and each Filter step is $k^{O(\log^{(3)} n)}$. Summing over all such steps, the total time required is $o(\log n)$.

The work required by the Filter procedure and *each* Borůvka step is linear in the number of edges. Since the number of edges is $O(m/(\log^{(i)} n)^2)$, and there are $O(\log^{(i)} n)$ Borůvka steps performed in this invocation, the work required in each invocation is $O(m/\log^{(i)} n)$. There are $2^{\log^* n - i}$ invocations with depth parameter i, hence the total work is given by $\sum_{i=3}^{\log^* n} 2^{\log^* n - i} O(m/\log^{(i)} n)$, which is $O(m)$.

4 Phase 2

Recall the Phase 2 portion of our overall algorithm **High-Level**:

(the number of vertices in G_s is $\leq n/k$)

G_s :=Sample edges of G' with prob. $1/\sqrt{k} = 1/\log^{(2)} n$
F_s :=Find-MSF(G_s)
G_f := Filter(G', F_s)
F := Find-MSF(G_f)

The procedure Filter(G, F) ([KPRS97]) returns the F-light edges of G in logarithmic time and linear work. The procedure Find-MSF(G_1), finds the MSF of G_1 in time $O((m_1/m) \cdot \log n \log^{(2)} n)$, where m_1 is the number of edges in G_1.

The graphs G_s and G_f each have expected $m/\sqrt{k} = m/\log^{(2)} n$ edges. G_s is derived by sampling each edge with probability $1/\sqrt{k}$, and by the sampling lemma of [KKT95], the expected number of edges in G_f is $(m/k)/(1/\sqrt{k}) = m/\sqrt{k}$. Because we call Find-MSF on graphs having expected size $O(m/\log^{(2)} n)$, each call takes $O(\log n)$ time.

4.1 The Find-MSF Procedure

The procedure Find-MSF(H) is similar to previous randomized parallel algorithms, except it uses no recursion. Instead, a separate *base case* algorithm is used in place of recursive calls. We also use slightly different Borůvka steps, in order to reduce the work. These modifications are inspired by [PR97] and [PR98] respectively.

As its Base-case, we use the algorithm of Chong, Han, and Lam [CHL99], which takes time $O(\log n)$ using $m + n$ processors. By guaranteeing that it is only called on graphs of expected size $O(m/\log n)$, the running time remains $O(\log n)$ with $(m + n)/\log n$ processors.

> **Find-MSF(H)**
> $H_c := \text{Borůvka-B}(H, \log^2 n, F)$
> $H_s := \text{Sample edges of } H_c \text{ with prob. } p = 1/\log n$
> $F_s := \text{BaseCase}(H_s)$
> $H_f := \text{Filter}(H_c, F_s)$
> $F' := \text{BaseCase}(H_f)$
> $\text{Return}(F \cup F')$

After the call to Borůvka-B, the graph H_c has $< m/\log^2 n$ vertices. Since H_s is derived by sampling the edges of H_c with probability $1/\log n$, the expected number of edges to the first BaseCase call is $O(m/\log n)$. By the sampling lemma of [KKT95], the expected number of edges to the second BaseCase call is $< (n/\log^2 n)/(1/\log n)$, thus the total time spent in these subcalls is $O(\log n)$. Assuming the size of H conforms to its expectation of $O(m/\log^{(2)} n)$, the calls to Filter and Borůvka-B also take $O(\log n)$ time, as described below.

The Borůvka-B(G_1, l, F) procedure returns a contracted version of G_1 with $\leq m/l$ vertices. If G_1 has m_1 edges, it requires time $O(\log n + \log^2 l)$ and work $O(m_1 \log l)$. Borůvka-B uses a simple growth control schedule similar to the one used in [JM92]. We omit the details.

5 Proof of Main Theorem

Proof. (Of Theorem 1.) The set of edges M returned by Find-k-Min is a subset of the MSF of G. By contracting the edges of M to produce G', the MSF of G is given by the edges of M together with the MSF of G'. The call to Filter produces graph G_f by removing from G' edges known not to be in the MSF. Thus the MSF of G_f is the same as the MSF of G'. Assuming the correctness of Find-MSF, the set of edges F constitutes the MSF of G_f, thus $M \cup F$ is the MSF of G.

Earlier we have shown that each step of High-Level requires $O(\log n)$ time and expected work $O(m)$. In the next section we show that w.h.p. the total work is linear in the number of edges.

6 High Probability Bounds

Consider a single invocation of Find-k-min(H, i), where H has m' edges and n' vertices. We want to place likely bounds on the number of edges in each recursive call to Find-k-min, in terms of m' and i.

For the first recursive call, the edges of H are sampled independently with probability $1/(\log^{(i-1)} n)^2$. Call the sampled graph H_1. By applying a Chernoff bound, the probability that the size of H_1 is less than twice its expectation is $1 - \exp(-\Omega(m'/(\log^{(i-1)} n)^2))$.

Before analyzing the second recursive call, we recall the sampling lemma of [KKT95] which states that the number of F-light edges conforms to the negative binomial distribution with parameters n' and p, where p is the sampling probability, and F is the MSF of H_1. As we saw in the proof of Lemma 1, every k-Min-light edge must also be F-light. Using this observation, we will analyze the size of the second recursive call in terms of F-light edges, and conclude that any bounds we attain apply equally to k-Min-light edges.

We now bound the likelihood that more than twice the expected number of edges are F-light. This is the probability that in a sequence of more than $2n'/p$ flips of a coin, with probability p of heads, the coin comes up heads less than n' times (since each edge selected by a coin toss of heads goes into the MSF of the sampled graph). By applying a Chernoff bound, this is $\exp(-\Omega(n'))$. In this particular instance of Find-k-min, $n' \leq m/(\log^{(i-1)} n)^4$ and $p = 1/(\log^{(i-1)} n)^2$, so the probability that fewer than $2m/(\log^{(i-1)} n)^2$ edges are F-light is $1 - \exp(-\Omega(m/(\log^{(i-1)} n)^4))$.

Given a single invocation of Find-k-min(H, i), we can bound the probability that H has more than $2^{\log^* n - i} m/(\log^{(i)} n)^2$ edges by $\exp(-\Omega(m/(\log^{(i)} n)^4))$. This follows from applying the argument used above to each invocation of Find-k-min from the initial call to the current call at depth $\log^* n - i$. Summing over all recursive calls to Find-k-min, the total number of edges (and thus the total work) is bounded by $\sum_{i=3}^{\log^* n} 2^{2\log^* n - 2i} m/(\log^{(i)} n)^2 = O(m)$ with probability $1 - \exp(-\Omega(m/(\log^{(3)} n)^4))$.

The analysis of Phase 2 is entirely analogous and much simpler as it does not have to address the effect of recursive calls. We omit the details.

The probability that our bounds on the time and total work performed by the algorithm fail to hold is exponentially small in the input size. However, this assumes perfect processor allocation. If we use one of the allocation schemes described in [HZ94, HZ96, PR99], the probability that work fails to be distributed evenly among the processors is less than $1/m^{\omega(1)}$. Thus the overall probability of failure is very small, and the algorithm runs in logarithmic time and linear work w.h.p.

References

[AS87] B. Awerbuch, Y. Shiloach. New connectivity and MSF algorithms for shuffle-exchange networks and PRAM. *IEEE Trans. Computers*, vol. C-36, 1987, pp. 1258-1263.

[Bor26] O. Borůvka . O jistém problému minimaálním. *Moravské Přírodovědecké Společnosti* 3, (1926), pp. 37-58. (In Czech).

[CHL99] K. W. Chong, Y. Han, T. W. Lam. On the parallel time complexity of undirected connectivity and minimum spanning trees. In *Proc. SODA* 1999, pp. 225-234.

[CKT94] R. Cole, P.N. Klein, R.E. Tarjan. A linear-work parallel algorithm for finding minimum spanning trees. In *Proc. SPAA*, 1994, pp. 11–15.

[CKT96] R. Cole, P.N. Klein, R.E. Tarjan. Finding minimum spanning trees in logarithmic time and linear work using random sampling. In *Proc. SPAA*, 1996, pp. 213–219.

[Gaz91] H. Gazit An optimal randomized parallel algorithm for finding connected components in a graph. *SICOMP*, vol. 20, 1991, pp. 1046-1067.

[GMR94] P.B. Gibbons, Y. Matias, V. Ramachandran. The QRQW PRAM: Accounting for contention in parallel algorithms. In *Proc. SODA*, 1994, pp. 638–648 (*SICOMP* 1999, to appear.)

[GMR97] P. B. Gibbons, Y. Matias, V. Ramachandran. Can a shared-memory model serve as a bridging model for parallel computation? In *Proc. SPAA*, 1997, pp. 72–83. (*Theory of Comp. Sys.*, 1999, to appear.)

[HZ94] S. Halperin, U. Zwick. An optimal randomized logarithmic time connectivity algorithm for the EREW PRAM. In *Proc. SPAA*, 1994, pp. 1-10.

[HZ96] S. Halperin, U. Zwick. Optimal randomized EREW PRAM algorithms for finding spanning forests and for other basic graph connectivity problems. In *Proc. SODA*, 1996, pp. 438–447, 1996.

[JM92] D. B. Johnson, P. Metaxas. Connected components in $O(\log^{3/2} n)$ parallel time for CREW PRAM. *JCSS*, vol. 54, 1997, pp. 227–242.

[King97] V. King. A simpler minimum spanning tree verification algorithm. *Algorithmica*, vol. 18, 1997, pp. 263-270.

[KKT95] D. R. Karger, P. N. Klein, R. E. Tarjan. A randomized linear-time algorithm to find minimum spanning trees. *JACM*, 42:321-328, 1995.

[KPRS97] V. King, C. K. Poon, V. Ramachandran, S. Sinha. An optimal EREW PRAM algorithm for minimum spanning tree verification. *IPL*, 62(3):153–159, 1997.

[PR97] C. K. Poon, V. Ramachandran. A randomized linear work EREW PRAM algorithm to find a minimum spanning forest. *Proc ISAAC*, 1997, pp. 212.-222.

[PR98] C. K. Poon, V. Ramachandran. Private communication, 1998.

[PR99] S. Pettie, V. Ramachandran. A Randomized Time-Work Optimal Parallel Algorithm for Finding a Minimum Spanning Forest Tech. Report TR99-13, Univ. of Texas at Austin, 1999.

[Val90] L. G. Valiant. A bridging model for parallel computation. *CACM*, 33(8):103–111, 1990.

Fast Approximate PCPs for Multidimensional Bin-Packing Problems *

Tuğkan Batu[1], Ronitt Rubinfeld[1]**, and Patrick White[1]

Department of Computer Science, Cornell University, Ithaca, NY 14850
{batu,ronitt,white}@cs.cornell.edu

Abstract. We consider approximate PCPs for multidimensional bin-packing problems. In particular, we show how a verifier can be quickly convinced that a set of multidimensional blocks can be packed into a small number of bins. The running time of the verifier is bounded by $O(T(n))$, where $T(n)$ is the time required to test for *heaviness*. We give heaviness testers that can test heaviness of an element in the domain $[1, \ldots, n]^d$ in time $O((\log n)^d)$. We also also give approximate PCPs with efficient verifiers for recursive bin packing and multidimensional routing.

1 Introduction

Consider a scenario in which the optimal solution to a very large combinatorial optimization problem is desired by a powerful corporation. The corporation hires an independent contractor to actually find the solution. The corporation then would like to trust that the *value* of the solution is feasible, but might not care about the structure of the solution itself. In particular they would like to have a quick and simple test that checks if the contractor has a good solution by only inspecting a very small portion of the solution itself. Two hypothetical situations in which this might occur are:

- A major corporation wants to fund an international communications network. Data exists for a long history of broadcasts made over currently used networks, including bandwidth, duration, and integrity of all links attempted. The corporation wants to ensure that the new network is powerful enough to handle one hundred times the existing load.
- The services of a trucking company are needed by an (e-)mail-order company to handle all shipping orders, which involves moving large numbers of of boxes between several locations. The mail-order company wants to ensure that the trucking company has sufficient resources to handle the orders.

In both cases, large amounts of typical data are presented to the consulting company, which determines whether or not the load can be handled. The probabilistically checkable-proof (PCP) techniques (cf. [3, 4, 1]) offer ways of verifying such solutions quickly. In

* This work was partially supported by ONR N00014-97-1-0505, MURI, NSF Career grant CCR-9624552, and an Alfred P. Sloan Research Award. The third author was supported in part by an ASSERT grant.
** Part of this work was done while on sabbatical at IBM Almaden Research Center

these protocols a proof is written down which a verifier can trust by inspecting only a constant number of bits of the proof. The PCP model offers efficient mechanisms for verifying any computation performed in NEXP with an efficient verifier. We note that the verifiers in the PCP results all require $\Omega(n)$ time. Approximate PCPs were introduced in [7] for the case when the input data is very large, and even linear time is prohibitive for the verifier. Fast approximate PCPs allow a verifier to ensure that the answer to the optimization problem is at least *almost* correct. Approximate PCPs running in logarithmic or even constant time have been presented in [7] for several combinatorial problems. For example, a proof can be written in such a way as to convince a constant time verifier that there exists a bin-packing which packs a given set of objects into a small number of bins. Other examples include proofs which show the existence of a large flow, a large matching, or a large cut in a graph to a verifier that runs in sublinear time.

Our Results. We consider approximate PCPs for multidimensional bin packing. In particular, we show how a verifier can be quickly convinced that a set of multidimensional objects can be packed into a small number of bins. We also consider the related problems of recursive bin packing and multidimensional routing. Our results generalize the 1-dimensional bin packing results of [7]. The PCPs are more intricate in higher dimensions; for example, the placements and orientations of the blocks within the bin must be considered more carefully. In the 1-dimensional case, the approximate PCP of [7] makes use of a property called *heaviness* of an element in a list, introduced by [6]. Essentially, *heaviness* is defined so that testing if an element is heavy can be done very efficiently (logarithmic) in the size of the list and such that all heavy elements in the list are in monotone increasing order. We generalize this notion to the multidimensional case and give heaviness tests which determine the heaviness of a point $x \in [1, \ldots, n]^d$ in time $O((2 \log n)^d)$. Then, given a heaviness tester which runs in time $T(n)$, we show how to construct an approximate PCP for binpacking in which the running time of the verifier is $O(T(n))$.

In [9], multidimensional monotonicity testers are given which pass functions f that are monotone and fail functions f if no way of changing the value of f at at most ϵ fraction of the inputs will turn f into a monotone function. The query complexity of their tester is $\tilde{O}(d^2 n^2 r)$ where f is a function from $[n]^d$ to $[r]$. Our multidimensional heaviness tester can also be used to construct a multidimensional monotonicity tester which runs in time $O(T(n))$. However, more recently Dodis *et. al.* [5] have given monotonicity testers that greatly improve on our running times for dimension greater than 2, and are as efficient as ours for dimension 2. This gives hope that more efficient heaviness testers in higher dimensions can also be found.

2 Preliminaries

Notation. We use the notation $x \in_R S$ to indicate x is chosen uniformly and at random from the set S. The notation $[n]$ indicates the interval $[1, \ldots, n]$.

We define a partial ordering relation $\prec$ over integer lattices such that if x and y are d-tuples then $x \prec y$ if and only if $x_i \leq y_i$ for all $i \in \{1, \ldots, d\}$. Consider a function

$f : \mathcal{D}^d \to \mathcal{R}$, where $\mathcal{D}^d$ is a d-dimensional lattice. If $x, y \in \mathcal{D}^d$ are such that $x \prec y$ then if $f(x) \leq f(y)$ we say that x and y are in *monotone order*. We say f is *monotone* if for all $x, y \in \mathcal{D}^d$ such that $x \prec y$, x and y are in monotone order.

Approximate PCP. The approximate PCP model is introduced in [7]. The verifier has access to a written proof, Π, which it can query in order to determine whether the theorem it is proving is *close* to correct. More specifically, if on input x, the prover claims $f(x) = y$, then the verifier wants to know if y is close to $f(x)$.

Definition 1. *[7] Let $\Delta(\cdot, \cdot)$ be a distance function. A function f is said to have a $t(\epsilon, n)$-approximate probabilistically checkable proof system with distance function Δ if there is a randomized verifier V with oracle access to the words of a proof Π such that for all inputs ϵ, and x of size n, the following holds. Let y be the contents of the output tape, then:*

1. *If $\Delta(y, f(x)) = 0$, there is a proof, Π, such that V^{Π} outputs pass with probability at least 3/4 (over the internal coin tosses of V);*
2. *If $\Delta(y, f(x)) > \epsilon$, for all proofs Π', $V^{\Pi'}$ outputs fail with probability at least 3/4 (over the internal coin tosses of V); and*
3. *V runs in $O(t(\epsilon, n))$ time.*

The probabilistically checkable proof protocol can be repeated $O(\lg 1/\delta)$ times to get confidence $\geq 1 - \delta$. We occasionally describe the verifier's protocol as an interaction with a prover. In this interpretation, it is assumed that the prover is bound by a function which is fixed before the protocol begins. It is known that this model is equivalent to the PCP model [8]. The verifier is a RAM machine which can read a word in one step.

We refer to PCP using the distance function $\Delta(y, f(x)) = \max\{0, 1 - f(x)/y\}$ as an *approximate lower bound* PCP : if $f(x) \geq y$ then Π causes V^{Π} to pass; if $f(x) < (1 - \epsilon)y$ then no proof Π' convinces $V^{\Pi'}$ with high probability. This distance function applied to our bin-packing protocol will show that if a prover claims to be able to pack all of the n input objects, the verifier can trust that at least $(1 - \epsilon)n$ of the objects can be packed.

It also follows from considerations in [7] that the protocols we give can be employed to prove the existence of suboptimal solutions. In particular, if the prover knows a solution of value v, it can prove the existence of a solution of value at least $(1 - \epsilon)v$. Since v is not necessarily the optimal solution, these protocols can be used to trust the computation of approximation algorithms to the NP-complete problems we treat. This is a useful observation since the prover may not have computational powers outside of deterministic polynomial time, but might employ very good heuristics. In addition, since the prover is much more powerful than V it may use its computational abilities to get surprisingly good, yet not necessarily optimal, solutions.

Heaviness Testing. Our methods all rely on the ability to define an appropriate *heaviness* property of a function f. In the multidimensional case, heaviness is defined so that testing if a domain element is heavy can be done very efficiently in the size of the domain, and such that all heavy elements in the domain which are comparable according to $\prec$ are in monotone order.

We give a simple motivating example of a heaviness test for $d = 1$ from [6]. This one-dimensional problem can be viewed as the problem of testing whether a list $L = (f(1), f(2), \ldots, f(n))$ is mostly sorted. Here we assume that the list contains distinct elements (a similar test covers the nondistinct case). Consider the following for testing whether such a list L is mostly sorted: pick a point $x \in L$ uniformly and at random. Perform a binary search on L for the value x. If the search finds x then we call x *heavy*. It is simple to see that if two points x and y are heavy according to this definition, then they are in correct sorted order (since they are each comparable to their common ancestor in the search tree). The definition of a heaviness property is generalized in this paper. We can call a property a *heaviness property* if it implies that points with that property are in monotone order.

Definition 2. *Given a domain* $D = [1, \ldots, n]^d$, *a function* $f : D \to R$ *and a property* H, *we say that* H *is a* heaviness property *if*

1. $\forall x < y \; H(x) \wedge H(y)$ *implies* $f(x) \leq f(y)$
2. *In a monotone list all points have property* H

If a point has a heaviness property H then we say that point is *heavy*. There may be many properties which can be tested of points of a domain which are valid heaviness properties. A challenge of designing heaviness tests is to find properties which can be tested efficiently. A heaviness test is a probabilistic procedure which decides the heaviness property with high probability. If a point is not heavy, it should fail this test with high probability, and if a function is perfectly monotone, then every point should pass. Yet it is possible that a function is not monotone, but a tested point is actually heavy. In this case the test may either pass or fail.

Definition 3. *Let* $\mathcal{D}_{\prec} = [1, \ldots, n]^d$ *be a domain, and let* $f : \mathcal{D} \to \mathcal{R}$ *be a function on* D. *Let* $S(\cdot, \cdot)$ *be a randomized decision procedure on* $\mathcal{D}$. *Given security parameter* δ, *we will say* S *is a* heaviness test *for* x *if*

1. *If for all* $x \prec y$, $f(x) \leq f(y)$ *then* $S(x, \delta) = \text{Pass}$
2. *If* x *is not heavy then* $\Pr(S(x, \delta) = \text{Fail}) > 1 - \delta$

The heaviness tests we consider enforce, among other properties, local multidimensional monotonicity of certain functions computed by the prover. It turns out that multidimensional heaviness testing is more involved that the one dimensional version considered in earlier works, and raises a number of interesting questions.

Our results on testing bin-packing solutions are valid for any heaviness property, and require only a constant number of applications of a heaviness test. We give sample heaviness properties and their corresponding tests in Section 4, yet it is an open question whether heaviness properties with more efficient heaviness tests exist. Such tests would immediately improve the efficiency of our approximate PCP verifier for bin-packing.

Permutation Enforcement. Suppose the values of a function f are given for inputs in $[n]$ in the form of a list $y_1, \ldots, y_n$. Suppose further that the prover would like to convince the verifier that the y_i's are distinct, or at least that there are $(1 - \epsilon)n$ distinct y_i's. In [7], the following method is suggested: The prover writes array A of length n. $A(j)$

should contain i when $f(i) = j$ (its preimage according to f). We say that i is *honest* if $A(f(i)) = i$. Note that the number of honest elements in $[n]$ lower bounds the number of distinct elements in $y_1, \ldots, y_n$ (even if A is written incorrectly). Thus, sampling $O(1/\epsilon)$ elements and determining that all are honest suffices to ensure that there are at least $(1 - \epsilon)n$ distinct y_i's in $O(1/\epsilon)$ time. We refer to A as the *permutation enforcer*.

3 Multidimensional Bin-Packing

We consider the d-dimensional bin-packing problem. We assume the objects to be packed are d-dimensional rectangular prisms, which we will hereafter refer to as blocks. The blocks are given as d-tuples (in $\mathbb{N}^d$) of their dimensions. Similarly, the bin size is given as a d-tuple, with entries corresponding to the integer width of the bin in each dimension. When we say a block with dimensions $w = (w_1, \ldots, w_d) \in \mathbb{N}^d$ is located at position $x = (x_1, \ldots, x_d)$, we mean that all the locations y such that $x \prec y \prec x + w$ are occupied by this block. The problem of multidimensional bin-packing is to try to find a packing of n blocks which uses the least number of bins of given dimension $\mathcal{D} = (N_1, \ldots, N_d)$.

It turns out to be convenient to cast our problem as a maximization problem. We define the d-dimensional bin-packing problem as follows: given n blocks, the dimensions of a bin, and an integer k, find a packing that packs the largest fraction of the blocks into k bins. It follows that if $1 - \epsilon$ fraction of the blocks can be packed in k bins, then at most $k + \epsilon n$ bins are sufficient to pack all of the blocks, by placing each of the remaining blocks in separate bins.

We give an approximate lower bound PCP protocol for the maximization version of the d-dimensional bin-packing problem in which the verifier runs in $O((1/\epsilon)T(N, d))$ time where $T(N, d)$ is the running time for a heaviness tester on $\mathcal{D} = [N_1] \times \cdots \times [N_d]$ and we take $N = \max_i N_i$. In all of these protocols, we assume that the block and bin dimensions fit in a word.

In this protocol, we assume that the prover is trying to convince the verifier that all the blocks can be packed in k bins. We address the more general version of this problem in the full version of this paper. In doing so we use the approximate lower bound protocol for set size from [7].

We require that the prover provides an encoding of a feasible packing of the input blocks in a previously agreed format. This format is such that if all the input blocks can be packed in the bins used by the prover, the verifier accepts. If less than $1 - \epsilon$ fraction of the input blocks can be simultaneously packed, the verifier rejects the proof with some constant probability. In the intermediate case, the verifier either accepts or rejects.

3.1 A First Representation of a Packing

We represent a bin as a d-dimensional grid with the appropriate length in each dimension. The prover will label the packed blocks with unique integers and then label the grid elements with the label of the block occupying it in the packing. In Figure 1, we illustrate one such encoding. The key to this encoding is that we can give requirements by which the prover can define a monotone function on the grid using these labels only if he knows a feasible packing. To show such a reduction exists, we first define a relation on blocks

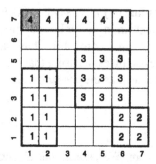

Fig. 1. A 2D Encoding

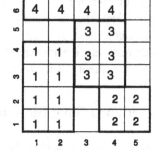

Fig. 2. Compressed Grid Encoding

Definition 4. *For a block b, the highest corner of b, denoted $h^{(b)}$, is the corner with the largest coordinates in the bin it is packed with respect to the $\prec$ relation. Similarly, the lowest corner of b, denoted $l^{(b)}$, is the corner with the smallest coordinates.*

In our figure, $l^{(1)} = (1, 1)$ and $h^{(1)} = (2, 4)$. We can order blocks by only considering the relative placement of these two corners.

Definition 5. *Let b_1 and b_2 be two blocks packed in the same bin. Block b_1 precedes block b_2 in a packing if $l^{(b_1)} \prec h^{(b_2)}$.*

Note that for a pair of blocks in dimension higher than 1 it may be the case that neither of the two blocks precedes the other. This fact along with the following observation makes this definition interesting: For two blocks, b_1 and b_2, such that b_1 precedes b_2, b_1 and b_2 overlap if and only if b_2 precedes b_1. Surely if b_1 precedes b_2 and this pair overlaps it must be the case that $l^{(b_2)} \prec h^{(b_1)}$. It follows that the precedence relation on blocks is a reflexive-antisymmetric ordering precisely when the packing of blocks is feasible. Given such an ordering, it is easy to construct a monotone function.

Lemma 1. *Given a feasible packing of a bin with blocks, we can label the blocks with distinct integers such that when we assign each grid element in the d-dimensional grid (of the bin) with the label of the block occupying it, we get a monotone function on this grid.*

Proof. Without loss of generality, assume that the bin is filled up completely. We know that by inserting extra "whitespace" blocks we can fill up the bin. It can be shown that the bin can be packed in such a way that $4n$ whitespace blocks are sufficient. The relation from Definition 5 gives a relation on the blocks that is reflexive and antisymmetric. Therefore we can label the blocks according to this relation such that a block gets a label larger than those of all its predecessors. This labeling gives us a monotone function on the grid.

Now we can describe the proof that the prover will write down. The proof will consist of three parts: the first one is a table which will have an entry for each block containing the label assigned to the block; a pointer to the bin where the object was assigned and the locations of the two (lowest and highest) corners of the block in this bin. The second part is a permutation enforcer on the blocks and the labels of the blocks.

Finally, the third part consists of a d-dimensional grid of size $\prod[N_j]$ for each bin used that numbers each grid element with the label of the block occupying it.

3.2 Testing Multidimensional Bin-Packing Using Heaviness

The heaviness test we have defined can be used to test that the prover's labeling agrees with a monotone function. By using Observation 1, we will be able to show if all the defining corners of a pair of blocks are heavy then they cannot overlap.

Protocol. We will define "good" blocks such that all "good" blocks can be packed together feasibly. Our notion of "good" should have the properties that (1) a good block is actually packed inside a bin, and it is not overlapping any other "good" block; and (2) we can efficiently test a block for being good. Then, the verifier will use sampling to ensure that at least $1 - \epsilon$ fraction of the blocks are "good" in the protocol.

Definition 6. *The block i with dimensions $w = (w_1, \ldots, w_d)$ is good with respect to an encoding of a packing if it has the following properties:*

- *Two corners defining the block in the proof have positive coordinates with values inside the bin, i.e., $1 \prec l^{(i)}, h^{(i)} \prec N$.*
- *The distance between these corners exactly fits the dimensions of the block, i.e., $w = h^{(i)} - l^{(i)} + 1$.*
- *The grid elements at $l^{(i)}$ and $h^{(i)}$ are heavy.*
- *The block is assigned a unique label among the good blocks, i.e., it is honest with respect to the permutation enforcer.*

Given this definition, we can prove that two good blocks cannot overlap.

Lemma 2. *If two blocks overlap in a packing, then both of the blocks cannot be good with respect to this packing.*

Proof. Note that when two blocks overlap, according to Definition 5, they must both precede each other. Without loss of generality, b_1 precedes b_2. Since these blocks overlap, the lowest corner of b_2, $l^{(b_2)}$, is smaller than the highest corner of b_1, $h^{(b_1)}$ ($l^{(b_2)} \prec h^{(b_1)}$). We know, by definition of a heaviness tester, that two comparable heavy points on the grid do not violate monotonicity. But, since both defining corners of a good block must have the same label, either $l^{(b_1)}$ and $h^{(b_2)}$, or $l^{(b_2)}$ and $h^{(b_1)}$ violates monotonicity.

Corollary 1. *There is a feasible packing of all the good blocks in an encoding using k bins.*

The verifier's protocol can be given as follows: The verifier chooses a block randomly from the input, and using the encoding described above, confirms that the block is good. Testing a block for being good involves $O(d)$ comparisons for the first two conditions in the definition, $O(1)$ time for checking the unique labeling using the permutation enforcer, and 2 heaviness tests for the third condition. The verifier repeats this $O(1/\epsilon)$ times to ensure at least $(1 - \epsilon)$ fraction of the blocks are good.

Theorem 1. *There is an $O((1/\epsilon)T(N, d))$-approximate lower bound PCP for the d-dimensional bin packing problem where $T(N, d)$ is the running time for a heaviness tester on $\mathcal{D} = [N_1] \times \cdots \times [N_d]$.*

3.3 A Compressed Representation of a Packing

The previous protocol requires the prover to write down a proof whose size depends on the *dimensions* of the bins to be filled, since the values N_i were based on the actual size of the bins given. We show here how the prover may write a proof which depends only on the number, n, of objects to be packed. In the protocol from the previous section the verifier calls the heaviness tester only on grid elements which correspond to the lowest or the highest corners of the blocks. We use this observation for a compressed proof.

The prover constructs a set of *distinguished coordinate* values S_k for each dimension $k = 1, \ldots, d$. Each set is initially empty. The prover considers each block i and does the following: for the lower corner, $l^{(i)} = (c_1, \ldots, c_d)$, and higher corner, $h^{(i)} = (e_1, \ldots, e_d)$, of block i, the prover computes $S_i \leftarrow S_i \cup \{c_i\} \cup \{e_i\}$. After all the blocks are processed, $|S_i| \leq 2n$. The *compressed grid* will be a sublattice of $\mathcal{D}$ with each dimension restricted to these distinguished coordinates, that is the set $\{\langle x_1, \ldots, x_d \rangle | x_i \in S_i\}$. This grid will contain in particular all the corners of all the blocks and the size of this proof will be at most $O((2n)^d)$. Note that although in the previous test we have added "whitespace" blocks to generate our monotone numbering, those blocks themselves were never tested, hence they do not affect the number of distinguished coordinates. The fact that this new compressed encoding is still easily testable does not trivially follow from the previous section. In particular, we must additionally verify that the prover's compression is valid.

The proof consists of four parts. First the prover implicitly defines the proof from the previous section, which we refer to as the *original grid*. The prover then writes down a table containing the *compressed grid*. In each axis, the prover labels the coordinates $[1, \ldots, 2n]$ and provides a *lookup-table* (of length $2n$) for each axis which maps compressed grid coordinates to original grid coordinates. Finally the prover writes down the list of objects with pointers to the compressed grid, and a permutation enforcer as before. In Figure 2, we give the compressed encoding of the packing from Figure 1.

Protocol. By making the prover write only a portion of the proof from the first protocol, we provide more opportunities for the prover to cheat. For example, even if the prover uses the correct set of hyperplanes for the compression, he may reorder them in the compressed grid to hide overlapping blocks. The conversion tables we introduced to our proof will allow the verifier to detect such cheating.

The definition of a good block is extended to incorporate the lookup tables. In a valid proof, the lookup tables would each define a monotone function on $[2n]$. We will check that the entries in the lookup tables which are used in locating a particular block are *heavy* in their respective lookup tables. Additionally we test a that a block is good with respect to Definition 6 in the compressed grid[1]. A block which passes both phases is a *good* block.

Our new protocol is then exactly as before. The verifier selects $O(1/\epsilon)$ blocks and tests that each is good and if so concludes that at least $(1 - \epsilon)$ fraction of the blocks are good.

---

[1] Except when we test the size of the block, for which we refer to the original coordinates via the lookup table.

Correctness. Any two good objects do not overlap in the compressed grid, by applying Lemma 2. Furthermore, since the labels of good objects in the lookup table are heavy, it follows that two good objects do not overlap in the original grid either. Certainly, since the corresponding values in the lookup table form a monotone sequence, the prover could not have re-ordered the columns during compression to untangle an overlap of blocks. It also follows from the earlier protocol that good blocks are the right size and are uniquely presented.

Theorem 2. *There is an $O((1/\epsilon)T(n,d))$-approximate lower bound PCP for the d-dimensional bin packing problem with proof size $O((2n)^d)$, where $T(n,d)$ is the running time for a heaviness tester on $\mathcal{D} = [n]^d$.*

3.4 Further Applications

Multidimensional Routing A graph G with edge-capacity constraints is given along with a set of desired messages which are to be routed between vertex pairs. Each message has a bandwidth requirement and a duration. If $\mathcal{P}$ knows how to route f of these messages, he can convince $\mathcal{V}$ that a routing of $\geq (1 - \epsilon)f$ exists. We sketch the method: The prover presents the solution as a 2D bin packing proof, with one bin for each edge: one dimension corresponds to the bandwidth, the other to the duration. The portion of a message routed along a particular bin is a 2D block. To verify that a routing is legal, $\mathcal{V}$ selects a message at random and the prover provides the route used as a list of edges. The verifier checks that sufficient bandwidth is allocated and that durations are consistent along all edges of the route and that the message *(block)* is "good" with respect to the packings of blocks in each of the edges *(bins)*. If we assume that the maximum length of any routing provided by the prover is length k, this yields a protocol with running time $O((k/\epsilon) \cdot \log^2(n))$, where n is the maximum number of calls ever routed over an edge. To achieve this running time we employ the heaviness tester in Section 4. Higher dimensional analogues of this problem can be verified by an extension of these methods.

Recursive Bin Packing At the simplest level the recursive bin packing problem takes as input a set of objects, a list of container sizes (of unlimited quantity), and a set of bins. Instead of placing the objects directly in the bins, an object must first be fit into a container (along with other objects) and the containers then packed in the bin. The goal is to minimize the total number of bins required for the packing. We can solve this problem by applying an extension of our multidimensional bin-packing tester. In particular, we define an object as *good* if it passes the goodness test (with respect to its container) given in Section 2 and furthermore if the container it is in passes the same goodness test (with respect to the bin). After $O(1/\epsilon)$ tests we can conclude that most objects are good and hence that $(1 - \epsilon)$ fraction of the objects can be feasibly packed. For a k-level instance of recursive bin packing, therefore, the prover will write k compressed proofs and $O(k/\epsilon)$ goodness tests will be needed.

3.5 Can Monotonicity Testing Help?

Given the conceptual similarities between heaviness testing and monotonicity testing, it may seem that a monotonicity test could be used to easily implement our multidimensional bin packing protocol. The obvious approach, though, does not seem to work. The complications arise because we are embedding n objects in a $(2n)^d$ sized domain. If a monotonicity tester can determine that the domain of our compressed proof is has $(1-\epsilon')$ of its points in a monotone subset, we can only conclude that at least $n-\epsilon' \cdot (2n)^d$ boxes are "good", by distributing the bad points among the corners of the remaining boxes. Thus monotonicity testing on this domain seems to need an error parameter of $O(\epsilon/(n^d))$. If the running time of the monotonicity tester is linear in ϵ then this approach requires at least $O((2n)^{d-1})$ time.

4   Heaviness Tests

We give two heaviness tests for functions on a domain isomorphic to an integer lattice. The domains are given as $\mathcal{D} = [1, \dots, n]^d$. The range can be any partial order, but here we use $\mathbb{R}$, reals. Both tests which follow determine that a point is heavy in $O((2 \log n)^d)$ time, yielding efficient bin packing tests for small values of d. In particular, the examples applications of bin packing which we have cited typically have dimension less than 3. For complete proofs, please consult the full version of the paper [2].

4.1   The First Algorithm

We extend the protocol of [6] to multidimensional arrays. On input x our test compares x to several random elements y selected from a set of carefully chosen neighborhoods around x. It is tested that x is in order with a large fraction of points in each of these neighborhoods. From this we can conclude that any two comparable heavy points a and b can be ordered by a mutually comparable point c such that $a < c < b$ and $f(a) < f(c) < f(b)$. The test is shown in Figure 4.

Proof of Correctness We consider a set of $\log^d n$ carefully chosen neighborhoods around a point x. We say that x is *heavy* if for a large fraction of points y in each of these neighborhoods, $f(x)$ and $f(y)$ are monotonically ordered. We are able to show from this that for any two heavy points x and y, two of these regions can be found whose intersection contains a point z with the property that $x < z < y$ and $f(x) < f(z) < f(y)$. Hence this defines a valid heaviness property. The efficiency of the test is bound by the fraction of points in each neighborhood which must be tested, which is given to us by Chernoff bounds. It follows that

Theorem 3. *Algorithm* Heavy-Test *is a heaviness tester performing* $O(\log(1/\delta)(2 \log(n))^d)$ *queries.*

4.2   The Second Algorithm

This algorithm is based on a recursive definition of heaviness. Namely a point x is heavy in dimension d if a certain set of projections of x onto hyperplanes are each heavy in dimension $d - 1$. We are able to use the heaviness of these projection points to conclude that d-dimensional heavy points are appropriately ordered.

Given a dimension d hypercube, C, consider a subdividing operation ϕ which maps C into 2^d congruent subcubes. This operation passes through the center d hyperplanes parallel to each of the axes of the hypercube. This is a basic function in our algorithm. For notational convenience, we extend ϕ to Φ which acts on sets of cubes such that $\Phi(\{x_1, \ldots, x_n\}) = \{\phi(x_1), \ldots, \phi(x_n)\}$. It is now possible to compose Φ with itself. We also define a function $\hat{\Phi}(x, C) = S \Rightarrow x \in S \in \Phi(C)$. This function is also a notational convenience which identifies the subcube a point lies in after such a division.

Now consider any two distinct points in the hypercube, x and y. We wish to apply Φ to the cube repeatedly until x and y are no longer in the same cube. To quantify this we define a new function $\varrho : C^2 \to \mathcal{Z}$ such that $\varrho(x, y) = r \Rightarrow \hat{\Phi}^r(x, C) = \hat{\Phi}^r(y, C)$ and $\hat{\Phi}^{r+1}(x, C) \neq \hat{\Phi}^{r+1}(y, C)$. That is, the $r + 1$st composition of Φ on C separates x from y.

Definition 7. *A point x is heavy in a domain $\mathcal{D} = [n]^d$ if the $2d$ perpendicular projections of x onto each cube in the series $\hat{\Phi}(x, \mathcal{D}), \ldots, \hat{\Phi}^{\log n}(x, \mathcal{D})$ of shrinking cubes are all heavy in dimension $d - 1$. The domains $\mathcal{D}'$ for these recurive tests are the respective faces of the cubes. When $d = 1$ we use the test of [6].*

We can now give the heaviness test for a point. Let C be a d-dimensional integer hypercube with side length n. Let x be some point in C. Construct the sequence $\{s_1, \ldots, s_k\} = \{\hat{\Phi}^1(x, C), \ldots, \hat{\Phi}^k(x, C)\}$ where $k = \lceil \log(n) \rceil$. Note that $s_k = x$. At each cube s_k perform the following test: (1) Compute the $2d$ perpendicular projections $\{p_1, \ldots, p_{2d}\}$ of x onto the $2d$ faces of s_k. (2) Verify that f is consistent with a monotone function on each of the $2d$ pairs (x, p_k). (3) If $d > 1$ recursively test that each of the points p_i is heavy over the reduced domain of its corresponding face on s_k. If $d = 1$, we use the heaviness test of [6]. This test is shown in Figure 3.

Theorem 4. *If x and y are heavy and $x < y$ then $f(x) < f(y)$*

Proof. (by induction on d). Let $r = \varrho(x, y)$. Let $S = \hat{\Phi}^r(x, C)$. Let $s_x = \hat{\Phi}^{r+1}(x, C)$ and $s_y = \hat{\Phi}^{r+1}(y, C)$. There is at least one plane perpendicular to a coordinate axes passing through the center of S which separates x and y. This plane also defines a face of s_x and of s_y, which we denote as f_x and f_y respectively. By induction we know the projections of x and y onto these faces are heavy. Since y dominates x in every coordinate, we know that $p_x < p_y$. Inductively we can conclude from the heaviness of the projection points that $f(p_x) < f(p_y)$. Since we have previously tested that $f(x) < f(p_x)$ and $f(p_y) < f(y)$ we conclude $f(x) < f(y)$.

Running time analysis If we let $H_d(n)$ be the number of queries made by our algorithm in testing that a point of the function $f : \mathcal{Z}_n^d \to S$ is heavy, then we can show

Lemma 3. *For all $d > 1$, for sufficiently large n, $H_d(n) \leq (d - 1) \log^d(n) \log(1/\delta)$*

Proof. By induction. For the case $d = 1$ we employ the spot checker algorithm from [6], which performs $\log(1/\delta) \log(n)$ queries to determine that a point is heavy.

```
RecursiveTest(f,ε,δ,D,d)
if d = 1
  δ' ← δ/(d! log^d n)
  return SpotCheckTest(f,ε,δ')
else
  for i = 1...log n
    Φ = Φ̂^i(x,C)
    {p_1,...,p_d} = projections
      of x onto Φ
    for k = 1...d
      C ← the face of Φ̂^i(x,D)
        containing p_k
      HeavyTest(f,ε,δ,D,d)
    end
  end
end
return PASS
```

```
Heavy-Test(f,x,ε,δ)
for k_1 ← 0...log x_1,
  ⋮
  k_d ← 0...log x_d do
    repeat t = O(2^d log(1/δ)) times
    choose h_i ∈_R [1,2^{k_i}] 1 ≤ i ≤ d
    h ← (h_1,...,h_d)
    if (f(x) < f(x − h)) return FAIL
for k_d ← 0...log(n − x_1),
  ⋮
  k_d ← 0...log(n − x_d) do
    repeat t times
    choose h_i ∈_R [1,2^{k_i}] 1 ≤ i ≤ n
    h ← (h_1,...,h_d)
    if (f(x) > f(x + h)) return FAIL
return PASS
```

Fig. 3. Algorithm RecursiveTest   **Fig. 4.** Algorithm Heavy-Test

Theorem 5. *Algorithm* RecursiveTest *is a heaviness tester performing* $O((d \log(d) + d \log \log(n) + log(1/\delta))(d − 1) \log^d(n))$ *queries.*

Proof. The confidence parameter $\delta' = \delta/(d! \log^d(n))$ which appears in Figure 3 arises because the probability of error accumulates at each recursive call. Now apply Lemma 3.

References

1. S. Arora, C. Lund, R. Motwani, M. Sudan, and M. Szegedy. Proof verification and hardness of approximation problems, *J. of the ACM*, 45(3):501–555, 1998.
2. T. Batu, R. Rubinfeld, P. White. Fast approximate PCPs for multidimensional bin-packing problems. *http://simon.cs.cornell.edu/home/ronitt/PAP/bin.ps*
3. L. Babai, L. Fortnow, and C. Lund. Non-deterministic exponential time has two-prover interactive protocols, *Computational Complexity*, pp. 3–40, 1991.
4. L. Babai, L. Fortnow, C. Lund, and M. Szegedy. Checking computations in polylogarithmic time. *Proc. 31st Foundations of Computer Science*, pp. 16–25, 1990.
5. Y. Dodis, O. Goldreich, E. Lehman, S. Raskhodnikova, D. Ron and A. Samorodnitsky Improved Testing Algorithms for Monotonicity. RANDOM '99.
6. F. Ergun, S. Kannan, R. Kumar, R. Rubinfeld, and M. Viswanathan. Spot-checkers. *Proc. 30th Symposium on Theory of Computing*, pp. 259–268, 1998.
7. F. Ergün, R. Kumar, R. Rubinfeld. Fast PCPs for approximations. *Proc. 31st Symposium on Theory of Computing*, 1999.
8. L. Fortnow, J. Rompel, and M. Sipser. On the power of multi-prover interactive protocols. *Theoretical Computer Science*, 134(2):545-557, 1994.
9. O. Goldreich, S. Goldwasser, E. Lehman, D. Ron. Testing Monotonicity *Proc. 39th Symposium on Foundations of Computer Science*, 1998.

Pfaffian Algorithms for Sampling Routings on Regions with Free Boundary Conditions*

Russell A. Martin[1] and Dana Randall[2]

[1] School of Mathematics
[2] College of Computing and School of Mathematics,
Georgia Institute of Technology, Atlanta GA 30332, USA
{martin, randall}@math.gatech.edu

Abstract. Sets of non-intersecting, monotonic lattice paths, or *fixed routings*, provide a common representation for several combinatorial problems and have been the key element for designing sampling algorithms. Markov chain algorithms based on routings have led to efficient samplers for tilings, Eulerian orientations [8] and triangulations [9], while an algorithm which successively calculates ratios of determinants has led to a very fast method for sampling fixed routings [12]. We extend Wilson's determinant algorithm [12] to sample *free routings* where the number of paths, as well as the endpoints, are allowed to vary. The algorithm is based on a technique due to Stembridge for counting free routings by calculating the Pfaffian of a suitable matrix [11] and a method of Colbourn, Myrvold and Neufeld [1] for efficiently calculating ratios of determinants. As an application, we show how to sample tilings on planar lattice regions with free boundary conditions.

1 Introduction

Physicists study combinatorial structures on lattices in order to understand various physical systems. For example, tilings on planar lattice regions model systems of diatomic molecules, or dimers. By studying statistics of random configurations on families of regions of finite size (such as the $n \times n$ square or the Aztec diamond), physicists gain insight into the behavior of these systems on the infinite lattice, the so-called thermodynamic limit.

It is well known that the boundary of the region plays a crucial role. There are two relevant boundary effects. The first is the *shape* of the family of finite regions; the second is the *type* of boundary conditions defined for the regions. So far sampling has primarily been done for *fixed* boundary conditions, where the configurations are forced to precisely agree with the boundary. In the case of domino tilings this means that tiles are forced to cover all of the squares inside, and only inside, the region. Another important type of boundary condition considered permits all configurations that can be seen within a *window* in the shape of the region. Returning to tilings, this means that tiles can overlap the boundary (as long as the configuration can be extended to a tiling of the plane). In the context of tilings, these are commonly referred to as *free* boundary conditions.

* Research suported in part by NSF Career Grant No. CCR-9703206

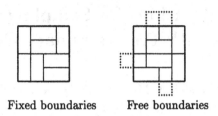

Fixed boundaries   Free boundaries

Fig. 1. Domino tilings on regions with fixed and free boundary conditions

One reason for studying free boundary conditions is that these eliminate the boundary effect due to the shape of the region (in the limit). This is not true for families of regions with fixed boundary conditions, where properties of random configurations can vary drastically according to their shape. Consider, for example, the entropy of the system, defined as $h(\Lambda) = \lim_{n\to\infty} \frac{\log \#(\lambda_n)}{\text{Area}(\lambda_n)}$, where $\Lambda = \{\lambda_n\}$ is a nested family of regions tending towards the infinite lattice and $\#(\lambda_n)$ is the number of tilings of λ_n. With fixed boundary conditions, the family of square regions has been proven to have *maximal* entropy over all finite families of regions Λ. In contrast, the family of Aztec diamonds is known to have lower entropy, which is related to the arctic circle phenomenon whereby frozen regions of the Aztec diamond emerge having a completely predictable local tiling [5]. On the other hand, with free boundary conditions, for any family of regions where the ratio of the length of the perimeter to the area of the region tends to zero, the entropy will converge to the same (maximal) value. In other words, statistics of tilings of square regions with free boundary conditions will agree with statistics of tilings of Aztec diamonds with free boundary conditions.

Several algorithms for sampling tilings on regions with fixed boundary conditions rely on a bijection between tilings and *fixed routings*, or sets of non-intersecting lattice paths where the number of paths and the position of their endpoints are fixed. The first is a Markov chain approach of Luby, Randall and Sinclair [8] which samples routings uniformly (and can be extended to the case where the paths are edge disjoint, but not necessarily vertex disjoint). A second approach, due to Wilson [12], uses the Gessel-Viennot method for enumerating routings by calculating a determinant [2] (and the close relationship between counting and sampling formalized by Jerrum, Valiant and Vazirani [4]). Wilson utilizes a technique introduced by Colbourn, Myrvold and Neufeld [1] which allows ratios of determinants of closely related matrices to be computed quickly without having to evaluate both determinants.

In this paper we sample *free routings*, or sets of non-intersecting lattice paths where the positions of the endpoints of the paths, as well as the number of paths, are allowed to vary. Our result relies on Stembridge's algorithm counting the number of free routings of a region by evaluating a Pfaffian [11]. We adapt the method of Colbourn, Myrvold and Neufeld to allow ratios of Pfaffians to be evaluated quickly, a special case of a technique of Kenyon for calculating statistics of random tilings [6]. The running time of our algorithm is $O(l^2 n)$, where n is the size of the region and l is the maximal number of paths in a routing. Typically $l = O(\sqrt{n})$, yielding an $O(n^2)$ algorithm. We apply this sampling method to generate random domino and lozenge tilings of hexagonal regions with free boundary conditions.

The remainder of the paper is organized as follows. In section 2 we review the counting techniques of Gessel-Viennot and Stembridge for fixed and free routings, respectively. In section 3 we present our algorithm for uniformly sampling free routings. Finally, in section 4 we show the bijections between free routings and tilings on regions with free boundary conditions which allow us to sample these tilings efficiently.

2 Background: Counting routings

First we begin with an overview of the method of Gessel and Viennot for counting fixed routings and that of Stembridge for counting free routings. Wilson shows how to sample fixed routings using self-reducibility and iterative applications of the Gessel-Viennot method. We give a similar method to sample free routings, utilizing Stembridge's method for counting free routings.

2.1 The Gessel-Viennot method

Gessel and Viennot[2, 3], and Lindström[7] introduce a method for finding the number of non-intersecting paths, with specified sources and sinks, in certain directed graphs by computing a determinant of a matrix. For their technique to work, the graph must be directed and acyclic. Furthermore, the sources and sinks must satisfy a condition known as *compatibility*. In this definition, we require that both the set of sources S and the set of sinks T be ordered.

Definition 1. *Let D be a directed acyclic graph. The ordered sets S and T are said to be* compatible *if $s < s'$ in S and $t < t'$ in T implies that every $s - t'$ path intersects every $s' - t$ path.*

Thus, if there is a set of l non-intersecting paths using sources $s_1 < s_2 < \cdots < s_l$ and sinks $t_1 < t_2 < \cdots < t_l$, then it must be the case that s_i is joined to t_i for all i. We call such a set of l non-intersecting paths a *fixed routing* of D.

Let D denote an acyclic directed graph with compatible sources $S = \{s_1, \ldots, s_l\}$ and sinks $T = \{t_1, \ldots, t_l\}$. Let p_{ij} denote the number of directed paths in D with source s_i and sink t_j. Since the graph is assumed to be acyclic, this number is finite for all i and j. Let P be the matrix with entries p_{ij}.

We have the following theorem [2, 11]:

Theorem 1. *With D, S, T, (S and T compatible) and P as above, the number of non-intersecting sets of l paths in D is equal to $\det(P)$.*

If D is not acyclic, or if S and T are not compatible, then the preceding theorem fails. (See [11] for an example for which the theorem fails.)

Theorem 2 (Wilson [12]). *Let D be a planar, acyclic digraph with n vertices, having compatible sources and sinks. Fixed routings of D can be uniformly sampled in $O(l^{1.688}n)$ time.*

2.2 Stembridge's extension

Stembridge[11] extends the Gessel-Viennot method to count free routings of a directed acyclic graph, D, with sources S and sinks T. In the case of free routings, the number of paths is no longer fixed, so S is really the set of *potential* sources and T is the set of *potential* sinks. Also, it is no longer always true that s_i will be joined to t_i, as was true in the case of fixed routings.

If $s_i \in S$ is a source in a free routing, we say that s_i is *used* in the routing; otherwise s_i is *unused*. Here we assume there are l sources and l sinks. First, we need a bit of linear algebra.

Definition 2. *Let B be a $2n \times 2n$ skew-symmetric matrix (i.e. $B^T = -B$), and let*

$$\pi = \{\{i_1, j_1\}, \{i_2, j_2\}, \ldots, \{i_n, j_n\}\}$$

be a partition of the set $\{1, \ldots, 2n\}$ into pairs. Let

$$b_\pi = \mathrm{sgn} \begin{pmatrix} 1 & 2 & 3 & 4 & \ldots & 2n-1 & 2n \\ i_1 & j_1 & i_2 & j_2 & \ldots & i_n & j_n \end{pmatrix} b_{i_1, j_1} b_{i_2, j_2} \cdots b_{i_n, j_n}.$$

The Pfaffian of B, denoted $\mathrm{Pf}(B)$, is defined by

$$\mathrm{Pf}(B) = \sum_\pi b_\pi.$$

Theorem 3. *If B is a skew-symmetric matrix of even size, then $\det(B) = \mathrm{Pf}(B)^2$.*

A skew-symmetric matrix, Q, will take the role of the matrix P in theorem 1, but instead of the determinant of Q, we look at its Pfaffian. For $1 \leq i < j \leq l$ and $1 \leq h < k \leq l$, let $\alpha_{ij}(h, k)$ denote the number of non-intersecting paths in D with sources s_i, s_j, and sinks t_h, t_k. We find $\alpha_{ij}(h, k) = \det \begin{pmatrix} p_{ih} & p_{ik} \\ p_{jh} & p_{jk} \end{pmatrix}$ using theorem 1, where, recall, p_{ih} is the number of paths from s_i to t_h.

Let $q_{ij} = \sum_{h<k} \alpha_{ij}(h, k)$. Then q_{ij} is the number of pairs of non-intersecting paths with sources s_i and s_j, where the sinks range over all pairs where t_h precedes t_k in the ordering of T. Finally, let q_i denote the number of paths with source s_i to any sink in T.

We assume that l is odd; if not, we can add an additional isolated vertex s_{l+1} to S. The following is due to Stembridge[11]:

Theorem 4. *Let $S = (s_1, \ldots, s_l)$ be an l-tuple of vertices in an acyclic digraph D, with l odd. Let T be an ordered subset of vertices that is compatible with S. Let Q be the skew-symmetric matrix where the upper triangular entries are given by*

$$[Q]_{ij} = (-1)^{i+j-1} + q_{ij}^*$$

for $1 \leq i < j \leq l+1$, where $q_{ij}^* = q_{ij}$ for $j \leq l$ and $q_{i,l+1}^* = q_i$. Then $\Phi = \mathrm{Pf}(Q)$ is the number of free routings of D.

The matrix Q looks like:

$$Q = \begin{pmatrix} 0 & 1+q_{12} & \cdots & -1+q_{1l} & 1+q_1 \\ -1-q_{12} & 0 & \cdots & 1+q_{2l} & -1+q_2 \\ \vdots & \vdots & \ddots & \vdots & \vdots \\ -1-q_1 & 1-q_2 & \cdots & -1-q_l & 0 \end{pmatrix}.$$

Stembridge uses this theorem to study shifted tableaux, plane partitions, and Schur's Q-functions. As we will see in section 4 it can be used to count and generate tilings with free boundary conditions. We first give an extension of Stembridge's result.

2.3 Fixing sources

We can extend theorem 4 to count the number of free routings where we specify that certain sources must be used (or unused) in a routing.

Informally, if s_i is used as a source then we replace terms $\pm 1 + q_{ij}$ in Q by q_{ij}, and if not used, by ± 1. The following theorem formalizes this:

Theorem 5. Let $S = (s_1, \ldots, s_l)$ be an l-tuple of vertices in an acyclic digraph D, with l odd. Let T be an ordered subset of vertices that is compatible with S. Suppose $S_{in}, S_{out} \subseteq S$ with $S_{in} \cap S_{out} = \emptyset$. Let Q be the skew-symmetric matrix where the upper triangular entries are given by

$$[Q]_{ij} = \begin{cases} 0 & \text{if } s_i \in S_{in} \text{ and } s_j \in S_{out}, \text{ (or vice-versa)} \\ q_{ij} & \text{if } j \leq l \text{ and } (s_i \text{ and/or } s_j \text{ in } S_{in}, \text{ neither in } S_{out}) \\ (-1)^{i+j-1} & \text{if } j \leq l \text{ and } (s_i \text{ and/or } s_j \text{ in } S_{out}, \text{ neither in } S_{in}) \\ q_i & \text{if } j = l+1 \text{ and } s_i \in S_{in} \\ (-1)^{i+l} & \text{if } j = l+1 \text{ and } s_i \in S_{out} \\ (-1)^{i+j-1} + q_{ij}^* & \text{otherwise, where } q_{ij}^* \text{ is as in theorem 4} \end{cases}$$

for $1 \leq i < j \leq l+1$. Then $\Phi = \text{Pf}(Q)$ is the number of free routings of D with S_{in} included in the set of used sources, and S_{out} in the set of unused sources.

Proof. For $J \subseteq \{1, 2, \ldots, n\}$, let A_J denote the square submatrix of A obtained by selecting the rows and columns indexed by J. We use the result (from [11, lemma 4.2]) that (for n even, A and B $n \times n$ in size) we can write

$$\text{Pf}(A+B) = \sum_J (-1)^{\sigma(J) - \frac{|J|}{2}} \text{Pf}[A_J]\text{Pf}[B_{J^C}] \tag{1}$$

where $\sigma(J) = \sum_{j \in J} j$, and the sum is taken over all partitions J, J^C of $\{1, \ldots, n\}$ with $|J|$ even.

Decompose Q into a sum of two matrices, A and B, where $[A]_{ij} \in \{0, 1, -1\}$ and $[B]_{ij} \in \{0, q_{ij}^*, -q_{ij}^*\}$. Now apply the above result for the Pfaffian of a sum. We have $\text{Pf}[A_J] = 0$ if $J \cap S_{in} \neq \emptyset$ since A will contain a row of zeros. Similarly, $\text{Pf}[B_{J^C}] = 0$ if $J^C \cap S_{out} \neq \emptyset$. So the only terms that survive in the sum (1) are those with $S_{in} \subseteq J^C$ and $S_{out} \subseteq J$. Note that if q_{ij} or q_i appears in one of the terms of Φ, that term corresponds to a set of paths (not necessarily non-intersecting) that uses s_i. Thus, if $s_i \in S_{in}$, by the choice of the entries of Q we ensure that s_i is used in every routing of D, as one of the q_{ij}'s or q_i will appear in each term of Φ. Similarly, if $s_i \in S_{out}$ then none of the q_{ij}'s or q_i appears in Φ, so that s_i is unused as a source in every routing of D. □

3 Generating random routings

We present an algorithm to uniformly generate a free routing of a planar acyclic digraph D with compatible sources and sinks. This algorithm is similar to the determinant algorithm of [12] for generating fixed routings. Once again, we assume that $|S| = |\mathcal{T}| = l$, with l odd.

We use self-reducibility to find the routing by building paths one edge at a time. We move through the graph, deciding (probabilistically) if a source s_i is used in the routing. Then, if used, we select one of its out-going edges, with appropriate probabilities, and add it to the routing, thereby starting a path using s_i. We push the source forward to w, the other end of the selected edge, and will eventually complete a path from s_i into $\mathcal{T}$. We use the fact that ratios of Pfaffians can be computed efficiently to determine the probability of using a particular source or edge. The following theorem is analogous to the result in [1] for ratios of determinants of matrices that differ by a single row.

Theorem 6. *Let A be an invertible, skew-symmetric matrix and let B be a skew-symmetric matrix which differs from A by only the ith row and column. Then*

$$\frac{\mathrm{Pf}(B)}{\mathrm{Pf}(A)} = [BA^{-1}]_{ii}.$$

Proof. The proof relies on a closely related fact that if A is an invertible matrix and C differs from A by only the ith row, then

$$\frac{\det(C)}{\det(A)} = [CA^{-1}]_{ii}$$

(which follows from Cramer's rule). Given A, an invertible, skew-symmetric matrix and B, a skew-symmetric matrix which differs from A by only the ith row and column, let C be the matrix formed by replacing the ith row of A with the ith row of B (so C differs from A by the ith row and from B by the ith column). Assume first that C is invertible. Then we have that $\det(C)/\det(A) = [CA^{-1}]_{ii}$ and $\det(B)/\det(C) = [B^T C^{-1\,T}]_{ii} = [C^{-1}B]_{ii}$. Finally, we use the fact that since A and C differ in only the ith row and $A_{ii} = C_{ii} = 0$ since A and B are skew-symmetric, then the ith rows of A^{-1} and C^{-1} must agree. Hence,

$$\left(\frac{\mathrm{Pf}(B)}{\mathrm{Pf}(A)}\right)^2 = \frac{\det(B)}{\det(A)} = \frac{\det(B)}{\det(C)} \cdot \frac{\det(C)}{\det(A)} = [C^{-1}B]_{ii}[CA^{-1}]_{ii}$$

$$= [A^{-1}B]_{ii}[BA^{-1}]_{ii} = \left([BA^{-1}]_{ii}\right)^2. \tag{2}$$

If C is not invertible, let B' be obtained from B by perturbing the ith row of B by $\varepsilon \cdot$ (random vector) and the ith column so that B' is skew-symmetric (and differs from A only in the ith row and ith column). With C' as the matrix formed by replacing the ith row of A by the ith row of B' we proceed as before (C' is invertible), then let $\varepsilon \to 0$ (so $B' \to B$) to get the same result as in (2).

Taking square roots, and recalling that $A^{-1} = \operatorname{adj}(A)/\det(A)$, where $\operatorname{adj}(A)$ is the (classical) adjoint of A, we can write (2) as

$$\operatorname{Pf}(B)\det(A) = \pm\operatorname{Pf}(A)[B\operatorname{adj}(A)]_{ii}. \tag{3}$$

Taking an invertible skew-symmetric matrix A and letting $B = A$, the sign in that case is $+$, and by continuity the sign for a whole neighborhood of the parameter values is also $+$. By taking partial derivatives and evaluating at 0, we see that the coefficients of the polynomials must be equal, so that the sign in (3) is everywhere $+$. $\qquad\square$

We use the Sherman-Morrison formula for updating A^{-1} after changing a single row or column of the matrix A. In our case we will be changing both a row and a column, but we can update the inverse by applying the Sherman-Morrison formula twice. Updating A^{-1} can be done in $\Theta(l^2)$ time using this method (for details see, e.g., [1, 12]). (The Sherman-Morrison formula for updating an inverse has shown some numerical instability in practice; we may achieve greater numerical stability by using other schemes for updating A^{-1} at a small cost in the running time of the algorithm.) Now we are ready to describe the algorithm.

The input to the algorithm is the planar digraph, D, having n vertices and m edges, and sets S and T, the sources and sinks, respectively. The variable x_i records the current position of source i, and the array R records the routing as it is constructed. We maintain a matrix Q, initially equal to the matrix Q of theorem 4, and a matrix U, initially the inverse of Q, which we use to compute probabilities of using sources or edges in the routing. Q and U are updated as we move through the digraph. We compute $P[v, i]$, which will be the number of paths from vertex v to sink t_i, and $\hat{P}[v, i]$, the number of paths from v to any of the sinks $t_i, t_{i+1}, \ldots, t_l$. (We use the $\hat{P}[v, i]$'s to help initialize the matrix Q in time $O(l^3)$ instead of $O(l^4)$, and later for updating entries of Q as we move through D.) With "$v \to w$" denoting that there is a directed edge from v to w the algorithm is:

$\mathcal{F}ree\mathcal{R}oute(D, \mathcal{S}, \mathcal{T})$

1. Do a topological sort on D, numbering the vertices 1 through n, so that $v \to w$ implies $v < w$. $\hfill$ ($O(n)$ time.)
2. For $v = 1$ to n, set $q_v = 0$. $\hfill$ ($O(n)$ time.)
3. For $i = l$ down to 1 (Dynamic programming step) $\hfill$ ($O(ln)$ time.)
   (a) Set $x_i = s_i$.
   (b) For $v = n$ down to 1
      i. If $v = t_i$, set $P[v, i] = 1$, else set $P[v, i] = \sum_{w:w \to v} P[w, i]$.
         ($P[v, i]$ now contains the number of paths from v to t_i.)
      ii. Set $q_v = q_v + P[v, i]$.

    iii. If $i = l$, set $\hat{P}[v, i] = P[v, i]$, else set $\hat{P}[v, i] = \hat{P}[v, i+1] + P[v, i]$.

4. For $i = 1$ to l ($O(l^3)$ time.)

    (a) Find v such that $v = s_i$.

    (b) Set $q_i = q_v$. (Initialize q_i's.)

    (c) For $j = i + 1$ to l

        i. Find w such that $w = s_j$.

        ii. Set $q_{ij} = \sum_{k=1}^{l-1} \det \begin{pmatrix} P[v, k] & \hat{P}[v, k+1] \\ P[w, k] & \hat{P}[w, k+1] \end{pmatrix}$. (Initialize q_{ij}'s.)

5. Initialize the matrix Q as in theorem 4 using the q_i's and q_{ij}'s, find $U = Q^{-1}$, and set $S_{in} = \emptyset$ and $S_{out} = \emptyset$.   ($O(l^3)$ time.)

6. For $v = 1$ to n, if $v = x_i$ for some i then

    (a) If $v \in S \backslash (S_{in} \cup S_{out})$ then decide if v is used as a source (see details below).
If it is, add v to S_{in}. If not, add v to S_{out} and set $R[v] = 0$.
In either case, update row and column i of Q and U.   ($O(n + l^3)$ time.)

    (b) If $v \in S_{in}$ then decide which edge leaving v to include in the path of the routing (see details and remarks). Let w be the other endpoint of this edge. If $w = s_k$ for some k, see remark 1 below.
Set $R[v] = w$, $x_i = w$, and add w to S_{in}.
Set $q_i = q_w$. Update the ith row and column of Q and update U.   ($O(l^2 n)$ time.)

Remark 1. In step 6(b), we may try to push $v = x_i$ forward to an (as yet) unused source $w = s_k \in S$. In this case, we want to add w to S_{out} so that it is not used during some later step to begin a different path. However, we also want to add w to S_{in} so that in later iterations of step 6(b) we push w forward to complete a full path into $\mathcal{T}$ that started from s_i. This conflicts with the condition of theorem 5 that $S_{in} \cap S_{out} = \emptyset$. We get around this difficulty as follows: Remove x_i from S_{in} and add it to S_{out}, then add $w = s_k$ to S_{in} so that it is pushed forward in later steps of the algorithm. Update row and column i of Q to reflect that x_i is unused, then row and column k so that s_k is used, and update U accordingly with successive applications of the Sherman-Morrison formula. Finally, set $R[v] = w$ to join the path between s_i and w to the path from w into $\mathcal{T}$. We will see examples of digraphs in which this situation might arise in section 4 where we consider tilings of reduced Aztec diamonds.

Remark 2. During step 6(b) it is possible that $x_i \in \mathcal{T}$ but we might still push x_i forward. This could occur if x_i has out-neighbors that are also in $\mathcal{T}$. Informally, in this situation we may consider that x_i is joined to a phantom sink by a single (phantom) edge. Pushing x_i forward to this phantom sink corresponds to terminating the path at x_i and not continuing to any of x_i's neighbors. In practice, we need not handle this situation as a special case, since we can examine all of the out-neighbors of x_i in turn and if we reject using any of them then terminate the path, i.e., x_i is not pushed.

Details for step 6(a): In this step, we determine if the source s_i is used in a routing The probability that s_i is used is given by $\frac{\mathrm{Pf}(Q')}{\mathrm{Pf}(Q)}$, where Q' is a skew-symmetric matrix differing from Q in the ith row and ith column. In particular, the ith row of Q' can be found using theorem 5, where we apply the theorem with s_i used in the set of current potential sources (the x_j's, restricting S_{in} and S_{out} to that set). We use theorem 6 to compute this probability as the dot product of the new ith row of Q' with column i of U. If v is used, we replace the ith row of Q by the ith row of Q' to reflect this (and then update the ith column of Q so that it remains skew-symmetric), and add v to S_{in}. If v is not used, we update row and column i of Q as appropriate in theorem 5, where v is now in S_{out}. In either case, we update U (so that it is still equal to Q^{-1}), using two successive applications of the Sherman-Morrison formula, once for changing row i of Q, and again for changing column i. Updating U takes time $\Theta(l^2)$, and hence the total time spent in step 6(a) is $O(n + l^3)$.

Details for step 6(b): Moving the source x_i forward in step 6(b) changes the ith row and column of Q. As before, the probability that the edge $v \rightarrow w$ is used is $\frac{\mathrm{Pf}(Q')}{\mathrm{Pf}(Q)}$, where Q' is the matrix with w used as a source in place of x_i. If this edge is taken, we update Q (and U) by replacing the ith row and column of Q with those of Q'. In the special case that $w \in S$, we proceed as outlined in remark 1. The time to update U (at any instance when Q is updated) is $\Theta(l^2)$, so the total time spent in step 6(b) is $O(l^2 n)$.

We have demonstrated the following theorem:

Theorem 7. *Let D be a planar acyclic digraph with n vertices, having compatible sources and sinks. $\mathcal{F}ree\mathcal{R}oute$ uniformly samples a free routing of D in time $O(l^2 n)$.*

4 Lattice paths

In this section, we demonstrate applications of the techniques from the previous section. We show how to generate random domino tilings of the reduced Aztec diamond with free boundary conditions and lozenge tilings of the hexagon with free boundary conditions. The key idea is the existence of a bijection between the set of tilings of this region and the set of free routings in a related digraph. For details of the analogous bijections in the case of fixed boundary conditions, we refer the reader to [8].

4.1 Domino tilings of the reduced Aztec diamond

The *reduced Aztec diamond of order n*, denoted Γ_n, is a region composed of $2n^2$ unit squares arranged as $2n$ centered rows of squares, where the kth row has $\min\{2k - 1, 4n - 2k + 1\}$ squares in it. A domino tiling is a cover of Γ_n using non-overlapping dominoes, where a domino covers two adjacent squares. A domino tiling with *free boundary conditions* is a tiling in which all the squares of Γ_n are covered, but the dominoes are allowed to "stick out" of (or overlap) the boundary of the region. We assume that we know the orientation of a domino that overlaps the boundary, i.e., a single square (or half-domino) is designated as the bottom, top, left or right half of a domino.

Given a tiling of Γ_n with free boundary conditions (or simply, a free tiling), we define a routing of a digraph, D_n. To get D_n, first color the left square of row n of Γ_n black, then extend the coloring to Γ_n using alternating black and white squares (as on the underlying infinite chessboard). Mark the midpoint of each vertical edge that has a black square to its right. Fix $(0,0)$ as the coordinates of the point on the left edge of row n. Add $n+1$ additional points at coordinates $(-1,-1),(0,-2),(1,-3),\ldots,(n-1,-n-1)$, and another $n+1$ points at $(n,n),(n+1,n-1),(n+2,n-2),\ldots,(2n,0)$. Join a point with coordinates (x,y) to the points $(x+1,y+1),(x+1,y-1)$, and $(x+2,y)$. Finally, delete edges that lie completely outside the boundary of Γ_n.

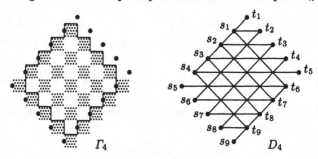

Fig. 2. The reduced Aztec diamond

The marked points form the vertex set of D_n, and the edges of D_n are those that remain between points after the deletion step. Direct edges from left to right. Starting at the source in the top square, label the sources $s_1, s_2, \ldots, s_{2n+1}$ in the counterclockwise direction, and then label each sink t_i where s_i is the last unmatched source. The left picture of figure 2 is Γ_4, the reduced Aztec diamond of order 4, along with the sources and sinks of D_4. The right picture is the digraph D_4.

Theorem 8. *There is a bijection between free boundary tilings of Γ_n and free routings of D_n.*

Proof. Given a free tiling of Γ_n we map it to a free routing of D_n as follows: Examine the sources in this order: $s_n, s_{n-1}, \ldots, s_1, s_{n+1}, s_{n+2}, \ldots, s_{2n+1}$. It's possible that no source lies on the edge of a domino, in which case the routing is empty. Otherwise, the routing consists of the paths constructed as follows: If s_n lies on the edge of a domino, this determines the first edge in a path starting at s_n (otherwise move onto s_{n-1}). Connect s_n to the unique vertex in D_n that lies on the right side of the domino. This new vertex must lie on the left side of another domino, so repeat this process. Stop when we reach a vertex in $\mathcal{T}$ that does not have a domino to its right. Choose the next source, in the prescribed order, that is not on a path already constructed, and repeat this procedure. The paths are non-intersecting since dominoes cannot overlap and because of the order in which the sources were examined. See figure 3 for an example of a free boundary tiling of Γ_4 and the corresponding routing. (An arrow in the tiling points to the location of the other half of a domino that overlaps the boundary.) The proof that this map forms a bijection is analogous to the proof given in [8] which establishes a similar bijection between domino tilings of regions with fixed boundary conditions and fixed routings of related regions. □

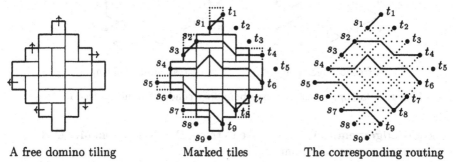

| A free domino tiling | Marked tiles | The corresponding routing |

Fig. 3. A domino tiling with free boundary conditions and its free routing

It follows from this connection between tilings of Γ_n and routings of D_n that we can generate free boundary tilings of Γ_n by using the algorithm given in section 3 for generating free routings of D_n.

4.2 Lozenge tilings of the hexagon

We use a similar approach as in the previous section to generate lozenge tilings of a hexagonal region of the triangular lattice with free boundary conditions. There is a bijection between the collection of free boundary tilings, and the set of free routings of a related digraph.

Let H_n denote a hexagonal region on the triangular lattice with n edges on each side. A *lozenge tiling* of H_n is a covering of the region with lozenges, where a lozenge covers two adjacent triangles, and lozenges do not overlap. As in the previous section, a lozenge tiling of H_n with *free boundary conditions* is a tiling in which lozenges may overlap the boundary of the region. We describe a digraph, G_n, associated with H_n, in which free routings correspond to free boundary tilings of H_n. First, augment H_n to get a region $\hat{H}_n$ by adding the triangles in the underlying lattice that share an edge with the boundary of H_n. Mark the midpoint of each vertical edge in $\hat{H}_n$. These marked points form the vertex set of G_n. Join two points if they lie on adjacent triangles. These are the edges of G_n. Direct these edges from left to right. A free boundary lozenge tiling of H_n corresponds to a free routing of G_n. Again, the proof of this bijection follows analogously to the proof given in [8] for establishing a bijection between fixed lozenge tilings and fixed routings. Figure 4 provides a pictorial illustration of this correspondence.

Applying the $\mathcal{F}ree\mathcal{R}oute$ algorithm of section 3 allows us to uniformly generate free routings of G_n, which we may then map to their corresponding free boundary tilings of H_n.

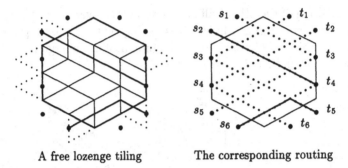

A free lozenge tiling   The corresponding routing

Fig. 4. A lozenge tiling with free boundary conditions and its free routing

Acknowledgements: We would like to thank Dan Klain, Tom Morley and the referees for many useful comments.

References

1. C. J. Colbourn, W. J. Myrvold, and E. Neufeld. Two algorithms for unranking arborescences. *J. Algorithms*, 20:268–281, 1996.
2. I. M. Gessel and X. G. Viennot. Binomial determinants, paths, and hook length formulae. *Advances in Mathematics*, 58:300–321, 1985.
3. I. M. Gessel and X. G. Viennot. Determinants, paths, and plane partitions. Preprint, 1989.
4. M. R. Jerrum, L. G. Valiant, and V. V. Vazirani. Random generation of combinatorial structures from a uniform distribution. *Theoretical Computer Science*, 43:169–188, 1986.
5. W. Jockusch, J. Propp, and P. Shor. Random domino tilings and the arctic circle theorem. Preprint, 1995.
6. R. Kenyon. Local statistics of dimers on a lattice. *Annales de l'Institut Henri Poincaré – Probabilités et Statistiques*, 33:591–618, 1997.
7. B. Lindström. On the vector representations of induced matroids. *Bulletin of the London Mathematical Society*, 5:85–90, 1973.
8. M. Luby, D. Randall, and A. Sinclair. Markov chain algorithms for planar lattice structures. *Proc. 36th IEEE Symposium on Foundations of Computer Science*, 150–159, 1995.
9. L. McShine and P. Tetali. On the mixing time of the triangulation walk and other Catalan structures. *Randomization Methods in Algorithm Design, DIMACS-AMS* 43, 1998.
10. J. Sherman and W. J. Morrison. Adjustment of an inverse matrix corresponding to changes in the elements of a given column or a given row of the original matrix. *The Annals of Mathematical Statistics* 20:621, 1949.
11. J. R. Stembridge. Non-intersecting paths, Pfaffians, and plane partitions. *Advances in Mathematics*, 83:96–131, 1990.
12. D. B. Wilson. Determinant algorithms for random planar structures. *Proc. 8th Annual ACM-SIAM Symposium on Discrete Algorithms*, 258–267, 1997.

Scheduling with Unexpected Machine Breakdowns

Susanne Albers[1] and Günter Schmidt[2]

[1] Max-Planck-Institut für Informatik, Im Stadtwald, 66123 Saarbrücken, Germany.
albers@mpi-sb.mpg.de, http://www.mpi-sb.mpg.de/~albers/
[2] Information and Technology Management, University of Saarland, 66041
Saarbrücken, Germany. Email: gs@itm.uni-sb.de

Abstract. We investigate an online version of a basic scheduling problem where a set of jobs has to be scheduled on a number of identical machines so as to minimize the makespan. The job processing times are known in advance and preemption of jobs is allowed. Machines are *non-continuously* available, i.e., they can break down and recover at arbitrary time instances *not known in advance*. New machines may be added as well. Thus machine availabilities change online.
We first show that no online algorithm can construct optimal schedules. We also show that no online algorithm can achieve a bounded competitive ratio if there may be time intervals where no machine is available. Then we present an online algorithm that constructs schedules with an optimal makespan of $C_{\max}^{OPT}$ if a *lookahead* of one is given, i.e., the algorithm always knows the next point in time when the set of available machines changes. Finally we give an online algorithm without lookahead that constructs schedules with a nearly optimal makespan of $C_{\max}^{OPT} + \epsilon$, for any $\epsilon > 0$, if at any time at least one machine is available. Our results demonstrate that not knowing machine availabilities in advance is of little harm.

1 Introduction

In scheduling theory the basic model assumes that a fixed set of machines is continuously available for processing throughout the planning horizon. This assumption might be justified in some cases but it does not apply if certain maintenance requirements, breakdowns or other constraints that cause the machines not to be available for processing have to be considered.

Machine availability constraints appear very often. Clearly, machines may be faulty and break down. Moreover, availability constraints arise on the operational level of production scheduling. Here some jobs are fixed in terms of starting and finishing times and resource assignment. When new jobs become available for processing, there are already jobs assigned to time intervals and corresponding machines while the new ones have to be processed using the remaining free processing intervals. A similar problem occurs in operating systems for single- and multi-processors when subprograms with higher priority have to be scheduled before subprograms with lower priority.

Thus, limited machine availability is common in practice. Knowledge about machine availabilities might be complete or incomplete. In an *online setting* machine availabilities are not known in advance. Machine breakdowns are a typical example of events that arise online. Sometimes a scheduler has partial knowledge of the availabilities, i.e, he has some *lookahead*. He might know of the next time interval where a machine requires maintenance or he might know when a broken machine will be available again. In an *offline setting* all machine availabilities are known prior to schedule generation.

In this paper we study a very basic scheduling problem with respect to limited machine availability: A set of jobs has to be scheduled on a set of identical machines so as to minimize the makespan. More specifically, let $\mathcal{J} = \{J_i | i = 1, \ldots, n\}$ be a set of independent jobs to be scheduled. Job J_i has a processing time of p_i time units known in advance, $1 \leq i \leq n$. The jobs have to be scheduled on a set of machines that operate with the same speed. Preemptions of jobs is allowed. Each machine may work only on one job at a time, and each job may be processed by only one machine at a time. We wish to minimize the *makespan*, i.e., the completion time of the last job that finishes. Machines may have different time intervals of availability. We emphasize here that we are interested in the online version of the problem where the machine availabilities are not known in advance. We also call an interval where a machine is not available a *machine break down*. Machines may break down or recover at arbitrary time instances. New machines may be added as well. We also consider the online problem with *lookahead one*, i.e., a scheduler always knows the next point in time where the set of available machines changes. However, he does not have to know which machines break down or become available. In the previous literature [4], this setting is also referred to as *nearly online*.

Given a scheduling algorithm A and a problem instance, let $C_{\max}^A$ denote the makespan of the schedule produced by A. In particular, $C_{\max}^{OPT}$ denotes the makespan of an optimal offline algorithm that knows the machine availabilities in advance. Following [8] we call an online scheduling algorithm A *c-competitive* if, for all problem instances, $C_{\max}^A \leq c \cdot C_{\max}^{OPT}$.

Related work: Schmidt [6] was the first who studied scheduling problems with limited machine availability. He concentrated on the offline version of the above problem when all the machine breakdown times are known in advance. Note that if the down times are identical for all the machines, then an optimal schedule can be constructed using McNaughton's algorithm [3]. The algorithm runs in $O(n)$ time and uses no more than $S - 1$ preemptions, where S is the total number of intervals where machines are available. Schmidt [6] studied the problem with arbitrary machine availabilities and gave an algorithm that always constructs an optimal schedule. His algorithm has a running time of $O(n + m \log m)$ and uses at most $S - 1$ preemptions if the intervals of availability are rearranged such that they form a staircase pattern. Again, S is the total number of intervals where machines are available.

There are results for nearly online problems, i.e., the next point in time when a machine breaks down or recovers is known. In [4], Sanlaville presents an algorithm for the problem variant that jobs have release and due dates and the goal is to minimize maximum lateness. At any point in time, the algorithm also has to know the next release date. The algorithm constructs optimal schedules for zigzag machine availability patterns (only m or $m-1$ machines are available at any point in time) but not for arbitrary patterns. The running time of the algorithm is $O(n^2 p_{max})$, where p_{max} is the longest processing time of the jobs. The running time does *not* include the cost incurred in repeatedly updating the set of available machines. Sanlaville [4] also reports that his algorithm constructs optimal schedules for arbitrary availability patterns if there are no release dates and the objective to minimize the makespan. However, neither the paper nor a private communication [5] contains any additional information regarding the optimality proof.

As for the online setting, scheduling with unexpected machine breakdowns was studied by Kalyanasundaram and Pruhs [1,2]. In [1] online algorithms with optimal competitive ratios are given for various numbers of faulty machines. The authors assume that if a machine breaks down, the job currently being processed has to be restarted later from the beginning. Also two specific types of breakdowns are considered. In a *permanent* breakdown a machine does not recover again; in a *transient* breakdown the machine is available again right after the breakdown. This is different from the problem setting we consider. In [2] Kalyanasundaram examine to which extent redundancy can help in online scheduling with faulty machines.

Our contribution: In this paper we study the scheduling problem defined above. As mentioned before we are mainly interested in the online version of the problem. In Section 2 we prove that no online algorithm can construct optimal schedules if machines can break down and recover at arbitrary time instances. We also show that no online algorithm can achieve a bounded competitive ratio if there may be time intervals where no machine is available. In Section 3 we present an online algorithm that constructs schedules with an optimal makespan of C_{max}^{OPT} if a *lookahead* of one is given, i.e., the algorithm always knows the next point in the when the set of available machines changes. However, the algorithm does not need to know which machines break down or become available. Our algorithm has a running time of $O(an + S)$, where a is the number of time instances where the set of available machines changes and S is the total number of intervals where machines are available. The additive term of S represents the cost of repeatedly updating the set of available machines. Note that our algorithm has a better running time than Sanlaville's if $a < n p_{max}$, which will be true in practical applications. If $a \geq n p_{max}$, then the set of available machines changes after each time unit. Finally, in Section 4 we give an online algorithm without lookahead that constructs schedules with a nearly optimal makespan of $C_{max}^{OPT} + \epsilon$, for any $\epsilon > 0$, if at any time at least one machine is available. This implies that not knowing machine availabilities does not really hurt the performance of an algorithm.

2 The performance of online algorithms

First note that if at any time at most one machine is available, an optimal online schedule is trivial to construct. In the following we concentrate on problems with an arbitrary set of machines.

Theorem 1. *No online algorithm can, in general, construct optimal schedules. If there may be time intervals where no machines are available, then no online algorithm can achieve a bounded competitive ratio.*

Proof. Let A be any online algorithm. Initially, at time $t = 0$ only one of m machines is available. We consider n jobs $J_1, \ldots, J_n$, each of which has a processing time of 1 time unit. We asume $n = m$. At time $t = 0$, algorithm A starts processing one job J_{i_0}. Let t' be the first time instance such that A first preempts J_{i_0} or A finishes processing J_{i_0}. At that time t' all machines become available. A's makespan is at least $t' + 1$ because none of the jobs J_i, $i \neq i_0$, has been processed so far. An optimal algorithm will divide the interval from 0 to t' evenly among the n jobs so that its makespan is $C_{\max}^{OPT} = t' + 1 - (t'/n)$. This proves the first part of the theorem. For the proof of the second part we modify the problem instance so that no machine is available during the interval $(C_{\max}^{OPT}, c \cdot C_{\max}^{OPT}]$, for any $c > 1$. The algorithm A cannot finish before $c \cdot C_{\max}^{OPT}$ because it has jobs left at time $C_{\max}^{OPT}$. $\square$

3 Optimal schedules

In this section we give an algorithm that constructs optimal schedules with a makespan of $C_{\max}^{OPT}$. The algorithm is online with a *lookahead of one*, i.e., the algorithm always knows the next point in time when the set of available machines changes. The algorithm does not need to know, however, which machines break down or become available.

Let $J_1, \ldots, J_n$ be the given jobs and let p_i, $1 \leq i \leq n$, denote the processing time of J_i. We assume that p_i is known in advance. Without loss of generality jobs are numbered such that $p_1 \geq p_2 \geq \ldots \geq p_n$. At any time during the scheduling process, r_i denotes the remaining processing time of J_i, $1 \leq i \leq n$. We will show later that the algorithm always maintains the invariant $r_1 \geq r_2 \geq \ldots \geq r_n$.

Starting at time $t = 0$, the algorithm repeatedly schedules time intervals $I = [t, t')$ in which the set of available machines remains the same. The availability changed at t and will next change at time t'. In each interval, the algorithm schedules as much load as possible while minimizing the length of the largest remaining processing time.

More specifically, suppose that the algorithm has already scheduled the interval $[0, t)$ and that the set of available machines changes at t. At time t, using lookahead information, the algorithm determines the next point in time $t' > t$ at which the machine availability changes. Let $\delta = t' - t$ and m_a be the number of available machines in $I = [t, t')$. Intuitively, the algorithm now tries to determine the largest possible ϵ, $r_1 \geq \epsilon > 0$, such that, for all jobs J_k, $1 \leq k \leq n$, the remaining processing time in excess to $r_1 - \epsilon$ can be scheduled in I. Thus, at the end of I, all jobs would have a remaining processing time of at most $r_1 - \epsilon$. Note that the total processing time in excess to $r_1 - \epsilon$ is

$$\sum_{k=1}^{n} \max\{0, r_k - (r_1 - \epsilon)\}$$

and that the total processing capacity available in I is $m_a \delta$. However, the algorithm has to satisfy the constraint that at most δ time units of each job can be scheduled in I. Thus, if $\sum_{k=1}^{n} \max\{0, r_k - (r_1 - \epsilon)\} = m_a \delta$ for some $\epsilon > \delta$, then the algorithm cannot schedule ϵ time units of J_1 in I. Only δ time units are permissible.

For this reason, the algorithm first determines a set of jobs that are scheduled for δ time units in I, see lines 6–8 of the code in Figure 1. Suppose that the algorithm has already scheduled δ time units of $J_1, \ldots, J_{i-1}$ in I. The available processing capacity $m_a \delta$ has reduced by $(i-1)\delta$. The algorithm also schedules δ time units of J_i in I if the total remaining processing time in excess to $r_i - \delta$ is not sufficient to fill the processing capacity still available and $r_i \geq \delta$. (Formally, if $\sum_{k=i}^{n} \max\{0, r_k - (r_i - \delta)\} < m_a \delta$ and $r_i \geq \delta$.)

Suppose that the while-loop in lines 6–8 terminates and $i > n$. Then, the algorithm can schedule no more jobs in I. If $i \leq n$, then there are two cases to consider.

(a) $\sum_{k=i}^{n} \max\{0, r_k - (r_i - \delta)\} \geq m_a \delta$

In this case, the algorithm determines the ϵ, $0 < \epsilon \leq \delta$ such that for all jobs J_k, $i \leq k \leq n$, the total remaining processing time in excess to $r_i - \epsilon$ is exactly equal to $m_a \delta$. Each of these jobs is scheduled in I to an extent of $\max\{0, r_k - (r_i - \epsilon)\}$.

(b) $\sum_{k=i}^{n} \max\{0, r_k - (r_i - \delta)\} < m_a \delta$ and $r_i < \delta$

In this case, the algorithm can schedule the rest of $J_i, \ldots, J_n$, if it exists, in I.

In each case, the scheduling of the jobs is done using McNaughton's algorithm. The formal algorithm is given in Figure 1.

We analyze the running time of the algorithm and first argue that within an iteration of the outer while-loop, all executions of lines 6–8 take $O(n)$ time. The critical part are the computations of the sums $S_i = \sum_{k=i}^{n} \max\{0, r_k - (r_i - \delta)\}$. Set $S_0 = 0$. We show that S_{i+1} can be easily derived from S_i. When computing S_i we determine the largest job index l_i such that $r_{l_i} - (r_i - \delta) \geq 0$. We will show below that $r_1 \geq r_2 \geq \ldots \geq r_n$, see Lemma 2. Given l_i, we can easily find l_{i+1} by going through the jobs starting at J_{l_i+1} and find the largest index l_{i+1} such that $r_{l_{i+1}} - (r_{i+1} - \delta) \geq 0$. Then $S_{i+1} = S_i - \delta + (l_i - i)(r_i - r_{i-1}) + \sum_{k=l_i+1}^{l_{i+1}} (r_k - (r_{i+1} - \delta))$. Thus all sums can be computed in $O(n)$ time. Similarly, in line 10, we can compute the desired ϵ in $O(n)$ time. Hence, the scheduling process in each interval $I = [t, t')$ can be done in $O(n)$ time. Whenever the set of available machines changes, we have to update the set of active machines. Each update in availability (a machine breaks down or recovers) can be implemented in constant

Algorithm Lookahead (LA)

1. $t := 0$;
2. $r_i := p_i$, for $1 \leq i \leq n$;
3. **while** there exist jobs with positive remaining processing time **do**
4. $t' :=$ next point in time when set of available machines changes;
5. $\delta := t' - t$; $i := 1$; $m_a :=$ number of machines available in $[t, t+\delta)$;
6.   **while** $i \leq n$ and $\sum_{k=i}^{n} \max\{0, r_k - (r_i - \delta)\} < m_a \delta$ and $r_i \geq \delta$ **do**
7.   Schedule δ time units of J_i in $[t, t+\delta)$;
8. $r_i := r_i - \delta$; $m_a := m_a - 1$; $i := i + 1$;
9.   **if** $i \leq n$ **then**
10.   Compute the maximum ϵ, $\epsilon \leq \min\{\delta, r_i\}$, such that
            $\sum_{k=i}^{n} \max\{0, r_k - (r_i - \epsilon)\} \leq m_a \delta$;
11.   For $k = i, \ldots, n$, schedule $\max\{0, r_k - (r_i - \epsilon)\}$ time units of J_k in
            $[t, t+\delta)$ using McNaughton's algorithm and set $r_k = \min\{r_k, r_i - \epsilon\}$;
12. $t := t'$;

Fig. 1. The online algorithm with a lookahead of one

time. We conclude that the total running time of our algorithm is $O(an + S)$, where a is the number of times instances where the set of available machines changes and S is the total number of intervals where machines are available.

In the analysis of the algorithm we consider the sequence of intervals in which LA schedules jobs. Within each interval, the set of available machines remains the same. Machine availability only changes at the beginning of an interval.

We first show that the algorithm works correctly. When the algorithm terminates, all jobs have a remaining processing time of zero, i.e. the scheduling process is complete. The condition in line 6 of the algorithm ensures that at most $\delta = t' - t$ time units of each job are scheduled in an interval. The update $m_a := m_a - 1$ in line 8 and the constraint $\sum_{k=i}^{n} \max\{0, r_k - (r_i - \epsilon)\} \leq m_a \delta$ in line 10 ensure that the total amount of processing time scheduled in an interval is not greater than the available processing capacity.

Next we prove two useful lemmas.

Lemma 2. *At the beginning of each interval, $r_1 \geq r_2 \geq \ldots \geq r_n$.*

Proof. The invariant holds at time $t = 0$ because initially $r_k = p_k$, for $1 \leq k \leq n$, and $p_1 \geq p_2 \geq \ldots \geq p_n$. Suppose that $r_1 \geq r_2 \geq \ldots \geq r_n$ holds at the beginning of some interval I. We show that the invariant is also satisfied at the end of I. Let $r'_1, \ldots, r'_n$ denote the remaining processing times at the end of I.

Suppose that while executing the while-loop in lines 6–8, the algorithm schedules δ time units of $J_1, \ldots, J_{i-1}$. The remaining processing time of each of these jobs decreases by δ and thus $r'_1 \geq \ldots \geq r'_{i-1}$. If $i > n$, we are done. Otherwise we have to consider two cases.

(a) $\sum_{k=i}^{n} \max\{0, r_k - (r_i - \delta)\} \geq m_a \delta$

    If $i > 1$, then in the last iteration of the while-loop, the condition in line 6 was satisfied, i.e. $\sum_{k=i-1}^{n} \max\{0, r_k - (r_{i-1} - \delta)\} < (m_a + 1)\delta$, which implies

$\sum_{k=i}^{n} \max\{0, r_k - (r_{i-1} - \delta)\} < m_a\delta$. In line 10, the algorithm chooses an ϵ such that $\sum_{k=i}^{n} \max\{0, r_k - (r_i - \epsilon)\} = m_a\delta$. Thus, if $i > 1$, $r'_{i-1} = r_{i-1} - \delta > r_i - \epsilon = r'_i$. For any $i \geq 1$, the invariant now follows because $r'_i = \ldots = r'_l$, where l is the largest job index such that $r_l - (r_i - \epsilon) \geq 0$, and $r'_k = r_k$ for $k > l$.

(b) $\sum_{k=i}^{n} \max\{0, r_k - (r_i - \delta)\} < m_a\delta$ and $r_i < \delta$

In this case, the rest of $J_i, \ldots, J_n$ is scheduled in I, i.e. $r'_i = \ldots = r'_n = 0$ and the invariant holds. $\square$

Now consider any other algorithm A for scheduling $J_1, \ldots, J_n$. In particular, A may be an optimal algorithm that knows the machine breakdowns in advance. At any time consider the sorted sequence $q_1 \geq q_2 \geq \ldots \geq q_n$ of remaining processing times maintained by A. That is, q_i is the i-th value in the sorted sequence, $1 \leq i \leq n$. Note that q_i is not necessarily the remaining processing time of J_i.

Lemma 3. *At the beginning of each interval, $r_1 \leq q_1$ and $\sum_{k=1}^{n} r_k \leq \sum_{k=1}^{n} q_k$.*

Proof. We show inductively that at any time

$$\sum_{k=1}^{j} r_k \leq \sum_{k=1}^{j} q_k \quad \text{for} \quad j = 1, \ldots, n. \tag{1}$$

The lemma follows from the special case $j = 1$ and $j = n$. The above inequalities hold at time $t = 0$. Suppose that they hold at the beginning of some interval I. We show that they are also satisfied at the end of I. Let $r'_1, \ldots, r'_n$ and $q'_1, \ldots, q'_n$ be the remaining processing times at the end of I. Recall that r'_k is the remaining processing time of J_k, $1 \leq k \leq n$. By Lemma 2, $r'_1 \geq \ldots \geq r'_n$. We have $q'_1 \geq \ldots \geq q'_n$ by the definition of the q-values. Note that q_k and q'_k can be the processing times of different jobs. However, $q'_k \leq q_k$ for $1 \leq k \leq n$.

Suppose that in lines 6–8, algorithm LA schedules δ time units of $J_1, \ldots, J_{i-1}$. Then $r'_k = r_k - \delta$, for $k = 1, \ldots, i - 1$. We have $q'_k \geq q_k - \delta$, for $1 \leq k \leq n$, because the processing times of jobs decrease by at most δ in I. Thus, inequality (1) holds for $j = 1, \ldots, i - 1$. Again, for $i \leq n$, we consider two cases.

(a) $\sum_{k=i}^{n} \max\{0, r_k - (r_i - \delta)\} < m_a\delta$ and $r_i < \delta$

The algorithm LA schedules the rest of $J_i, \ldots, J_n$ in I so that $r'_i = \ldots = r'_n = 0$. Inequality (1) also holds for $j = i, \ldots, n$.

(b) $\sum_{k=i}^{n} \max\{0, r_k - (r_i - \delta)\} \geq m_a\delta$

LA computes an ϵ, $0 < \epsilon \leq \delta$, such that $\sum_{k=i}^{n} \max\{0, r_k - (r_i - \epsilon)\} = m_a\delta$. It reduces the remaining processing times of $J_i, \ldots, J_l$ to $r_i - \epsilon$, where l is the largest job index such that $r_l - (r_i - \epsilon) \geq 0$.

Let $\overline{m}_a$ be the number of machines that were initially available in I. Since LA uses all of the available processing capacity, $\sum_{k=1}^{j} r'_k = \sum_{k=1}^{j} r_k - \overline{m}_a\delta$ for $j = l, \ldots, n$. Since $\sum_{k=1}^{j} q'_k \geq \sum_{k=1}^{j} q_k - \overline{m}_a\delta$ for $j = l, \ldots, n$, inequality (1) holds for $j = l, \ldots, n$. It remains to show that the inequality is also satisfied for $j = i, \ldots, l - 1$.

Let $R_1 = \sum_{k=1}^{i-1} r'_k$, $R_2 = \sum_{k=i}^{l} r'_k$ and similarly $Q_1 = \sum_{k=1}^{i-1} q'_k$, $Q_2 = \sum_{k=i}^{l} q'_k$. We have already shown (i) $R_1 \leq Q_1$ and (ii) $R_1 + R_2 \leq Q_1 + Q_2$. Suppose that $Q_1 = R_1 + x$ for some $x \geq 0$. Then (ii) implies $Q_2 + x \geq R_2$. Consider the $l - i + 1$ values $q'_i, \ldots, q'_l$. Since $q'_i \geq \ldots \geq q'_l$, the sum of the first μ values, for any $1 \leq \mu \leq l - i + 1$, is at least $\mu Q_2/(l - i + 1)$. Thus, for any j with $i \leq j \leq l$,

$$\sum_{k=1}^{j} q'_k \geq Q_1 + (j - i + 1)\frac{Q_2}{l - i + 1} = R_1 + x + (j - i + 1)\frac{Q_2}{l - i + 1}$$

$$\geq R_1 + (j - i + 1)\frac{Q_2 + x}{l - i + 1} \geq R_1 + (j - i + 1)\frac{R_2}{l - i + 1}$$

$$= \sum_{k=1}^{j} r'_k.$$

The last equation follows because $r_i - \epsilon = r'_i = r'_{i+1} = \ldots = r'_l = R_2/(l - i + 1)$. $\square$

Theorem 4. *For any problem instance, $C_{\max}^{LA} = C_{\max}^{OPT}$.*

Proof. Given a set of jobs $J_1, \ldots, J_n$, let $I = [t, t')$ be the last interval in which LA has scheduled jobs, i.e., $t \leq C_{\max}^{LA} \leq t'$. Consider the makespan $C_{\max}^{OPT}$ produced by an optimal offline algorithm. We distinguish two cases.

(1) In the online schedule, the interval from t to $C_{\max}^{LA}$ contains no idle machines Thus, in the online schedule all machines finish at the same time. Lemma 3 implies that at the beginning of I, the total remaining processing time $\sum_{k=1}^{n} r_k$ of LA is not greater than the total remaining processing time $\sum_{k=1}^{n} q_k$ of OPT. Thus, $C_{\max}^{OPT} \geq C_{\max}^{LA}$.

(2) In the online schedule, the interval from t to $C_{\max}^{LA}$ contains idle machines Since LA schedules job portions using McNaughton's algorithm, there must exist a job that spans the entire interval from t to $C_{\max}^{LA}$. Thus, at the beginning of I the largest remaining processing time r_1 equals $C_{\max}^{LA} - t$. By Lemma 3, the largest remaining processing time q_1 of OPT is not smaller. Thus OPT cannot finish earlier than LA. $\square$

4 Nearly optimal schedules

In this section we study the problem that an online algorithm has no information about the future machine availabilities. It does not know the next point in time when the set of available machines changes. We present an algorithm that always produces a makespan of $C_{\max}^{OPT} + \epsilon$, for any $\epsilon > 0$. It is assumed that at any time at least one machine is available since otherwise, by Theorem 1, no bounded performance guarantee can be achieved.

We number the jobs to be scheduled such that $p_1 \geq p_2 \geq \ldots \geq p_n$. Given a fixed $\epsilon > 0$, our online algorithm, called ON, computes $\delta = \epsilon/n^2$. Starting at time $t = 0$, the algorithm always schedules jobs within the time interval $[t, t + \delta)$. Let m_a be the number of machines available at time t. The algorithm determines the m_a jobs with the largest remaining processing times (ties are broken arbitrarily) and schedules them on the available machines. If a machine breaks down or becomes available at some time $t + \delta'$, $\delta' < \delta$, then the algorithm preempts the jobs currently being processed and computes a new schedule for the next δ time units from $t + \delta'$ to $t + \delta' + \delta$. Otherwise, if the set of available machines remains the same throughout $[t, t + \delta)$, the algorithm computes a new partial schedule at time $t + \delta$. A formal description of the algorithm is given in Figure 2. At any time r_i denotes the remaining processing time of J_i, $1 \leq i \leq n$.

Algorithm Online (ON)
1. $t := 0$; $\delta = \epsilon/n^2$;
2. $r_i := p_i$, for $1 \leq i \leq n$;
3. **while** there exist jobs with positive remaining processing time **do**
4. $m_a :=$ number of machines available at time t;
5. $n_a :=$ number of jobs with positive remaining processing time;
6. $S :=$ set of the $\min\{m_a, n_a\}$ jobs with the largest remaining processing time;
7.   Process the jobs J_i, $i \in S$, on the available machines;
8.   **if** machines break down or become available at some time $t + \delta'$, $\delta' < \delta$ **then**
9.   Set $r_i := \max\{0, r_i - \delta'\}$ for $i \in S$; $t := t + \delta'$;
10.   **else**
11.   Set $r_i := \max\{0, r_i - \delta\}$ for $i \in S$; $t := t + \delta$;

Fig. 2. The online algorithm ON

In the scheduling process, the algorithm repeatedly has to find jobs with the largest remaining processing time. If we keep a priority queue of the remaining processing times, each such job can be found in $O(\log n)$ time. Let a be the total number of time instances where the set of available machines changes and let m_i, $1 \leq i \leq a$, be the number of machines that are available right after the i-th change; m_0 is the number of machines that are available initially. Let $P = \sum_{i=1}^{n} p_i$. Note that the total number of job portions scheduled by the algorithm is no more than $P/\delta + \sum_{i=0}^{a} m_i$. This is because at the end of a scheduled job portion, δ time units have been processed or the set of available machines changes. Thus the total running time of the algorithm is $O((Pn^2/\epsilon + \sum_{i=0}^{a} m_i) \log n + S \cdot M)$, where S is the total number of intervals where machines are available and $M = \max_{0 \leq i \leq a} m_i$. The additive term accounts for the fact that the algorithm has to repeatedly update the set of available machines. Note that the number of preemptions is no more than $Pn^2/\epsilon + \sum_{i=0}^{a} m_i$.

For the analysis of the algorithm we partition the time into intervals such that at the beginning of an interval the online algorithm computed a new partial schedule, i.e., it executed lines 4–7. Note that intervals have a length of at most δ and that within each interval the set of available machines remains the same.

We start with a useful observation on the relative job lengths. The proof of the next lemma is omitted.

Lemma 5. *At the beginning of each interval, for any two jobs J_i and J_j with $i < j$, $r_i \geq r_j - \delta$.*

For the further analysis, we maintain a sequence of job sets $S_1, \ldots, S_l$, for some $1 \leq l \leq n$. Initially, at time 0, S_i contains J_i, $1 \leq i \leq n$. At the end of each interval I, the sets are updated as follows.

> Let i be the smallest job index such that J_i was not processed in I and let j be the largest job index such that J_j was processed in I. Suppose that $J_i \in S_{k_i}$ and $J_j \in S_{k_j}$. If $k_i < k_j$, then replace $S_{k_i}, S_{k_i+1}, \ldots, S_{k_j}$ by the union of these sets. Renumber the new sequence of sets so that the k-th set in the sequence has index k.

Note that at any time the sequence of sets forms a partitioning of the jobs $J_1, \ldots, J_n$. The update rule ensures that every set contains a sequence of consecutive jobs with respect to the job numbering. In the following, let n_k denote the number of jobs in S_k, and let $N_k = n_1 + \ldots + n_k$.

Lemma 6. *At the beginning of each interval, for every set S_k, $1 \leq k \leq l$, and jobs $J_i, J_j \in S_k$, $|r_i - r_j| \leq (n-1)\delta$.*

Proof. We prove inductively that at the beginning of each interval, for every set S_k and jobs $J_i, J_j \in S_k$,

$$|r_i - r_j| \leq (n_k - 1)\delta. \tag{2}$$

This holds initially because at time $t = 0$, every set contains exactly one job. Consider an interval $I = [t, t + \delta')$, for some $\delta' \leq \delta$, and suppose (2) holds at the beginning of I.

We first show that (2) is maintained while jobs are processed in I and before the update rule for the sets is applied. Given a set S_k, let $J_i, J_j \in S_k$ be any two jobs with $i < j$. Let r_i, r_i' and r_j, r_j' be the remaining processing times at the beginning and at the end of I. If $r_i' \leq r_j'$, then by Lemma 5, $|r_i' - r_j'| = r_j' - r_i' \leq \delta$.

If $r_i' > r_j'$, we have to consider several cases. If none of the two jobs was processed in I or if both jobs were processed for δ' time units, then there is nothing to show. Otherwise, let δ_i and δ_j be the number of time units for which J_i and J_j are processed in I. If only J_j is processed in I, then $r_j \geq r_i$ and thus $|r_i' - r_j'| = r_i' - r_j' = r_i - (r_j - \delta_j) \leq \delta_j \leq \delta$. The case that both J_i and J_j are scheduled in I, but J_j is processed for a longer period, cannot occur. This would imply that the processing of J_i is complete, i.e. $r_i' = 0$, which contradicts $r_i' > r_j'$. Finally suppose that J_i is processed as least as long as J_j in I, i.e. $0 \leq \delta_j \leq \delta_i$. Then $|r_i' - r_j'| = r_i - \delta_i - (r_j - \delta_j) = r_i - r_j + \delta_j - \delta_i \leq r_i - r_j \leq (n_k - 1)\delta$. Inequality (2) is satisfied.

We now study the effect when the set update rule is applied at the end of I. Suppose that a sequence of sets $S_{k_1}, \ldots, S_{k_2}$ is merged. Let $J_i \in S_{k_1}$ be a job not scheduled in I and let $J_j \in S_{k_2}$ be the job with the largest index scheduled in I. Let $J_{\max}$ be the job with the largest remaining processing time at time $t + \delta'$ and let $J_{\min}$ be the job with the smallest remaining processing time. We will show $|r'_{\max} - r'_{\min}| \leq (n_{k_1} + n_{k_2} - 1)\delta$. This completes the proof because the newly merged set contains $\sum_{k=k_1}^{k_2} n_k \geq n_{k_1} + n_{k_2}$ jobs. We have

$$|r'_{\max} - r'_{\min}| = r'_{\max} - r'_{\min} = (r'_{\max} - r'_i) + (r'_i - r'_j) + (r'_j - r'_{\min}).$$

If $J_{\max} \in S_{k_1}$, then $r'_{\max} - r'_i \leq (n_{k_1} - 1)\delta$. If $J_{\max} \notin S_{k_1}$, then $r'_{\max} - r'_i \leq \delta$ by Lemma 5 because $J_{\max}$ has a higher index than J_i. In any case $r'_{\max} - r'_i \leq (n_{k_1} - 1)\delta$. Similarly, if $J_{\min} \in S_{k_2}$, then $r'_j - r'_{\min} \leq (n_{k_2} - 1)\delta$. If $J_{\min} \notin S_{k_2}$, then $r'_j - r'_{\min} \leq \delta$ by Lemma 5. In any case $r'_j - r'_{\min} \leq (n_{k_2} - 1)\delta$. Since J_i was not scheduled in I but J_j was scheduled, $r_j \geq r_i$. Job J_j was scheduled for at most δ time units, which implies $r'_i = r_i \leq r_j \leq r'_j + \delta$ and hence $r'_i - r'_j \leq \delta$. In summary we obtain

$$
\begin{aligned}
|r'_{\max} - r'_{\min}| &= (r'_{\max} - r'_i) + (r'_i - r'_j) + (r'_j - r'_{\min}) \\
&\leq (n_{k_1} - 1)\delta + \delta + (n_{k_2} - 1)\delta \\
&= (n_{k_1} + n_{k_2} - 1)\delta. \quad \square
\end{aligned}
$$

At any time let $l_{\max}$ denote the maximum index such that $S_1, \ldots, S_{l_{\max}}$ contain only jobs with positive remaining processing times. If there is no such set, then let $l_{\max} = 0$. Let A be any other scheduling algorithm. In particular, A may be an optimal offline algorithm. At any time consider the sequence of remaining processing times maintained by A, sorted in non-increasing order. Let q_i be the i-th value in this sorted sequence.

Lemma 7. *At the beginning of each interval, for $k = 1, \ldots, l_{\max}$, $\sum_{i=1}^{N_k} r_i \leq \sum_{i=1}^{N_k} q_i$.*

Proof. The lemma holds initially because at time $t = 0$, $r_i = q_i = p_i$ for $1 \leq i \leq n$. Suppose that the lemma holds at the beginning of an interval $I = [t, t + \delta')$, for some $\delta' \leq \delta$. Let $S_1, \ldots, S_l$ and $S'_1, \ldots, S'_{l'}$ be the sequences of job sets at the beginning and at the end of I. Furthermore, let j be the largest index such that all jobs in $S'_1, \ldots, S'_j$ were scheduled in I and still have a positive remaining processing time. These sets were not involved in a merge operation at the end of I and, hence, each S'_k contains the same jobs as S_k, $1 \leq k \leq j$. Since the jobs of these sets have a positive remaining processing time, all of them were scheduled for exactly δ' time units in I. Let r_i, r'_i and q_i, q'_i denote the remaining processing times at the beginning and at the end of I. Since $q'_i \geq q_i - \delta'$, for $1 \leq i \leq n$, we obtain

$$\sum_{i=1}^{N_k} r'_i = \sum_{i=1}^{N_k} (r_i - \delta') \leq \sum_{i=1}^{N_k} (q_i - \delta') \leq \sum_{i=1}^{N_k} q'_i,$$

for $k = 1, \ldots, j$. If $j = l'_{\max}$, then we are done.

Suppose that $j < l'_{\max}$. By the definition of $l'_{\max}$, the set S'_{j+1} does not contain jobs with zero remaining processing time. Also, by the definition of j,

S'_{j+1} contains jobs not scheduled in I. The update rule for job sets ensures that S'_{j+1} contains all jobs J_i, $i > N_j$, that were scheduled in I. Let N be the number of jobs in S'_{j+1} scheduled in I. All of these jobs were scheduled for δ' time units because they all have positive remaining processing time. The total number of available machines in I is $N_j + N$ since, otherwise, the algorithm ON would have scheduled more jobs of S'_{j+1} in I. Thus any other algorithm cannot process more than $(N_j + N)\delta$ time units in I. We conclude

$$\sum_{i=1}^{N_k} r'_i = \sum_{i=1}^{N_k} r_i - (N_j + N)\delta' \leq \sum_{i=1}^{N_k} q_i - (N_j + N)\delta' \leq \sum_{i=1}^{N_k} q'_i,$$

for $k = j + 1, \ldots, l'_{\max}$. $\square$

Theorem 8. *For any fixed $\epsilon > 0$ and any problem instance, $C^{ON}_{\max} \leq C^{OPT}_{\max} + \epsilon$.*

Proof. Let $I = [t, t + \delta')$, $\delta' \leq \delta$, be the last interval such that $l_{\max} > 0$ at the beginning of I. Consider the total remaining processing time of the jobs in $S_1, \ldots, S_{l_{\max}}$ at time t. By Lemma 7, the value of ON is not larger than the value of an optimal offline algorithm. Thus $C^{OPT}_{\max} \geq t$. We analyse ON's makespan. At time $t + \delta'$, S_1 contains a job J_i with zero remaining processing time. By Lemma 6, all jobs belonging to the first set have a remaining processing time of at most $(n - 1)\delta$. All jobs not belonging to the first set have a higher index than J_i and, by Lemma 5, they have a remaining processing time of at most δ. Thus at time $t + \delta'$, we are left with at most $n - 1$ jobs having a remaining processing time of at most $(n - 1)\delta$, i.e., the total remaining processing time of ON is at most $(n - 1)^2\delta$. Since at any time at least one machine is available $C^{ON}_{\max} \leq t + \delta' + (n - 1)^2\delta \leq C^{OPT}_{\max} + n^2\delta \leq C^{OPT}_{\max} + \epsilon$. $\square$

References

1. B. Kalyanasundaram and K.P. Pruhs. Fault-tolerant scheduling. In *Proceedings of the 26th Annual ACM Symposium on the Theory of Computing*, pages 115–124, 1994.
2. B. Kalyanasundaram and K.P. Pruhs. Fault-tolerant real-time scheduling. In *Proc. 5th Annual European Symposium on Algorithms (ESA)*, Springer Lecture Notes in Computer Science, 1997.
3. R. McNaughton. Scheduling with deadlines and loss functions. *Management Science*, 6:1-12, 1959.
4. E. Sanlaville. Nearly on line scheduling of preemptive independent tasks. *Discrete Applied Mathematics*, 57:229–241, 1995.
5. E. Sanlaville. Private communication, 1998.
6. G. Schmidt. Scheduling on semi-identical processors. *Z. Oper. Res.*, 28:153–162, 1984.
7. G. Schmidt. Scheduling independent tasks with deadlines on semi-identical processors. *J. Oper. Res. Soc.*, 39:271–277, 1988.
8. D.D. Sleator and R.E. Tarjan. Amortized efficiency of list update and paging rules. *Communications of the ACM*, 28:202–208, 1985.

Scheduling on a Constant Number of Machines

F. Afrati[1], E. Bampis[2], C. Kenyon[3], and I. Milis[4]

[1] NTUA, Division of Computer Science, Heroon Polytechniou 9, 15773, Athens, Greece
[2] LaMI, Université d'Evry, Boulevard François Mitterrand, 91025 Evry Cedex, France
[3] LRI, Bât 490, Université Paris-Sud, 91405 Orsay Cedex, France
[4] Athens University of Economics, Dept. of Informatics, Patission 76, 10434 Athens, Greece

Abstract. We consider the problem of scheduling independent jobs on a constant number of machines. We illustrate two important approaches for obtaining polynomial time approximation schemes for two different variants of the problem, more precisely the multiprocessor-job and the unrelated-machines models, and two different optimization criteria: the makespan and the sum of weigthed completion times.

1 Introduction

In the past few years, there have been significant developments in the area of approximation algorithms for $\mathcal{N}P$-hard scheduling problems, see *e.g.* [8] and the references at the end of this paper. Our current, admittedly optimistic, opinion is the following: for any scheduling problem where the schedules can be stretched without unduly affecting the cost function, and in which each job is specified by a constant number of parameters, there should be a way to construct a PTAS by a suitable combination of known algorithmic and approximation techniques.

We present here two approaches for obtaining efficient PTAS that we illustrate by two examples dealing eachone with a different optimization criterion. In the first case, we consider the problem of minimizing the makespan for a *multiprocessor job system* [12]. In this model, we are given a set of jobs J such that each job requires to be processed simultaneously by several processors. In the *dedicated* variant of this model, each job requires the simultaneous use of a prespecified set of processors fix_j. Since each processor can process at most one job at a time, jobs that share at least one resource cannot be scheduled at the same time step and are said to be *incompatible*. Hence, jobs are subject to compatibility constraints. Thus, every job is specified by its execution time p_j and the prespecified subset of processors fix_j on which it must be executed. By t_j we denote the starting time of job j and the completion time of j is equal to $C_j = t_j + p_j$. The objective is to find a feasible schedule minimizing the makespan $C_{\max}$, i.e. the maximum completion time of any job. Using the standard three field notation of Graham *et al.* [14], this problem is classified as $Pm|fix_j|C_{\max}$. In the second case, we are given n independent jobs that have to be executed on m unrelated machines. Each job j is specified by its execution times $p_j^{(i)}$ on each machine M_i, $i = 1, 2, \ldots, m$, and by its positive weight w_j. The jobs must be processed without interruption, and each machine can execute at most one job at a time. Hence, if t_j is the starting time of job j executed on machine M_i in a particular schedule then the completion time of j is $C_j = t_j + p_j^{(i)}$. The objective is to minimize the weighted sum of job completion times $\sum_{j \in J} w_j C_j$. Using the standard three-field notation, the considered problem is denoted as $Rm| \mid \sum w_j C_j$. For example, one may think of an application where each job is associated with specific deliverables for customers: then the sum of completion times measures the quality of a schedule constructed.

In the next section, we present the most important known results for the dedicated variant of the multiprocessor job problem, and we briefly sketch the principle of the method used in [2] in order to obtain a PTAS. In the third section, we also start with a brief state of the art about the unrelated machines problem and we give the rough idea of the PTAS proposed in [1].

2 Minimizing the makespan

2.1 State of the art

The problem of scheduling independent jobs on dedicated machines has been extensively studied in the past in both the preemptive and the non-preemptive case. In the *preemptive* case, each job can be at no cost interrupted at any time and completed later. In the *non-preemptive* case, a job once started has to be processed (until completion) without interruption.

The non-preemptive three-processor problem, i.e. $P3|\text{fix}_j|C_{\max}$, has been proved to be strongly NP-hard by a reduction from 3-partition [4,17]. The best constant-factor approximation algorithm is due to Goemans [13] who proposed a $\frac{7}{6}$-algorithm improving the previous best performance guarantee of $\frac{5}{4}$ [11]. A linear time PTAS has been proposed in [2]. For unit execution time jobs, *i.e.* $Pm|\text{fix}_j, p_j = 1|C_{max}$, the problem is solvable in polynomial time through an integer programming formulation with a fixed number of variables [17]. However, if the number of processors is part of the problem, *i.e.* $P|\text{fix}_j, p_j = 1|C_{\max}$, the problem becomes NP-hard. Furhermore, Hoogeveen et al. [17] showed that, for $P|\text{fix}_j, p_j = 1|C_{\max}$, there exists no polynomial approximation algorithm with performance ratio smaller than 4/3, unless $P = NP$.

2.2 The principle of the algorithm

The PTAS presented in [2] is based on the transformation of a preemptive schedule to a non-preemptive one. A simple approach to obtain a feasible non-preemptive schedule from a preemptive schedule S is to remove the preempted jobs from S and process them sequentially in a naïve way at the end of S. The produced schedule has a makespan equal to the makespan of S plus the total processing time of the delayed jobs. However, even if S is an optimal preemptive schedule, the schedule thus produced is almost certainly not close to optimal, since there can be a large number of possibly long preempted jobs. The algorithm in [2] does use a preemptive schedule to construct a non-preemptive schedule in the way just described. However, to ensure that the non preemptive solution is close to optimal, only "short" jobs are allowed to be preempted.

The intuitive idea of the algorithm is the following: first, partition the set of jobs into "long" jobs and "short" jobs. This is similar in spirit to Hall and Shmoys' polynomial time approximation scheme for single-machine scheduling [16]. For each possible schedule of the long jobs, complete it into a schedule of J, **assuming that the short jobs are preemptable**[1]. From each such preemptive schedule, construct a non-preemptive schedule. Finally, among all the schedules thus generated, choose the one that has the minimum makespan. The details are given in [2].

Theorem 1. [2] *There is an algorithm A which given a set J of n independent multiprocessor jobs, a fixed number m of dedicated processors and a constant $\epsilon > 0$, produces, in time at most $O(n)$, a schedule of J whose makespan is at most $(1 + \epsilon)C_{\max}^{opt}$.*

This algorithmic paradigm is appropriate for problems in which the objective function is the makespan. Indeed, the same approach has been used in order to obtain PTASs for the parallel variant of the multiprocessor job problem [3], and for scheduling malleable parallel tasks [20].

3 Minimizing the weigthed completion time

3.1 State of the art

The problem of scheduling jobs to minimize the total weighted job completion time is one of the most well-studied problems in the area of scheduling theory. The basic situation is the single-machine case with no release dates; in that case, in 1956 Smith designed a very easy greedy algorithm: sequencing in order of non-decreasing p_j/w_j ratio produces an optimal schedule [41]. The problem $R|\ |\sum C_j$ is also polynomial [18]. However with general weights and $m \geq 2$ machines the problem is $\mathcal{N}P$-hard, even for fixed m and identical machines (*i.e.* when $p_j^{(i)} = p_j$ is independent of the machine) [5] [24].

There has been a lot of recent progress on scheduling on *identical* parallel machines: Kawagachi and Kyan [21] showed an 1.21 approximation algorithm. For any fixed number of identical parallel machines Sahni proposed a PTAS [32]. Woeginger [43] extended this result in the case of uniform parallel machines which run at different but not job dependent speeds. Moreover, Skutella and Woeginger [38] proposed a PTAS for the case where the number of identical parallel machines is an input of the problem ($P|\ |\sum w_j C_j$).

---

[1] A job is *preemptable* if it can be at no cost interrupted at any time and completed later.

The situation is not so good for scheduling on *unrelated* machines. A sequence of papers has proposed various approximation algorithms [29,19,36,37]. When the number of machines is a parameter of the problem the last of these papers, due to Schultz and Skutella, provides an $(3/2-\epsilon)$-approximation algorithm. More recently Skutella gave an approximation algorithm with performance ratio 1.5 for a constant number m of machines, $m > 2$, and 1.2752 for the two-machine problem [39]. In [1], an $O(n \log n)$-time approximation scheme (PTAS) for scheduling on general unrelated machines, when the number m of machines is fixed, *i.e.* for the problem $Rm| \ |\sum w_j C_j$, and an $O(n \log n)$-time approximation scheme (PTAS) for the single-machine-release dates case $1|r_j|\sum w_j C_j$, have been presented. Independently, Chekuri, Karger, Khanna, Stein and Skutella [6] designed a PTAS for scheduling on a constant number of unrelated machines with release dates.

3.2   The principle of the algorithm

The approach of the previous section cannot be used when the considered objective function is the sum of the weighted completion times since each job is now specified by its execution(s) time(s) and its weight, and thus the partition into short and long jobs based on the execution times does not help. The general principle used in [1] in order to obtain a PTAS is that the only really well-understood minsum scheduling problem is Smith's setting (single-machine-no-release-dates), and so the problem is simplified until it closely resembles that setting. Two well-known ingredients have been used:

1. grouping and rounding, as in the classical bin-packing approximation schemes, in order to simplify the problem; and
2. dynamic programming to design an algorithm, once we are close to but not exactly in Smith's setting (Skutella and Woeginger's paper [38] gave us a strong belief that jobs should only interact with jobs that had similar ratios, and hence dynamic programming should work).

For simplicity of notation, we present the principle of the algorithm when there are just two machines, but it should be clear that this holds for any constant number of machines.

We start from a simple but fundamental observation. Consider the restriction of the optimal schedule to each of the two machines: then Smith's ratio rule applies, i.e. the jobs executed on machine M_i are processed in order of non-increasing ratios $r_j^{(i)} = w_j/p_j^{(i)}$. In particular, observe that if two jobs executed on the same machine have the same ratio on that machine, then their relative order does not matter.

The algorithm [1] is based on the following two results:

The first result states that unrelated machines are not all that different from identical machines: if the processing times of a job are very different on the two machines (i.e. if $p_j^{(1)} < \epsilon p_j^{(2)}$ or $p_j^{(2)} < \epsilon p_j^{(1)}$), then we can schedule it on the machine on which it has the shorter processing time. Thus, if a job j satisfies

$p_j^{(1)} < \epsilon p_j^{(2)}$ then we say that j is M_1-**decided**, and similarly if $p_j^{(2)} < \epsilon p_j^{(1)}$ then we say that j is M_2-**decided**.

The second result states that if the ratios of two jobs are very different, then we can neglect their interaction.

Exploiting these two results we define windows of ratios with appropriate size such that any undecided job has its two ratios in the same window or in adjacent windows, and that we only need to worry about interactions between jobs which ratios are in the same window or in adjacent windows. This is the key to the dynamic program.

Theorem 2. [1] *There is a polynomial time approximation scheme for* $Rm|\ |\sum w_j C_j$.

With additional ideas one can extend this approach to deal with single-machine scheduling in the presence of release dates.

Theorem 3. [1] *There is a polynomial time approximation scheme for* $1|r_j|\sum w_j C_j$.

The details of the algorithm can be found in [1].

References

1. F. AFRATI, E. BAMPIS, C. KENYON, I. MILIS, *Scheduling to minimize the weighted sum of completion times, submitted.*
2. A.K. AMOURA, E. BAMPIS, C. KENYON, Y. MANOUSSAKIS, *How to schedule independent multiprocessor tasks, Proc. ESA'97, LNCS No 1284, pp. 1-12, 1997.*
3. A.K. AMOURA, E. BAMPIS, C. KENYON, Y. MANOUSSAKIS, *Scheduling independent multiprocessor tasks, submitted.*
4. J. BLAZEWICZ, P. DELL'OLMO, M. DROZDOWSKI, AND M.G. SPERANZA, *Scheduling Multiprocessor Tasks on Three Dedicated Processors, Information Processing Letters, 41:275-280, 1992.*
5. J.L. BRUNO, E.G. COFFMAN JR., R. SETHI, *Scheduling independent tasks to reduce mean finishing time, Communications of the Association for Computing Machinery, 17, 382-387, 1974.*
6. C. CHEKURI, D. KARGER, S. KHANNA, C. STEIN, M. SKUTELLA, *Scheduling jobs with release dates on a constant number of unrelated machines so as to minimize the weighted sum of completion times, submitted.*
7. C. CHEKURI, R. MOTWANI, B. NATARAJAN, C. STEIN, *Approximation techniques for average completion time scheduling, in Proc. SODA'97, 609-618, 1997.*
8. B. CHEN, C.N. POTTS, G.J. WOEGINGER, *A review of machine scheduling: Complexity, algorithms and approximability, Report Woe-29, TU Graz, 1998.*
9. S. CHAKRABARTI, C.A. PHILLIPS, A. S. SHULZ, D. B. SHMOYS, C. STEIN, J. WEIN, *Improved scheduling algorithms for minsum criteria, in F. Meyer auf der Heide and B. Monien, editors, Automata, Languages and Programming, volume 1099 of LNCS, pp. 646-657, Springer, Berlin, 1996.*
10. F. A. CHUDAK, D. B. SHMOYS, *Approximation algorithms for precedence-constrained scheduling problems on parallel machines that run at different speeds, in Proceedings of the 8th Annual ACM-SIAM Symposium on Discrete Algorithms, pp. 581-590, 1997.*

11. P. DELL'OLMO, M. G. SPERANZA, AND Z. TUZA, *Efficiency and Effectiveness of Normal Schedules on Three Dedicated Processors, Submitted.*

12. J. DU, AND J. Y-T. LEUNG, *Complexity of Scheduling Parallel Task Systems, SIAM J. Discrete Math., 2(4):473-487, 1989.*

13. M. X. GOEMANS, *An Approximation Algorithm for Scheduling on Three Dedicated Processors. Disc. App. Math., 61:49-59, 1995.*

14. R.L. GRAHAM, E.L. LAWLER, J.K. LENSTRA, A.H.G. RINNOOY KAN, *Optimization and approximation in deterministic sequencing and scheduling , Ann. Discrete Math., Vol. 5, 287-326, 1979.*

15. L. A. HALL, A. S. SHULZ, D. B. SHMOYS, J. WEIN, *Scheduling to minimize average completion time: Off-line and on-line approximation algorithms, Mathematics of Operations Research, 22: 513-544, 1997.*

16. L.A. HALL AND D.B. SHMOYS, *Jackson's rule for single-machine scheduling: making a good heuristic better, Mathematics of Operations Research, 17:22-35, 1992.*

17. J. HOOGEVEEN, S. L. VAN DE VELDE, AND B. VELTMAN, *Complexity of Scheduling Multiprocessor Tasks with Prespecified Processors Allocation, Disc. App. Math., 55:259-272, 1994.*

18. W.A. HORN, *Minimizing average flow time with parallel machines, Operations Research 21, pp. 846-847, 1973.*

19. L.A. HALL, A.S. SCHULTZ, D.B. SHMOYS, J. WEIN, *Scheduling to minimize average completion time, Mathematics of Operations Research, 22, 513-544, 1997.*

20. K. JANSEN, L. PORKOLAB, *Linear-Time approximation schemes for malleable parallel tasks, Technical Report MPI-I-98-1-025, Max Planck Institute, also in Proceedings of SODA'99.*

21. T. KAWAGUCHI, S. KYAN, *Worst case bound of an LRF schedule for the mean weighted flow-time problem, SIAM J. on Computing, 15, 1119-1129, 1986.*

22. M. X. GOEMANS, *Improved approximation algorithms for scheduling with release dates, in Proceedings of the 8th Annual ACM-SIAM Symposium on Discrete Algorithms, pp. 591-598, 1997.*

23. J. LABETOULLE, E.L. LAWLER, J.K. LENSTRA, A.H.G. RINNOOY KAN, *Preemptive scheduling of uniform machines subject to release dates, in W.R. Pulleyblanck (ed.) Progress in Combinatorial Optimization, Academic Press, 245-261, 1984.*

24. J.K. LENSTRA, A.H.G. RINNOOY KAN, P. BRUCKER, *Complexity of machine scheduling problems, Annals of Discrete Mathematics, 1, 343-362, 1977.*

25. J. K. LENSTRA, D. B. SHMOYS, E. TARDOS, *Approximation algorithms for scheduling unrelated parallel machines, Mathematical programming 46:259-271, 1990.*

26. R. H. MÖHRING, M. W. SCHÄFFTER, A. S. SHULZ, *Scheduling jobs with communication delays: Using unfeasible solutions for approximation, in J. Diaz and M. Serna, editors, Algorithms-ESA'96. volume 1136 of LNCS, pp. 76-90, Springer, Berlin 1996.*

27. A. MUNIER, M. QUEYRANNE, A. S. SHULZ, *Approximation bounds for a general class of precedence-constrained parallel machine scheduling problems, in R. E. Bixby, E. A. Boyd, R. Z. Rios-Mercado, editors, Integer Programming and Combinatorial Optimization, volume 1412 of LNCS, pp. 367-382, Springer, Berlin, 1998.*

28. C. PHILIPS, C. STEIN, J. WEIN, *Scheduling job that arrive over time in Proc. 4th Workshop on Algorithms and Data Structures (1995), to appear in Mathematical Programming.*

29. C. PHILIPS, C. STEIN, J. WEIN, *Task scheduling in networks, SIAM J. on Discrete Mathematics, 10, 573-598, 1997.*

30. C. PHILIPS, C. STEIN, J. WEIN, *Minimizing average completion time in the presence of release dates*, Mathematical Programming 82:199-223, 1998.

31. C. A. PHILLIPS, A. S. SHULZ, D. B. SHMOYS, C. STEIN, J. WEIN, *Improved bounds on relaxations of a parallel machine scheduling problem*, Journal of Combinatorial Optimization, 1:413-426, 1998.

32. S. SAHNI, *Algorithms for scheduling independent tasks*, Journal of the Association for Computing Machinery, 23, 116-127, 1976.

33. M. W. P. SAVELSBERGH, R. N. UMA, J. M. WEIN, *An experimental study of LP-based approximation algorithms for scheduling problems*, In Proceedings of the 9th Annual ACM-SIAM Symposium on Discrete Algorithms, pp. 453-462, 1998.

34. L. SCHRAGE, *A proof of the shortest remainig processing time processing discipline*, Operations Research 16, pp. 687-690, 1968.

35. A. S. SHULZ, *Scheduling to minimize total weighted completion time: Performance guarantees of LP-based heuristics and lower bounds*, in W. H. Cunningham, S. T. McCormick, and M. Queyranne, editors, Integer Programming and Combinatorial Optimization, volume 1084 of LNCS, pp. 301-315, Springer, Berlin, 1996.

36. A.S. SCHULZ, M. SKUTELLA, *Random-based scheduling: New approximations and LP lower bounds*, In J. Rolimmm ed., Randomization and Approximation Techiques in Computer Science, LNCS 1269, 119-133, 1997.

37. A.S. SCHULZ, M. SKUTELLA, *Scheduling-LPs bear probabilities: Randomized approximations for min-sum criteria*, In R. Burkard and G.J. Woeginger eds, ESA'97, LNCS 1284, 416-429, 1997.

38. A.S. SCHULZ, G.J. WOEGINGER, *A PTAS for minimizing the weigthed sum of job completion times on parallel machines*, In Proceedings of STOC'99.

39. M. SKUTELLA, *Semidefinite relaxations for parallel machine scheduling*, In Proc. FOCS, 472-481 1998.

40. M. SKUTELLA, *Convex quadratic programming relaxations for network scheduling problems*, In Proc. ESA'99, 1999, to appear.

41. W. SMITH, *Various optimizers for single-stage production*, Naval Res. Logist. Quart. Vol. 3, 59-66, 1956.

42. D. SHMOYS, E. TARDOS, *An approximation algorithm for the generalized assignment problem*, Mathematical Programming, 62, 461-474, 1993.

43. G.J. WOEGINGER, *When does a dynamic programming formulation guarantee the existence of an FPTAS*, Report Woe-27, Institut für Mathematik B, Technical University of Graz, Austria, April 1998.

Author Index

F. Afrati 281
Susanne Albers 269
Noga Alon 16
Uri Arad 16
Yonatan Aumann 109
Yossi Azar 16

Andreas Baltz 138
E. Bampis 281
Tuğkan Batu 245
Andrei Z. Broder 1

Christine T. Cheng 209
Andrea E.F. Clementi 197
Anne E. Condon 221

Yevgeniy Dodis 97
Benjamin Doerr 39

Uriel Feige 189
Dimitris A. Fotakis 156

Oded Goldreich 97, 131
Joachim Gudmundsson 28

Johan Håstad 109
Magnús M. Halldórsson 73

Csanád Imreh 168

Klaus Jansen 177

Richard M. Karp 221
Hans Kellerer 51
C. Kenyon 281
Guy Kortsarz 73

Eric Lehman 97
Christos Levcopoulos 28

Russell A. Martin 257
Milena Mihail 63
I. Milis 281
Michael Mitzenmacher 1
Rolf H. Möhring 144
Benjamin J. Morris 121

Giri Narasimhan 28
John Noga 168

Michal Parnas 85
Paolo Penna 197
Seth Pettie 233

Michael O. Rabin 109
Vijaya Ramachandran 233
Dana Randall 257
Sofya Raskhodnikova 97
Dana Ron 85, 97
Ronitt Rubinfeld 245

Michael Saks 11
Alex Samorodnitsky 97
Günter Schmidt 269
Tomasz Schoen 138
Andreas S. Schulz 144
Riccardo Silvestri 197
Roberto Solis-Oba 177
Paul G. Spirakis 156
Aravind Srinivasan 11
Anand Srivastav 39, 138
Madhu Sudan 109
Maxim Sviridenko 177

Marc Uetz 144

Patrick White 245
Avi Wigderson 130, 131

Shiyu Zhou 11
David Zuckerman 11

Springer
and the
environment

At Springer we firmly believe that an international science publisher has a special obligation to the environment, and our corporate policies consistently reflect this conviction.

We also expect our business partners – paper mills, printers, packaging manufacturers, etc. – to commit themselves to using materials and production processes that do not harm the environment. The paper in this book is made from low- or no-chlorine pulp and is acid free, in conformance with international standards for paper permanency.

Springer

Lecture Notes in Computer Science

For information about Vols. 1–1574
please contact your bookseller or Springer-Verlag

Vol. 1575: S. Jähnichen (Ed.), Compiler Construction. Proceedings, 1999. X, 301 pages. 1999.

Vol. 1576: S.D. Swierstra (Ed.), Programming Languages and Systems. Proceedings, 1999. X, 307 pages. 1999.

Vol. 1577: J.-P. Finance (Ed.), Fundamental Approaches to Software Engineering. Proceedings, 1999. X, 245 pages. 1999.

Vol. 1578: W. Thomas (Ed.), Foundations of Software Science and Computation Structures. Proceedings, 1999. X, 323 pages. 1999.

Vol. 1579: W.R. Cleaveland (Ed.), Tools and Algorithms for the Construction and Analysis of Systems. Proceedings, 1999. XI, 445 pages. 1999.

Vol. 1580: A. Včkovski, K.E. Brassel, H.-J. Schek (Eds.), Interoperating Geographic Information Systems. Proceedings, 1999. XI, 329 pages. 1999.

Vol. 1581: J.-Y. Girard (Ed.), Typed Lambda Calculi and Applications. Proceedings, 1999. VIII, 397 pages. 1999.

Vol. 1582: A. Lecomte, F. Lamarche, G. Perrier (Eds.), Logical Aspects of Computational Linguistics. Proceedings, 1997. XI, 251 pages. 1999. (Subseries LNAI).

Vol. 1583: D. Scharstein, View Synthesis Using Stereo Vision. XV, 163 pages. 1999.

Vol. 1584: G. Gottlob, E. Grandjean, K. Seyr (Eds.), Computer Science Logic. Proceedings, 1998. X, 431 pages. 1999.

Vol. 1585: B. McKay, X. Yao, C.S. Newton, J.-H. Kim, T. Furuhashi (Eds.), Simulated Evolution and Learning. Proceedings, 1998. XIII, 472 pages. 1999. (Subseries LNAI).

Vol. 1586: J. Rolim et al. (Eds.), Parallel and Distributed Processing. Proceedings, 1999. XVII, 1443 pages. 1999.

Vol. 1587: J. Pieprzyk, R. Safavi-Naini, J. Seberry (Eds.), Information Security and Privacy. Proceedings, 1999. XI, 327 pages. 1999.

Vol. 1589: J.L. Fiadeiro (Ed.), Recent Trends in Algebraic Development Techniques. Proceedings, 1998. X, 341 pages. 1999.

Vol. 1590: P. Atzeni, A. Mendelzon, G. Mecca (Eds.), The World Wide Web and Databases. Proceedings, 1998. VIII, 213 pages. 1999.

Vol. 1592: J. Stern (Ed.), Advances in Cryptology – EUROCRYPT '99. Proceedings, 1999. XII, 475 pages. 1999.

Vol. 1593: P. Sloot, M. Bubak, A. Hoekstra, B. Hertzberger (Eds.), High-Performance Computing and Networking. Proceedings, 1999. XXIII, 1318 pages. 1999.

Vol. 1594: P. Ciancarini, A.L. Wolf (Eds.), Coordination Languages and Models. Proceedings, 1999. IX, 420 pages. 1999.

Vol. 1595: K. Hammond, T. Davie, C. Clack (Eds.), Implementation of Functional Languages. Proceedings, 1998. X, 247 pages. 1999.

Vol. 1596: R. Poli, H.-M. Voigt, S. Cagnoni, D. Corne, G.D. Smith, T.C. Fogarty (Eds.), Evolutionary Image Analysis, Signal Processing and Telecommunications. Proceedings, 1999. X, 225 pages. 1999.

Vol. 1597: H. Zuidweg, M. Campolargo, J. Delgado, A. Mullery (Eds.), Intelligence in Services and Networks. Proceedings, 1999. XII, 552 pages. 1999.

Vol. 1598: R. Poli, P. Nordin, W.B. Langdon, T.C. Fogarty (Eds.), Genetic Programming. Proceedings, 1999. X, 283 pages. 1999.

Vol. 1599: T. Ishida (Ed.), Multiagent Platforms. Proceedings, 1998. VIII, 187 pages. 1999. (Subseries LNAI).

Vol. 1601: J.-P. Katoen (Ed.), Formal Methods for Real-Time and Probabilistic Systems. Proceedings, 1999. X, 355 pages. 1999.

Vol. 1602: A. Sivasubramaniam, M. Lauria (Eds.), Network-Based Parallel Computing. Proceedings, 1999. VIII, 225 pages. 1999.

Vol. 1603: J. Vitek, C.D. Jensen (Eds.), Secure Internet Programming. X, 501 pages. 1999.

Vol. 1604: M. Asada, H. Kitano (Eds.), RoboCup-98: Robot Soccer World Cup II. XI, 509 pages. 1999. (Subseries LNAI).

Vol. 1605: J. Billington, M. Diaz, G. Rozenberg (Eds.), Application of Petri Nets to Communication Networks. IX, 303 pages. 1999.

Vol. 1606: J. Mira, J.V. Sánchez-Andrés (Eds.), Foundations and Tools for Neural Modeling. Proceedings, Vol. I, 1999. XXIII, 865 pages. 1999.

Vol. 1607: J. Mira, J.V. Sánchez-Andrés (Eds.), Engineering Applications of Bio-Inspired Artificial Neural Networks. Proceedings, Vol. II, 1999. XXIII, 907 pages. 1999.

Vol. 1608: S. Doaitse Swierstra, P.R. Henriques, J.N. Oliveira (Eds.), Advanced Functional Programming. Proceedings, 1998. XII, 289 pages. 1999.

Vol. 1609: Z. W. Raś, A. Skowron (Eds.), Foundations of Intelligent Systems. Proceedings, 1999. XII, 676 pages. 1999. (Subseries LNAI).

Vol. 1610: G. Cornuéjols, R.E. Burkard, G.J. Woeginger (Eds.), Integer Programming and Combinatorial Optimization. Proceedings, 1999. IX, 453 pages. 1999.

Vol. 1611: I. Imam, Y. Kodratoff, A. El-Dessouki, M. Ali (Eds.), Multiple Approaches to Intelligent Systems. Proceedings, 1999. XIX, 899 pages. 1999. (Subseries LNAI).

Vol. 1612: R. Bergmann, S. Breen, M. Göker, M. Manago, S. Wess, Developing Industrial Case-Based Reasoning Applications. XX, 188 pages. 1999. (Subseries LNAI).

Vol. 1613: A. Kuba, M. Šámal, A. Todd-Pokropek (Eds.), Information Processing in Medical Imaging. Proceedings, 1999. XVII, 508 pages. 1999.

Vol. 1614: D.P. Huijsmans, A.W.M. Smeulders (Eds.), Visual Information and Information Systems. Proceedings, 1999. XVII, 827 pages. 1999.

Vol. 1615: C. Polychronopoulos, K. Joe, A. Fukuda, S. Tomita (Eds.), High Performance Computing. Proceedings, 1999. XIV, 408 pages. 1999.

Vol. 1616: P. Cointe (Ed.), Meta-Level Architectures and Reflection. Proceedings, 1999. XI, 273 pages. 1999.

Vol. 1617: N.V. Murray (Ed.), Automated Reasoning with Analytic Tableaux and Related Methods. Proceedings, 1999. X, 325 pages. 1999. (Subseries LNAI).

Vol. 1618: J. Bézivin, P.-A. Muller (Eds.), The Unified Modeling Language. Proceedings, 1998. IX, 443 pages. 1999.

Vol. 1619: M.T. Goodrich, C.C. McGeoch (Eds.), Algorithm Engineering and Experimentation. Proceedings, 1999. VIII, 349 pages. 1999.

Vol. 1620: W. Horn, Y. Shahar, G. Lindberg, S. Andreassen, J. Wyatt (Eds.), Artificial Intelligence in Medicine. Proceedings, 1999. XIII, 454 pages. 1999. (Subseries LNAI).

Vol. 1621: D. Fensel, R. Studer (Eds.), Knowledge Acquisition Modeling and Management. Proceedings, 1999. XI, 404 pages. 1999. (Subseries LNAI).

Vol. 1622: M. González Harbour, J.A. de la Puente (Eds.), Reliable Software Technologies – Ada-Europe'99. Proceedings, 1999. XIII, 451 pages. 1999.

Vol. 1625: B. Reusch (Ed.), Computational Intelligence. Proceedings, 1999. XIV, 710 pages. 1999.

Vol. 1626: M. Jarke, A. Oberweis (Eds.), Advanced Information Systems Engineering. Proceedings, 1999. XIV, 478 pages. 1999.

Vol. 1627: T. Asano, H. Imai, D.T. Lee, S.-i. Nakano, T. Tokuyama (Eds.), Computing and Combinatorics. Proceedings, 1999. XIV, 494 pages. 1999.

Col. 1628: R. Guerraoui (Ed.), ECOOP'99 - Object-Oriented Programming. Proceedings, 1999. XIII, 529 pages. 1999.

Vol. 1629: H. Leopold, N. García (Eds.), Multimedia Applications, Services and Techniques - ECMAST'99. Proceedings, 1999. XV, 574 pages. 1999.

Vol. 1631: P. Narendran, M. Rusinowitch (Eds.), Rewriting Techniques and Applications. Proceedings, 1999. XI, 397 pages. 1999.

Vol. 1632: H. Ganzinger (Ed.), Automated Deduction – Cade-16. Proceedings, 1999. XIV, 429 pages. 1999. (Subseries LNAI).

Vol. 1633: N. Halbwachs, D. Peled (Eds.), Computer Aided Verification. Proceedings, 1999. XII, 506 pages. 1999.

Vol. 1634: S. Džeroski, P. Flach (Eds.), Inductive Logic Programming. Proceedings, 1999. VIII, 303 pages. 1999. (Subseries LNAI).

Vol. 1636: L. Knudsen (Ed.), Fast Software Encryption. Proceedings, 1999. VIII, 317 pages. 1999.

Vol. 1637: J.P. Walser, Integer Optimization by Local Search. XIX, 137 pages. 1999. (Subseries LNAI).

Vol. 1638: A. Hunter, S. Parsons (Eds.), Symbolic and Quantitative Approaches to Reasoning and Uncertainty. Proceedings, 1999. IX, 397 pages. 1999. (Subseries LNAI).

Vol. 1639: S. Donatelli, J. Kleijn (Eds.), Application and Theory of Petri Nets 1999. Proceedings, 1999. VIII, 425 pages. 1999.

Vol. 1640: W. Tepfenhart, W. Cyre (Eds.), Conceptual Structures: Standards and Practices. Proceedings, 1999. XII, 515 pages. 1999. (Subseries LNAI).

Vol. 1642: D.J. Hand, J.N. Kok, M.R. Berthold (Eds.), Advances in Intelligent Data Analysis. Proceedings, 1999. XII, 538 pages. 1999.

Vol. 1643: J. Nešetřil (Ed.), Algorithms – ESA '99. Proceedings, 1999. XII, 552 pages. 1999.

Vol. 1644: J. Wiedermann, P. van Emde Boas, M. Nielsen (Eds.), Automata, Languages, and Programming. Proceedings, 1999. XIV, 720 pages. 1999.

Vol. 1645: M. Crochemore, M. Paterson (Eds.), Combinatorial Pattern Matching. Proceedings, 1999. VIII, 295 pages. 1999.

Vol. 1647: F.J. Garijo, M. Boman (Eds.), Multi-Agent System Engineering. Proceedings, 1999. X, 233 pages. 1999. (Subseries LNAI).

Vol. 1648: M. Franklin (Ed.), Financial Cryptography. Proceedings, 1999. VIII, 269 pages. 1999.

Vol. 1649: R.Y. Pinter, S. Tsur (Eds.), Next Generation Information Technologies and Systems. Proceedings, 1999. IX, 327 pages. 1999.

Vol. 1650: K.-D. Althoff, R. Bergmann, L.K. Branting (Eds.), Case-Based Reasoning Research and Development. Proceedings, 1999. XII, 598 pages. 1999. (Subseries LNAI).

Vol. 1651: R.H. Güting, D. Papadias, F. Lochovsky (Eds.), Advances in Spatial Databases. Proceedings, 1999. XI, 371 pages. 1999.

Vol. 1652: M. Klusch, O.M. Shehory, G. Weiss (Eds.), Cooperative Information Agents III. Proceedings, 1999. XI, 404 pages. 1999. (Subseries LNAI).

Vol. 1653: S. Covaci (Ed.), Active Networks. Proceedings, 1999. XIII, 346 pages. 1999.

Vol. 1654: E.R. Hancock, M. Pelillo (Eds.), Energy Minimization Methods in Computer Vision and Pattern Recognition. Proceedings, 1999. IX, 331 pages. 1999.

Vol. 1661: C. Freksa, D.M. Mark (Eds.), Spatial Information Theory. Proceedings, 1999. XIII, 477 pages. 1999.

Vol. 1662: V. Malyshkin (Ed.), Parallel Computing Technologies. Proceedings, 1999. XIX, 510 pages. 1999.

Vol. 1663: F. Dehne, A. Gupta. J.-R. Sack, R. Tamassia (Eds.), Algorithms and Data Structures. Proceedings, 1999. IX, 366 pages. 1999.

Vol. 1666: M. Wiener (Ed.), Advances in Cryptology – CRYPTO '99. Proceedings, 1999. XII, 639 pages. 1999.

Vol. 1671: D. Hochbaum, K. Jansen, J.D.P. Rolim, A. Sinclair (Eds.), Randomization, Approximation, and Combinatorial Optimization. Proceedings, 1999. IX, 289 pages. 1999.